Texts in Theoretical Computer Science
An EATCS Series

Springer
Berlin
Heidelberg
New York
Barcelona
Budapest
Hong Kong
London
Milan
Paris
Tokyo

Max Garzon

Models of
Massive Parallelism

Analysis of Cellular Automata
and Neural Networks

With 14 Figures and 10 Tables

 Springer

Author

Prof. Dr. Max Garzon, The University of Memphis, Department of Mathematical Sciences, Winfield Dunn Building, Room 373, Memphis, Tennessee 38152, USA

Series Editors

Prof. Dr. Wilfried Brauer, Fakultät für Informatik, Technische Universität München, Arcisstraße 21, D-80333 München, Germany

Prof. Dr. Grzegorz Rozenberg, Institute of Applied Mathematics and Computer Science, University of Leiden, Niels-Bohr-Weg 1, P.O. Box 9512, 2300 RA Leiden, The Netherlands

Prof. Dr. Arto Salomaa, Turku Centre for Computer Studies, Data City, 4th Floor, FIN-20520 Turku, Finland

Library of Congress Cataloging-in-Publication Data

Garzon, Max, 1953–
    Models of massive parallelism : analysis of cellular automata and
neural networks / Max Garzon.
        p.   cm. -- (Texts in theoretical computer science)   (EATCS
monographs on theoretical computer science)
    Includes bibliographical references and index.
    ISBN-13: 978-3-642-77907-7

    1. Parallel processing (Electronic computers) 2. Cellular
automata. 3. Neural networks (Computer science)   I. Title.
II. Series. III. Series: EATCS monographs on theoretical computer
science.
QA76.58.G37   1995
004'.35--dc20
                                                    95-18962
                                                    CIP

CR Subject Classification (1991): F.1–2, G.2, I.2

ISBN-13: 978-3-642-77907-7   e-ISBN-13: 978-3-642-77905-3
DOI: 10.1007/978-3-642-77905-3

© Springer-Verlag Berlin Heidelberg 1995
Softcover reprint of the hardcover 1st edition 1995

The use of general descriptive names, trademarks, etc., in this publication does not imply, even in the absence of a specific statement, that such names are exempt from the relevant protective laws and regulations and therefore free for general use.

Cover Design: MetaDesign plus GmbH
Typesetting: Camera ready by author
Copy and Production Editing: J. Andrew Ross

SPIN 10083937   45/3142 – 5 4 3 2 1 0

To my parents, Maximiliano and Carmen,
for all the wonder they sowed in me

Preface

Locality is a fundamental restriction in nature. On the other hand, adaptive complex systems, life in particular, exhibit a sense of permanence and timelessness amidst relentless constant changes in surrounding environments that make the global properties of the physical world the most important problems in understanding their nature and structure. Thus, much of the differential and integral Calculus deals with the problem of passing from local information (as expressed, for example, by a differential equation, or the contour of a region) to global features of a system's behavior (an equation of growth, or an area). Fundamental laws in the exact sciences seek to express the observable global behavior of physical objects through equations about local interaction of their components, on the assumption that the continuum is the most accurate model of physical reality. Paradoxically, much of modern physics calls for a fundamental discrete component in our understanding of the physical world.

Useful computational models must be eventually constructed in hardware, and as such can only be based on local interaction of simple processing elements. This may perhaps best explain why massively parallel processors have held such fascination for researchers and practitioners since the early days of the information age. The theme of this volume is the understanding of what is meaningfully achievable, in practice or in principle, by computational devices consisting of a large number of simple units evolving in time, lacking a central executive that would provide global coordination and control, and spread across space carrying a tiny program without the ability of instant remote communication.

We present a systematic exposition of the most important *analytical* results on the three better studied such models of fine-grained massively parallel computation, namely cellular automata, neural networks, and random boolean networks (also known as automata networks). The first five chapters emphasize the cellular automaton model, while the remaining ones focus on its generalizations, discrete neural and automata networks. The main topics include linear cellular automata, semi/totalistic rules, complexity of decision problems, the relationship between neural nets and cellular automata, the topological aspects of their global long-term behavior (including classification schemes for both), basic results about their dynamical behavior, and an exploration of the reach and significance of neural networks (perhaps with an infinite number of neurons) as computational tools in the realms of the discrete and of the continuum.

Given the growing interest in neural networks, we hasten to disclaim what material a casual reader might expect but will *not* find herein. The text does not formally include learning or cognitive-related issues in neural networks, or any experimental results where no theoretical framework or analytic proof has been established. The exposition is written mainly from the point of view of theoretical computer science, not formal mathematics, physics, cognitive science, or the like. However, useful analogies and links are drawn from and references made, when appropriate, to topics in these fields.

The presentation, while rigorous and mathematical, has been kept motivated and intuitively driven. A reader with a basic course in discrete mathematics can follow the text. Upper undergraduate and graduate students should be able to take full advantage. Experts should be able to quickly find references to the literature that may have escaped them. Early versions of the material have been class tested at The University of Memphis (formely Memphis State University). They have have also been used as notes for a seminar on the subject at the Laboratorie de L'Informatique du Parallélisme de l'Ecole Normale Superiéure de Lyon during the year 1991-1992, and at several other places.

In a new age, when data from computer experiments and observations fill up volumes and volumes of conference proceedings and memory archives, we hope the reader will find in this text a refreshing vantage point from which to contemplate the kaleidoscope of parallel virtual universes into which research in the subject has exploded.

The University of Memphis, February 1995   *Max Garzon*

Acknowledgements

This work would not have materialized without the help and supportive environment provided by a number of people in several places. I'd like to record my explicit appreciation of some of them. To Paul Schupp for introducing me to automata and Cayley graphs at The University of Illinois, Urbana. To a few generations of students at The University of Memphis who endured preliminary classnote versions for a course in fine-grained computation from which this volume evolved. To NSF for grants which supported a lot of our research included herein. To Michel Cosnard, Jacques Mazoyer, colleagues and the staff of the CNRS-LIP at the École Normale Superiéure de Lyon for providing me with a very supportive atmosphere where this project could finally "get over the bump". To my coworkers in the dynamical systems and models groups at UofM and LIP, especially Stan Franklin, Fernanda Botelho, Michel Cosnard, Pascal Koiran, and Ron Bartlett, for fruitful collaborations that led to results which, in some cases, even substantially comprise sections in several chapters. To Jarkko Kari, Arto Salomaa, Kenishi Morita, Andreas Weber, Wilfried Brauer, Tom Head and others whose interest in the notes and/or feedback on early drafts kept the project going. To Hans Wössner, Andy Ross, and the other staff of Springer-Verlag for their patience and proofreading on an ever expanding camera-ready form. And last, but by no means least, to my children Catalina and Ana María, for their understanding, encouragement, and, in the latter case, for many of the pictures drawn using xfig that illustrate the following pages.

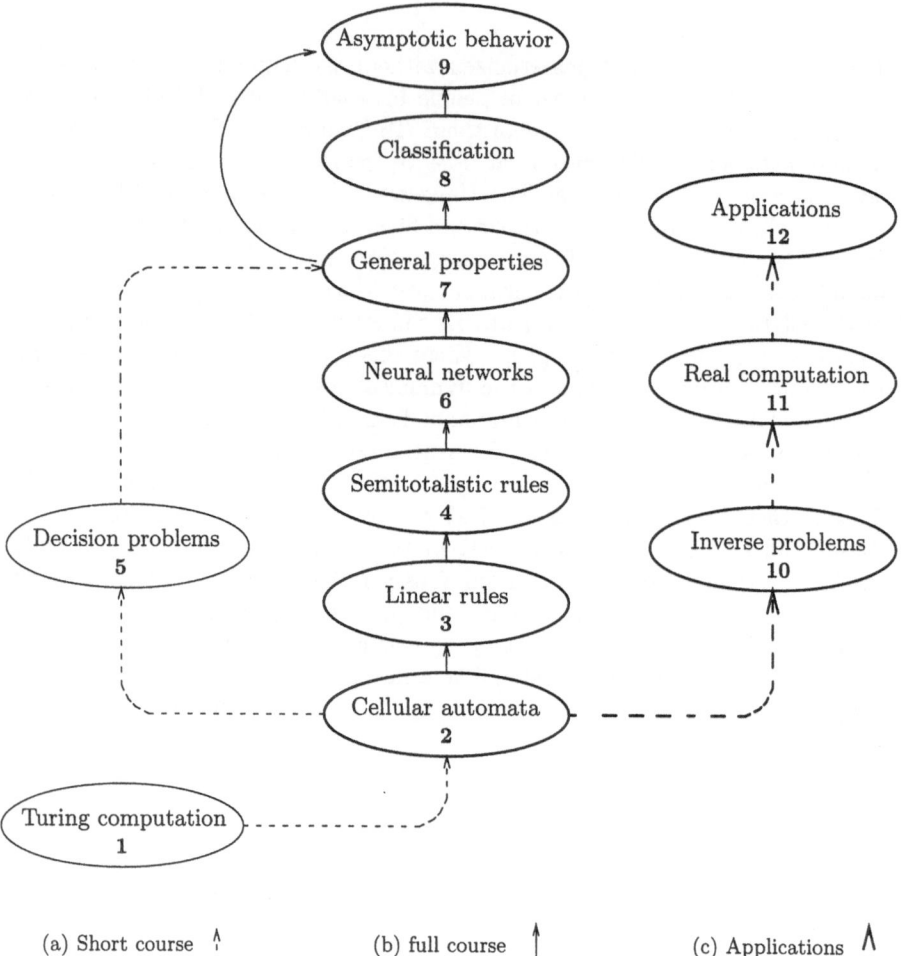

(a) Short course   (b) full course   (c) Applications

Chapter interdependence and reading options

Table of Contents

1. Turing Computability and Complexity

Computing is normally done by writing symbols on paper [···] We may suppose there is a bound to the number of symbols or squares that the computer can observe at one moment [···] If we admitted an infinity of states of [the computer's] mind, some of them will be "arbitrarily close" and will be confused [···]

Alan M. Turing

Ancient documents, like the Rhind Papyrus, dating as far back as 3600 B.C. attest to the fact that men have been performing computations for thousands of years, and for very practical reasons such as farming, land measurements (Egyptian geometry), or simply wishful thinking about the powers of human prediction (astrology). Years later, after Europe surmounted the laggish period of the Middle Ages, a reblooming in the arts during the Renaissance slowly led the way to the dawn of the scientific method, specifically, in the physical-mathematical sciences, with Galileo and the crowning work of Newton on celestial mechanics. In particular, from a modern perspective, their monumental work provides us with algorithms to establish facts about heavenly or earthly objects related to their position and/or their motion. Further research on electricity and magnetism by J.C. Maxwell, and on thermodynamics and heat theory by L. Boltzmann and other 19th century physicists may be regarded in a similar way.

It is thus striking to notice that only very recently, in this century, did we begin to wonder with the same eagerness for a 'deep' explanation of a mathematical character concerning the nature of computations and algorithms. The leading investigations of Alan Turing and Emile Post, most experts believe, paved the way for the later development of electromechanical, and later electronic computing machines, of which personal computers and workstations are increasing familiar examples nowadays.

The purpose of this first chapter is to present a summary description of current mathematical models of sequential computers. The basic reason for a computer model is not only to gain insight into what computers *can* do but also to discover tasks that they *cannot* do. It is a curious fact that men did not require a *formal* definition of an algorithm to design and use them, but that, on the other hand, it is a logical necessity to previously *define* what is meant by an *algorithm* (or a *computer*) in a rigorous way in order to establish that there are tasks beyond their power. Otherwise we will be just playing a futile hide-and-seek game which at the crucial point, upon a new puzzle, says "Ahh ... but *that* is just what I meant by *a finite sequence of steps that can be executed mechanically by somebody who is not knowledgeable about the meaning of the symbols involved.*"

The classical theory of computability attempts to characterize the power of perfect computers (in a sense to be specified below, such as Turing machines) that do not share the practical limitations of real machines. The usefulness of these models can be compared to that of the equations of free fall or the orbit of a planet. Just as no object falls according to these equations due to frictional forces or other forces that need to be neglected to make the problems tractable, these models also serve as idealized models of computational devices whose properties can be approximated to within any degree of accuracy by real machines.

1.1 Models of Sequential Computation

There is a phenomenon that implicitly pervades the computing field, but a satisfactory explanation of which remains as yet hidden to us. Despite substantial efforts to elucidate how it actually works, the mechanism of *memory* surely holds the key to understanding a number of central issues in computing. In particular, questions such as *What does it mean to remember?* and *How it is possible for humans to remember?* no longer admit the simplistic explanation that they have been traditionally given, especially in view of the fact that current storage-retrieval mechanisms used on computing machines perform so poorly when compared to human memory. Yet, memory seems to be an inherent property of a brain, human or not. Life in general requires some sort of learning and adaptation, and it seems impossible to argue that these qualities are possible without some form of memory capability. Therefore any attempt at defining precisely what a computer is must in some way or other involve some sort of speculation and modeling on the nature of memory.

Alan Turing, a British logician and mathematician who devoted a great deal of his efforts to deciphering the German secret military code 'Enigma' during World War II, implicitly proposed a mathematical model of memory when he introduced his now well-known Turing machine in 1936. Let us assume, Turing says, that a black box (the calculator's head or a computer's central processing unit) is capable of *being* in, or assuming a number of internal configurations. Further, he continues, assume that this box dynamically switches among these configurations, in such a way that only one of a fixed *finite* number of them is unmistakingly identifiable at any time as the one the box is "in" at the moment. Each one of these identifiable possible "patterns" or "configurations" or "snapshots" is called a *state*, and a box like this is called a *finite control*. We can, if we wish, imagine the finite control as a switching board with as many bulbs as states, which are to turn on and off to indicate that the corresponding memory "configuration" is being remembered or not, and thus that a certain definite action is taking place, at a particular instant of time. Incidentally, instead of arguing about its nature, he assumes that time flows discretely, at least as far as the machine is concerned, that is, that we can use the integer numbers $0, 1, 2, 3, \ldots$ as our time coordinate.

Another essential ingredient in a computing device is some form of input/output. After all, a computing device needs to be given the description of the task to be performed in some way or other. Turing further assumed that this task would be described by a *string* of symbols of a fixed alphabet, the way humans communicate in writing with the help of a standard set of symbols (say, the roman characters plus the numerical and special symbols in the English alphabet). Here an *alphabet* can be defined as an arbitrary finite set that contains a special symbol that is not visibly written (such as the 'blank' symbol). Further, assume these symbols are written to the finite control on a linear tape divided off into squares, each capable of holding one of these alphabet symbols, the same way they are written for humans just about everywhere (except that they are 'folded' into a book or a diskette for practical reasons). Thus a finite control can be given the further capability of detecting which symbol is on the square it is currently scanning and taking appropriate action about it. What specific action the finite control can take upon detecting a symbol on the tape will determine to a large extent, in the classical theory of computability, the computing power of the resulting device.

1.1.1 A Simple Model: the Finite-State Machine

Let us now look at the restricted case in which the computer cannot modify the input but just look at it and compute. Let us look at an example to fix ideas before spelling out the definition of a finite-state machine. The basic ingredient is a finite control reading problem instances on an input tape (in the form of strings of symbols) and "computing" according to a prewired program that drives the machine from one memory configuration (state) to the next depending on the symbol that it has just seen on the tape. We assume that the reader is familiar with the binary and, more generally, arbitrary representations of the natural numbers in ternary, decimal, etc. For instance, 1110 represents 'eleven hundred and ten' in decimal, but 14 in binary and 39 in ternary.

Example 1.1 We want to design a finite-state machine that recognizes divisibility by 3 in every integer.

This may sound like a precise task, because we are all familiar with the terms involved. We may even possess very sophisticated methods to solve the problem. The situation is very different, however, when we want to build an *independent* device that will execute the task *correctly*, "on its own" so to speak. This requires a much more explicit description of parameters (say, how is the machine given the number), which we assumed by default to be the ways that work for us. The reason is that different formulations of the problem may allow different solutions of it, or even none at all. For instance, if the integer is given in decimal notation, we have learned a quick way of checking divisibility by 3: add up the digits and thereby reduce the problem to a similar one of a smaller size to which the same rule can be reapplied, and so on down until we get to the familiar range 1 through 9 where the answer can be easily given. But if the number is given in ternary notation this test is not that easy to implement by *us* because we are not usually as familiar with ternary addition as decimal; more seriously, the add-up-the-digits test fails for it rarely provides right answers in ternary. Instead, we can simply look for a 0 in the last digit since in ternary divisibility by 3 is like divisibility by 10 in decimal.

Of course we can always say, 'Get your calculator/computer and do it!' but a little reflection will make us realize that this is plain begging the question because that is exactly what we are trying to explain: how can *we* design a device that performs such a task? So, let's fully specify the problem. We want to design a finite-state machine that will perform as follows:

> Upon being started in a certain *initial state*, say state 0, it will read, from left to right, the digits of *any* positive integer while switching from one state to the next depending on the symbol just read. If the number represented by the input symbols is an integer divisible by 3, then the machine will end up in a special *final* state, say 0, to signal this fact; otherwise, it will do so by finishing the reading of the integer in any other of its internal states.

If we want to design a finite-state machine to do any task, we must first answer the question: What are the bits and pieces of information that *need* to be manipulated to solve this problem? In the example, we realize that since the input digits are read one by one, we need to have the machine keep track somehow or other of the *remainders of the successive numbers represented by the digits already scanned when divided by 3*. On the other hand, we have the constraint that no machine with an infinite number of parts can be built, so we need to achieve this with a *fixed finite number* of components. These components can be realized in practice as voltage differences in a circuit, or the concentration of a certain chemical in some stream, or simply as the position of a certain piece in a wooden computer. If we manage to build a device like that, then we simply make the state corresponding to remainder 0 the final state and we are done. That is the desired finite-state automaton that the problem is asking for.

Therefore we decide on a finite control which has 3 states, conveniently labelled 0, 1, and 2, for these are exactly the possible remainders upon division by 3. We can draw 3 circles to represent them. And now comes the crucial moment, when we design the 'dynamics' of these states. What will make the machine work, that is, 'compute' out of a sequence of digits the *abstract* idea that *we* have of divisibility by 3?

The plot is very simple indeed. We will have the input symbols switch the machine from state to state to reflect the fact that this remainder changes as more input digits are read in. But as is the case with the remainders, the switch from one of the three states to the next in *not* random. We will do it in such a way that the machine will 'remember' the current remainder r by 'being in state' r. So it is easy to read the answer off after all digits are in by making the state **0** the special 'final state' (double-circled in Fig. 1.1). The other states **1**,

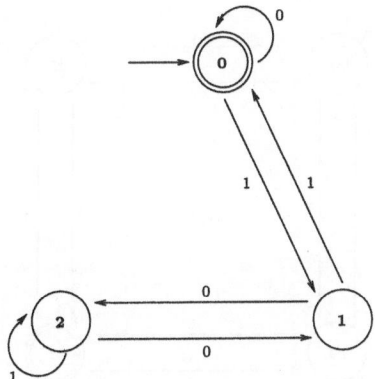

Fig. 1.1. A finite-state automaton for divisibility by 3

2 are intermediate states used in the course of the calculation or to represent 'failure' of the property of being divisible by 3.

Of course a finite control that has to be opened and switched to the proper state after *we* recognize the new remainder would not be of much use. If it has to do the task independently after it has been designed, its construction demands a definite, specific wiring. Yet, the same design has to answer correctly and succesfully a multitude, in fact, an infinite number of questions: check if *each* string of binary digits –of which there is an infinite number– is divisible by 3. And it has to do it by itself, with no other external wisdom than what has been 'blown' into it by the designer. How are the transitions to be wired up to achieve that? That's the question.

The answer is not difficult at all at this point. Just observe that if a numeral (i.e., a sequence of digits 0 or 1, like the long string in equation (1.2) below) represents a number x which has remainder r when divided by 3, that is, if

$$x = 3q + r, \quad \text{where } r = 0, 1, \text{ or } 2 \,,$$

then adding a 0 at the end of x multiplies it by 2 (the same way that adding a 0 at the right end of a decimal string multiplies it by 10), and adding a 1 at the end of x further adds 1 to twice x. In summary, adding a digit d at the end of x modifies x according to the equation

$$xd = 2x + d = 3(2q) + (2r + d) \,. \tag{1.1}$$

This equation says that the remainder of the new number represented by adding the extra digit d is precisely $2r + d$ mod 3. (An integer can be reduced modulo 3 by subtracting 3 as many times as necessary until it becomes **0**, **1** or **2**). Thus we see how the machine should switch between states to "remember" the right

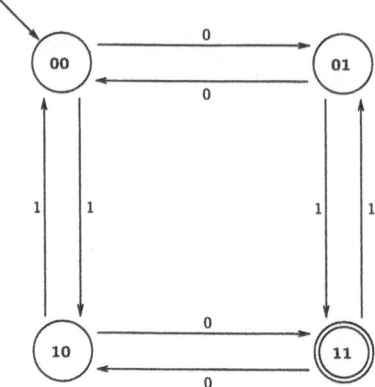

Fig. 1.2. Another finite-state automaton. What does it do?

remainders. Equation (1.1) says that, for instance, if the machine is in the initial state **0** ($r = 0$) then an incoming 1 should make it go into state 1, and if further a 1 comes in, it should go into state $2r + d = 2(1) + 1$ mod 3. If this was the last digit, the machine will end up in state 0 and thereby have computed that the binary string 11 representing the number 3 is divisible by 3. This may not sound like a big accomplishment, but now the reader can imagine the similar sequence of moves that the machine would go through from state 0 upon input

$$100101100000101001110011001 \tag{1.2}$$

while we just sip some coffee somewhere else. When we get back we will *know* that these bits represent a number divisible by 3 (for we will find the machine in state 0). We get this answer by chasing these bits along the arrows of the diagram of Fig. 1.1 from initial state 0 (which is indicated by an incoming arrow). The circle into which an arrow leads is the next state of the machine after reading the arrow label. It corresponds to the current remainder. A string checks all right if it leads into a final (double-circled) state.

This little example is representative of what happens in an arbitrary finite-state (machine) automaton. It is able to "remember" fragmentary pieces of information, corresponding to its states. It has some initial (start) state and some final states by which it can communicate the results of its computations to its environment. Finally the input is made up of symbols from a fixed alphabet (i.e., a character set) that drive the automaton from one memory configuration to another while trying to recognize or identify some common feature about the 'concepts' represented by strings of input symbols. And best of all, all this amounts to just a little diagram like Fig. 1.1. Figure 1.2 shows another example of a finite-state machine.

The reader may be already wondering about a crucial point: can you design a finite-state automaton to recognize other interesting sets, say *prime numbers*? The problem can be posed in decimal or binary, or any other base. The answer is no. The reason is that it is impossible to manipulate a *fixed* finite number of pieces of information to check if a number is prime (one needs, for example, arbitrarily long quotients and remainders upon division by a large number of divisors). The finite-state machine model therefore cannot possibly be accepted as the most general model of a computing device that would formalize the idea of an algorithm.

1.1.2 Turing Machines

The fundamental problem of a finite-state machine is that it attempts to remember everything in its limited primary memory. Its power can be greatly enhanced if the finite control can, in addition, write and rewrite symbols on its input tape. This is the model originally proposed by Turing. The Turing machine, as it is known today, can write on its input or also use additional work tapes (scratch paper if you will). It was suggested as a satisfactory solution to the problem of arriving at a rigorous definition of the notion of "computing procedure." This notion attempts to embody in precise terms our intuitive idea of "algorithm," that is, a list of instructions that do not require any ability on the part of the agent executing them other than an ability to read and rigidly interpret symbols in the data to the problem being solved or auxiliary to its solution. The instructions in the algorithm should not admit ambiguous interpretation and should, at least in principle, be executable by a piece of machinery, properly constructed.

Briefly, a *Turing machine* consists of a finite-state control (like a head with a limited ability to remember "isolated facts") with the added capability of reading and writing symbols from a fixed character set Σ onto a doubly-infinite tape divided into squares, each capable of holding one character at a time. The machine computes by performing successive atomic moves in discrete time steps. In each atomic move, a Turing machine M, keeping track in its internal memory of certain "states of mind" from a set Q, finite in number, changes its state and the contents of the cell being scanned under the read-write head, and moves along the tape, always as a function

$$\delta : Q \times \Sigma \longrightarrow Q \times \Sigma \times \{-1, 0, 1\} \tag{1.3}$$

of the current state of the finite control and the symbol on the tape currently under it. In addition, M comes with two special states, the *initial* and *final* halting state, and a special tape symbol called *blank*. The most important parameter thus determined is the transition function δ. It is indeed a *finite* compact description of a possibly very long (but finite) sequence of prescribed actions on an infinite number of possible inputs.

This idealized model of computation has become one of the most popular definitions of the (sequential) computer. Among specific reasons for this success are:

- its simplicity and its accuracy in reflecting so closely, although ideally, our paradigm for most of the physical machines currently in operation;

- more importantly, each Turing machine is made of *finitely* many *controllable* parts that can be physically implemented; in particular, this means that each machine can be completely described by a *finite* string of symbols (as any other truly *finitary* object) and that, moreover, time can be modeled by a discrete clock that ticks at integer time steps;

- a *single* machine of this type can work and solve, *exactly* and *correctly*, all (usually an *infinite* number of) instances of the same problem, each in a finite amount of time and using a finite number of tape cells. Either the time or the number of cells may not be uniformly bounded as the inputs vary.

- Further, the model has withstood the test of time and proven to be just as powerful as a number of other less natural devices suggested since Turing's paper [T] in 1936 as solutions to the same problem.

By their rendering the notion of *algorithm* precise and unambiguous, Turing machines afford researchers in computability a valuable tool to determine whether given tasks can or *prove* they cannot be completely solved by any physical machine or other models, regardless of how much memory or computing time we may be willing to provide it. This belief that this rigorous concept does capture the intuitive notion of an "algorithm" is usually stated as:

The Church-Turing thesis
Every algorithm can be implemented by a Turing machine.

A huge theory, the *theory of computability* and *recursion function theory*, has been developed during the last 50 years on the Church-Turing thesis in attempting to answer the ultimate question: What can a computer, with an unlimited number of tape cells and unlimited computing time, eventually do? The answer to this question has been completely transversal to what one would have expected, and even today, one can be surprised by regular discoveries that certain problems are (un)solvable in this sense.

1.2 Complexity

Church's thesis is all nice and well so far as we are only looking for a definition to tell *solvable* from *unsolvable* computational tasks, and so are willing to provide an unlimited amount of resources (say, tape and computing time). In reality, however, an algorithmic solution to a problem that requires resources

Table 1.1. Small and large growth rates

n	1000nlog n	100n²	n^{log n}	...	2^{n/3}	2^n
20	.09 sec	.04 sec	4 sec	...	.0001 sec	1 sec.
50	.3 sec	.25 sec	1.1 hr	...	.1 sec	35 years
100	.6 sec	1 sec	220 days	...	2.7 hr	10^{10} cent.
200	1.5 sec	4 sec	125 cent.	...	3×10^4 cent.	?
1000	10 sec	2 min	?	...	?!	??!!

that increase too fast (say exponentially, as 2^n) in relation to the size n of the problem cannot be realistically expected to be of any practical use, as the reader can ponder by examining Table 1.1. The last column of this table indicates the time it will take the standard truth-table algorithm to verify whether a given formula built from propositional symbols using the logical connectives *negation, conjunction* and *disjunction* is true in "all possible worlds", that is, true regardless of the specific truth value of each of its component propositions. It is assumed that the machine that executes the algorithm can read a symbol (propositional or connective) and perform a basic move at a rate of one per microsecond (that is, a millionth of a second.)

Thus, for example, to determine whether a formula containing 100 symbols, a small size in typical applications of the algorithm, is true or false would take

$$2^{100} sec/10^6 \times 3600 \times 24 \times 365 \times 100 \ \geq \ 10^{10} \text{ centuries}.$$

This is obviously unacceptable since it it would probably take longer than the age of the universe. The trouble is, however, that satisfiability (or just SAT) is not just an intellectual curiosity of interest only to academicians. The hard fact of the matter is, literally hundreds of problems of immediate importance in such diverse fields as scheduling, combinatorial optimization (e.g., linear programming), secure secret communication (e.g., factoring and primality testing) and map coloring can all be shown to be just as hard as (if not harder than) SAT. These considerations explain, at least in part, why one of the most important problems in the theory of computing today is precisely to find new computational strategies that do overcome the complexity barrier posed by this "combinatorial explosion." Computational complexity is the area of computation theory concerned with questions of this kind.

There is another issue raised by the above considerations. Using as yardstick the time (number of steps) taken by a Turing machine to measure the difficulty of a computational task, clearly there are tasks that are practically unfeasible (although theoretically possible) while others are not only possible but also feasible (for example, a task taking at most n^2 steps on a Turing machine, such as sorting). Thus the question arises, where do we draw a line to separate problems that are *efficiently computable* and so practically implementable from those that take so much resource (time in this case) that they must be considered *intractable*. For a number of reasons, S. Cook [C] and R. Karp [K] suggested in the early 1970s that we consider efficient, or computationally feasible, those tasks that take at most a polynomial number of steps $O(n^d)$ in the size n of the input, while problems such as theoremhood in additive number theory (i.e., checking whether a formula about addition and order holds in the natural numbers) on a Turing machine should be considered unsolvable *in practice* because it provably takes at least $2^{kn} = \lambda^n$ steps on an infinite number of instances (for some constant $\lambda > 1$ and various sizes n). The class of tractable problems thus defined is known as **P** (for polynomial time).

Faced with the fact that by these standards important problems are intractable, computer scientists have come up with a number of possibilities to approach these computationally "hard" problems. Most important among them are (a) *nondeterministic computation*, (b) *randomization*, and (c) *parallel computation*. The purpose of the following sections is to present some examples to convey the general idea in each of these alternatives.

1.2.1 Nondeterministic Computations

If one does not want to abandon the Turing model altogether and start from scratch in modeling computers, the obvious first approach is to enhance its computational power in order to reduce the amount of resource it requires to implement known algorithms. Let us illustrate the idea with two problems, our new friend SAT and PRIMALITY testing of integer numbers.

Suppose we are given a formula to verify satisfiability. Instead of plunging head on into the construction of the truth-table prescribed by the algorithm described in its definition (see Sect. 1.1), you may take the lazy approach: sit back and instead take a guess at the truth values of each of the component propositions that might lead to a true composite after evaluation of the connectives that appear in it. The key to the success of this method lies in the fact that regardless of how you come up with truth values for the atomic propositions, once you have them the evaluation of the whole proposition becomes a very simple task that can be done efficiently. In other words, if you had a potential answer it would be an easy matter to verify that indeed it is correct. If the evaluation yields *true*, you can skip the construction of *all* of the other lines in the truth table. Otherwise, you may guess again and repeat the procedure, until you get lucky (which may never happen).

There are two objections to this approach. First, there may be a single assignment of truth values that makes the whole proposition *true* but this assignment may never be guessed in the process. More seriously, we want algorithms that are to be performed by physical machines into which we may not be able, in biblical terms, to breathe the ability that *we* humans use when taking a guess. The second objection is not a stumbling block, however, because there are "random number generators" that make it possible to produce nearly "random integers" that can be used by machines in lieu of human guesses. The end result is that a nondeterministic algorithm such as this may find the answer very quickly or not at all. The hard fact remains, however, that if the original proposition is indeed satisfiable it is *possible* to hit the solution the first time you try it. Likewise in the case of compositeness testing. The class of problems for which a *given* solution can be *verified* (but not found) efficiently is called **NP** (nondeterministic polynomial time), as opposed to **P**, the class of problems for which some Turing machine can actually *find* a solution in polynomial time by a systematic procedure. Detailed accounts of the known properties of the classes **P** and **NP** can be found in [G-J, B-D-G].

One of the most striking properties of the class **NP** was discovered by S. Cook himself in 1971 [C]. We explain with SAT again. If there is a way of *systematically* cutting out large portions of the truth-table (say by trying to be very systematic about the guesswork in the nondeterministic algorithm) one could then, *deterministically* as well, evaluate the few lines left and thereby obtain an efficient **P**-time algorithm for SAT. Cook showed that this is very unlikely, by providing us instead with a systematic recipe to convert *every* nondeterministic algorithm for a problem in **NP** into a **P**-time solution that would use as a subroutine a **P**-time solution to just SAT. Specifically, he showed how to build an efficient program (machine) that converts *any* instance of an **NP**-problem into an instance of SAT, to which the clever deterministic algorithm could be applied. A problem of this sort is called **NP**-complete. R. Karp [K] proved a year later that many other combinatorial problems arising in important practical situations are **NP**-complete as well. In the same year, W. Savitch [S] gave a result that illustrated what might be happening with the other most common resource, by proving that if we would rather use *space* (i.e., the number of cells in the tape used by the machine in the course of its computation) as the basic resource, then one can eliminate guessing at the expense of just squaring the amount of space utilized in the computation. The best *known* trade-off in the case of **NP** time is an exponential amount of additional time to eliminate guesswork. Today, a quarter of a century later, these results stand as just about the only genuine leaps of insight into the nature of nondeterministic computations.

1.2.2 Randomized Algorithms

Another approach is suggested by the idea that the success in guesswork depends on the odds of hitting a right solution (guessing the right line in the truth-table, or the right factor in primality testing when the given integer is really composite). If a number has a large number of factors, it is not going to be very hard to guess one. Solovay–Strassen [S-S] and Rabin [R], independently, used this idea in 1976–77 to give an algorithm for PRIMALITY which, unlike nondeterministic algorithms, returns the right answer only with probability greater than $\frac{1}{2}$, although it may also return wrong answers (with probability less than $\frac{1}{2}$). To describe the procedure we need to recall Fermat's "little" theorem that asserts that if an integer b is coprime to m, then

$$b^{\phi(m)} \equiv 1 \pmod{m};$$

where $\phi(m)$ stands for the number of integers between 1 and m which are coprime to m (so-called Euler's totient function). In particular, if p is prime,

$$b^{p-1} \equiv 1 \pmod{p}$$

The tests proceeds in two steps on a given input m:

1. choose a number b at random between 1 and $m-1$ (uniform distribution);

2. perform a good pseudoprimality testing on the pair (m, b) and answer accordingly.

A *pseudoprimality test* is a **P**-time algorithm that produces exactly one of two answers: *composite* or *inconclusive*, so that if it answers *composite* then m is indeed composite but m may not really be prime if it says so. The test is *good* if a positive fraction of all possible choices of b yields the answer *composite* for every composite integer. For instance, checking the equation in Fermat's theorem is a pseudoprimality test, but it is not good. Here is an example of a good one.

Pseudoprimality Test(b,m: integer);
Let t be the largest odd factor in $m - 1$, i.e., $m - 1 = 2^s t$. Answer *inconclusive* if either of the following two conditions holds. Answer *composite* otherwise.

(a) $b^t \equiv 1 \pmod{m}$, or

(b) there is an integer k between 0 and $s - 1$ such that

$$b^{2^k t} \equiv -1 \pmod{m}. \quad \square$$

The insight of Solovay–Strassen and Rabin was to find the extra condition (b) to ensure that a pair (b, m) will cause this test to return *inconclusive* with probability less than $\frac{1}{2}$. Thus, if the test is performed independently twice on an integer m and still both answers are *inconclusive*, the odds of m being composite are no more than $\frac{1}{4} = 2^{-2}$. If t times it answers *inconclusive*, the probability that m is still composite is down to 2^{-t}. By the fourth test (all which can be done in **P**-time) the probability that m is still composite is so dim that we can, with a great deal of confidence, declare n a (pseudo-!)prime. Using this method, Rabin impressed the computer science community with integers about 200 digits long (precisely the order of magnitude needed for secret communication in cryptography to be considered secure) that could be declared prime in a matter of minutes with the greatest degree of confidence, while a deterministic test would not be completed any sooner than something like the age of the universe. This pioneering technique was soon applied to many other hard problems with varying degrees of success. Today, there are plenty of problems of this kind (such as factoring integers) which exhibit a similar phenomenon (see textbooks in complexity theory, sucb as Bovet-Crescenzi's [B-C] or Papadimitriou's [P]). The ability to guess is indeed a powerful gift!

1.2.3 Parallel Computation

Another important alternative is based on the idea that it may take a long time for a single man to build a house, but if a whole team gets on the job it may speed up the total construction time considerably. Turing machines are good models of *sequential* computation because only one processor is operating on the data. It may happen, however, that certain subtasks in the problems can be performed all at once because their results are not needed until they are all completed. Finding the maximum of a sequence of n values is a simple example. The sequential approach has to carry a value (initially equal to the first) with the successive ones, swapping it with any larger one it sees until it arrives at the other end of the input list. This takes roughly n comparisons to arrive at the result. If in place we had $n(n-1)/2$ smaller computers arranged in a network that resembles a tree so that they can communicate with the immediate neighbors, the problem can be solved in much faster $log\,n$ time by making each processor pick up the larger value residing in the two processors below. By using the same strategy, problems like SORTING, i.e. arranging a list of values in increasing order, can be given $log\,n$–time solutions that break the optimal sequential $n\log n$ barrier. This is another speedup that one hopes may overcome the combinatorial explosion of **NP**-complete problems. Several textbooks survey this type of approach –see for example [L, Sm].

The question will remain, however, in the case of any new model of computation: What are the computational tasks that can be considered efficient if solved on a computer built on that model? In the case of parallel computation, researchers have observed a correlation between sequential and parallel time. The *parallel computation thesis* asserts that the problems solvable efficiently in parallel using a reasonable (polynomial) number of relatively powerful processors equals **PSPACE**, the class of problems that require polynomial space on a Turing machine. In addition to further evidence of how robust the Turing machine model really is, here again, although one knows that **PSPACE** contains all problems in **NP**, it is not known whether some clever scheme will allow one to perform in **P**-time the computing necessary to do a problem that is known to be solvable in polynomial space.

1.3 Cellular Machines

Finally we come to the current scenario in computing. Parallel computation seems to be the only way to render deterministically tractable problems that are intractable by Turing complexity standards. We can distinguish three main streams in parallel computing. The first one, *distributed computing*, follows the classical theory in that no radical attempts are made to change the machines being used but rather they are to be interconnected together in a network. Different *relatively powerful* machines solve different parts of the input problem independently and simultaneously, and then the various partial solutions are somehow put together to produce the solution to the original problem. The second stream is motivated by the idea that there are already very powerful "computers" around, namely biological brains, and that efficient computation can be accomplished by imitating their most relevant functions. The models dealt with here are the so-called *connectionist models* or (artificial) *neural networks*. The atomic "neurons" of these machines are supposed to be very simple cells, by themselves lacking any significant computational ability. They are only capable of performing primitive operations similar to those of neurons in the human brain (basically firing excitations to neighbor cells only if their level of excitation rises over a certain threshold level, peculiar to the cell). The third stream is the cellular automaton model. A *cellular automaton* is, in a way, a neural network except that the atomic cells have the power of an arbitrary finite-state machine but are, on the other hand, restricted to be identical to each other. In the last two streams, the computing power of the network arises from the pattern of interconnections and the highly degree of cooperation exhibited by the individual cells. These last two models constitute examples of so-called *fine-grained* models of parallel computation, as opposed to the *coarse-grained*, distributed models of the first stream.

Parallel computing is still in its infancy and only a thorough investigation of their theoretical possibilities and limitations will allow us to decide which model is best suited to particular tasks. Despite their recent development, the literature on cellular automata and neural networks is already enormous (international conferences on neural networks began taking place in 1987 and have alone produced proceedings exceeding 40,000 pages, many of them applications). As a result, a survey of the subject is of necessity highly selective. This survey focuses on the important results on the same fundamental questions posed by the Turing model in regard to cellular automata and discrete neural networks. Only principled results with a rigorous and general flavor are included here. In particular, what many people consider the most important aspect of the models, namely their applications in the solution of a number of specific problems, has been left untouched, although some idea of fundamental techniques has been presented in Chapters 10, 11, and 12. The interested reader can initially consult the references in Chapters 2, 10, and 11 for a variety of applications and the references in Chapters 7, 8, and 9 for other aspects in the theory of cellular automata and neural networks.

1.4 Prerequisites

We would like to say that the only prerequisite to read these notes is some basic mathematical culture and the ever-elusive 'mathematical maturity'. But since this is a very old trick I will have to be more specific. If a reader wants to be able to just follow the arguments, a good course in discrete mathematics and calculus will do. S/he can get to the point of some creative thinking with elementary courses in discrete mathematics and analysis (maybe as far as elementary topology) at the undergraduate level. Full understanding of some sections and some problems will appear a lot easier armed with knowledge of certain topics usually covered in graduate courses, although no such specific knowledge has been assumed. Nevertheless, the book is intended to be a rigorous development of the material, although the underlying ideas are simple and explained often enough that they can be intuitively grasped easily. Group discussions should help clear them up ("You only understand something when you can explain it to the first wo/man on the street"). This is true even of the most demanding Chapters 7, 9, and 11 on the analytical and topological foundations of cellular automata and discrete neural networks. The flow of chapter logical interdependencies is shown in the diagram on page X.

References

[B-D-G] J.L Balcázar, J. Díaz, J. Gabarró: Structural complexity. Springer-Verlag, Berlin, vols. **I** (1988, 2nd. ed. 1995), **II** (1990)

[B-C] D.P. Bovet, P. Crescenzi: Introduction to the theory of complexity. Prentice-Hall, Hertfordshire, 1994

[C] S.A. Cook: The complexity of theorem proving procedures. In: Proc. 3rd Annual Symp. on the Theory of Computing STOC, Assoc. Comput. Mach. New York, 1971, pp. 151-158

[G-J] M. Garey and D.S. Johnson: Computers and intractability: a guide to the theory of NP-completeness. W.H. Freeman, San Francisco, CA, 1979

[H] D. Harel: Algorithmics (The spirit of computing). Addison-Wesley, Reading MA, 1992.

[L] F.T. Leighton: Introduction to parallel algorithms and architectures: arrays, trees, hypercubes. Morgan Kaufmann, San Mateo CA, 1992

[K] R.M. Karp: Reducibility among combinatorial problems. In: Complexity of computer computations. Plenum Press, New York 1972, pp. 86-103

[P] C. Papadimitriou: Computational complexity. Addison-Wesley, Reading MA, 1994

[R] M. Rabin: Probabilistic algorithms. In: Algorithms in complexity, New directions and results. J. Traub (ed.), Academic Press, New York, 1976, pp. 151-158

[S] W. Savitch: Relationships between nondeterministic and deterministic tape complexities. J. Comput. Syst. Sci. 4 (1970), 177-192

[Sm] J. Smith: The design and analysis of parallel algorithms. Oxford University Press, Oxford, 1993

[S-S] R. Solovay and V. Strassen: A fast Monte-Carlo test for primality. SIAM J. of Computing **6** (1977) 84-85. Erratum *ibid* **7** (1978), 118

[T] A.M. Turing: On computable numbers, with an application to the Entscheidungsproblem. Proc. London Math. Soc. **42** (1936) 230–265. A correction, *ibid* **43** (1936) 544–546

2. Cellular Automata

> *It always bothers me that according to the laws [of physics] as we*
> *understand them today, it takes a computing machine an infinite*
> *number of logical operations to figure out what goes on in no mat-*
> *ter how tiny a region of space and no matter how tiny a region of*
> *time. How can all that be going on in that tiny space? Why should*
> *it take an infinite amount of logic to figure out what one tiny piece*
> *of space-time is going to do?*
>
> Richard Feynman

In this chapter we define the fundamental concepts and establish our notation. One of the problems in surveying the literature of cellular automata is the great variety of notation and terminology that has been developed over the years by various authors to describe similar concepts and results. As a consequence, many results have been (and continue to be) rediscovered a number of times in a number of guises. Although the definitions, concepts, and notations given in this chapter may not be, in few instances, the most common current use, they at least provide a consistent framework to deal with general types of cellular automata. Moreover, this notational framework can be easily extended to include neural and other types of networks.

2.1 Finite-State Automata

An *alphabet* is a finite nonempty set Σ containing a special blank symbol '_'. A *word* (or *string*) on Σ is a finite (possibly empty) sequence of elements of Σ. The set of all words over an alphabet Σ is denoted Σ^*. There is a natural binary operation of *concatenation* on Σ^* that juxtaposes two strings x, y together into a string denoted xy and called the (con)catenation of x and y. Catenation is associative and has a neutral element, namely the empty word, denoted λ (only) in this chapter, and thus makes Σ^* an algebraic *monoid*.

A (formal) *language* L is any set of words (possibly empty), that is, a subset of Σ^*. Union and catenation induce three basic operations on the set of all languages. Given languages $K, L \subseteq \Sigma^*$,

1. the *sum* of K and L is the ordinary set-theoretic union

$$K + L := \{x : x \in K \text{ or } x \in L\}.$$

2. the *catenation* of K and L is given by

$$KL := \{xy : x \in K \text{ and } y \in L\}.$$

Catenation of languages remains associative and it distributes over union. The sets LL, $LLL, \ldots$ are denoted, respectively, L^2, $L^3, \ldots$.

3. the *Kleene closure* of L is given by

$$L^* := \{\lambda\} + L \cup L^2 + L^3 + \ldots = \bigcup_{n \geq 0} L^n .$$

The finite-state automaton, or finite-state machine, (for short, fsm) has been introduced in Chapter 1 as a model of sequential computers. We rehearse it briefly for parallel computation. A fsm M is a 5-tuple

$$M := \langle Q, \Sigma, \delta, q_0, F \rangle$$

consisting of an alphabet Q of 'memory configurations' with special symbol q_0, an alphabet Σ of input symbols, a transition function

$$\delta : Q \times \Sigma \to Q ,$$

(which will be called the *dynamics* of M for reasons made clear below), and, finally, a distinguished subset $F \subseteq Q$ of *final* states. A finite-state machine is intended to be a mathematical model of a primitive (sequential) computer that can read but not write on or change its input. An input string $x \equiv x_1 \ldots x_n$ is given to M on a linear tape divided into squares, one symbol per square. The fsm M also possesses a *finite control* or reading head that allows it to sense symbols on the input tape, one at a time, and which, at any instant of (discrete) time, is always found in one of the states of Q (i.e., M is 'thinking' of that particular 'memory configuration'), say p.

Remark. Some authors require that a fsm give an output in order to be a *machine*. One then has finite automata and Mealey and Moore machines. The distinction is immaterial for our purposes.

An fsm M is started in its initial state q_0 at the leftmost symbol x_1 of its input. Thereafter, M computes by performing a series of atomic moves on x until it has exhausted the input string past its rightmost symbol. An atomic move consists of sensing the current input symbol a under the finite control, shifting the state of the finite control to state $\delta(p, a)$ and moving on to the next symbol in x (or the end of x).

Naturally, the dynamics δ extends by induction on the length $|x|$ to an action

$$\begin{aligned} \delta : Q \times \Sigma^* &\to Q \\ (p, xa) &\mapsto \delta(\delta(p, x), a) \end{aligned}$$

of the (discrete) semigroup Σ^* on the state set Q. Thus, M can be regarded as a discrete dynamical system acting on the state set of the automaton.

An fsm can be used to *discriminate* inputs by their action on the state set. Specifically, if the action of x (i.e., the sequence of states of q_0 in Q) drives M into a final state of F at the end of x, M is said to *accept* the input x. The set of all words accepted by M is called the *language* of M, and it is denoted $L(M)$. Languages accepted by fsm's are called *regular* languages. Examples of the kind of pattern recognition that fsm's can (or cannot) perform as well as graphical representations by labeled diagrams have been given in Chapter 1.

The basic problem in the study of fsm's is the following: given a *local rule* that defines atomic moves of a fsm M (namely, the dynamics δ), characterize the *global* behavior of M (namely, identify the set L of words accepted by M). Conversely, given a (global) property that defines a formal language $L \subseteq \Sigma^*$ (e.g., binary strings divisible by 3), decide if there exists (and if so, find) a fsm M such that $L = L(M)$. This automaton M provides a very simple *algorithm* (namely, a sequence of local decisions embodied in δ) to solve the *recognition problem* of L.

Just what kinds of languages are (or are not) regular is therefore a natural and interesting question. A necessary condition for regularity involves the notion of *pumping words*. A language L is said to *pump* a word x in L if x can be expressed as the catenation of three subwords (blocks of consecutive symbols) u, v, w such that if $x \equiv uvw \in L$ then $uv^n w \in L$ for every $n \geq 0$. In this case L is also said to pump the subword v.

Lemma 2.1 (Pumping Lemma) *If L is regular language then L pumps all sufficiently long words in L.* □

This lemma is enough to show that a number of languages (such as the set of primes in binary) is not regular, that is, they do not admit an easy algorithm for their recognition problem. (The reader is referred to a textbook on sequential computation for a proof, for example, the text by Hopcroft–Ullman [H-U].) However, this condition is only necessary.

Theorem 2.2 [S. Kleene] *A set $L \subseteq \Sigma^*$ is regular if and only if it can be constructed from the one-element languages $\{a\}$ ($a \in \Sigma$) by a finite number of applications of the operations of union, catenation, and Kleene closure.*

For example, $(00 + 11)^*$ consists of all strings without an isolated 0 or an isolated 1. Thus regular languages admit a 'formula' (called a *regular expression*) of *finite* type that describes their elements exactly. Kleene's construction to pass from an fsm to regular expression and vice versa is algorithmic and can found in an introductory text in computability or automata theory. Finite automata and regular languages have very many applications (which, as stated above, are not our concern here).

In regard to cellular machines, one is only interested in a special kind of fsm where the input alphabet Σ is of the form $\Sigma := Q^d$ for some $d \geq 0$, as we will see in Sect. 2.2. A precise definition requires us to introduce first the underlying hardware of the automaton.

2.2 Regular Graphs

A *digraph* is a pair $\Gamma = (V, \mathcal{A})$ consisting of a (possibly countably infinite) set of vertices V and a subset of *arcs* $\mathcal{A} \subseteq V \times V$ of the set of ordered pairs of elements of V. If $\mathcal{A}$ is a symmetric relation, that is, if $\mathcal{A}$ contains the pair (j, i) whenever $(i, j) \in \mathcal{A}$, Γ is called a *graph* and its arcs are then called *edges*. In this case $\mathcal{A}$ is just a subset $E \subseteq V^{[2]}$ of 2-element subsets of V. Two digraphs $\Gamma = (V, \mathcal{A})$ and $\Gamma' = (V', \mathcal{A}')$ are *isomorphic* if it is possible to establish a 1-1 correspondence between their vertex sets $f : V \to V'$ that preserves adjacencies, that is, $(i, j) \in \mathcal{A} \Leftrightarrow (f(i), f(j)) \in \mathcal{A}'$. We will not distinguish between isomorphic (di)graphs. If $(i, j) \in \mathcal{A}$ we say that vertex i is a *nearest neighbor* of vertex j. The neighborhood of a given vertex i is the set $\Gamma(i)$ (sometimes denoted $\Gamma^+(i)$) of all vertices j adjacent to i. In general, this set depends on the vertex i. The cardinality $|\Gamma(i)|$ of the neighborhood of a (di)graph is called the (outer) *degree* of i. A *path* (or *walk*) in a digraph is a sequence of arcs $(i_0, i_1), (i_1, i_2), \ldots, (i_{n-1}, i_n)$. The path *connects* two of its vertices i_l, i_m if $0 \le l < m \le n$. A graph is *connected* if any two of its vertices can be connected by a path.

A (di)graph is (respectively, outer-) *regular* if all vertices have the same (outer-) degree. We will consider finite and infinite regular digraphs which are, moreover, edge colored. The colors come from a finite set X which is supposed to be symmetrized, in the sense that each color A in X has an *inverse* A^{-1} so that $(A^{-1})^{-1} = A$. A regular edge-colored (di)graph Γ has a *consistent* (or *strongly regular*) (edge-) coloring if there exists a finite edge-colored digraph X, called the *fundamental neighborhood* of Γ, which has a distinguished origin λ so that the neighborhood $\Gamma(i)$ of each vertex $i \in \Gamma$ is isomorphic to X by an isomorphism f_i which (a) maps i to λ; (b) preserves the edge colors; and (c) the color of the arc (i, j) in Γ under f_i is exactly the inverse of the color of the arc (j, i) under f_j. A regular coloring on a graph, in fact, induces an orientation of the graph Γ and makes it again a digraph.

A (regular) graph with a consistent coloring of this type is *homogeneous* in the sense that there is an edge-color-preserving automorphism of the graph that maps any vertex to any other vertex. More importantly, it admits a *coordinate system* that provides a one-one correspondence between the vertex set and the elements of a group (in the algebraic sense) as coordinates.

Theorem 2.3 *A connected graph has a strongly regular coloring if and only if it is the Cayley graph of a group.*   □

This means that the underlying graph of a cellular automaton can be described in a more geometric and algebraic way in the language of group theory as

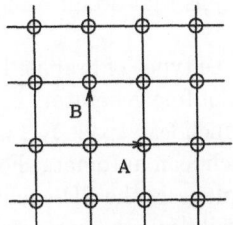 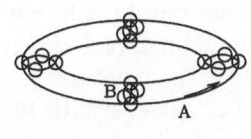

(a) 2D-euclidean tesselation and torus

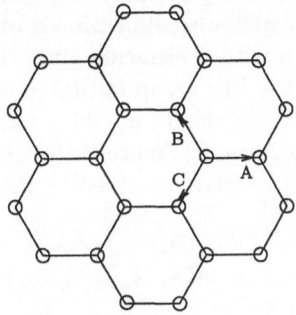

(b) Triangular tesselation

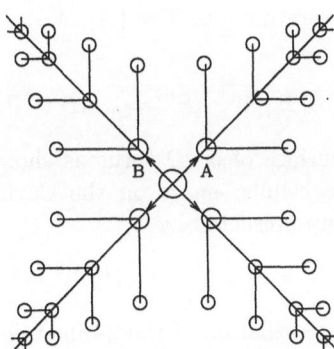

(c) Tree tesselation

Fig. 2.1. Common types of Cayley graphs

follows. Let G be a (countable) group and let X be a set of generators of G (none of which will be assumed trivial). A generator A (or its inverse A^{-1}) from X, connects vertices i and iA (or iA^{-1}, respectively), the product being taken as elements of G. Obviously, every Cayley graph is regular and has a regular coloring. Theorem 2.3 says that these are the only graphs that admit a regular coloring.

We will be concerned chiefly with automata on three types of graphs. First, the free abelian group of rank d with the standard set of free generators is (isomorphic to) the lattice of integer coordinate points of euclidean space $\mathbf{R}^d$, which will be denoted $\mathbf{Z}^d$. They will be referred to as dD-euclidean automata. For instance, the grid of integer-coordinate points in the plane is the 2D-euclidean grid. This is the most common type of digraph for cellular automata. A similar type of Cayley graph is the Cayley graph of finite commutative n-groups of rank d with the standard generating set illustrated in Fig. 2.1(a), which in the boolean case is a finite hypercube of dimension d on 2^d vertices, denoted $\mathbf{B}^d$. For example, a cyclic group with n elements gives rise to a ring of nodes. Secondly, the Cayley graph of the free group of rank d with the standard free set of generators is a uniform tree of degree d, which will be denoted $\mathbf{F}_d$. This is a less common type of Cayley graph. Figure 2.1(c) shows a portion of this graph. These groups can be respectively described by the presentations

$$\mathbf{B}^d := \langle A_1, \ldots, A_d \,|\, \forall m, n(A_m^2 = 1, A_m A_n = A_n A_m) \rangle$$
$$\mathbf{Z}^d := \langle A_1, \ldots, A_d \,|\, \forall m, n(A_m A_n = A_n A_m) \rangle$$
$$\mathbf{F}_d := \langle A_1, \ldots, A_d \,| \qquad\qquad \rangle .$$

Many other common tessellations can be obtained from Cayley graphs of other groups with suitable generating sets $X = \{A_1, A_2, \ldots\}$. The Cayley graph of the group presented by

$$\mathbf{Z}_n \times \mathbf{Z}_m = \langle A, B \,|\, A^n = B^m = 1, AB = BA \rangle,$$

is a discrete lattice on the surface of a 2D torus as shown in Fig. 2.1(a). In general, *torus* will refer to a cellular space on the Cayley graph of a finite commutative group. The group presented by

$$\langle A, B, C \,|\, A^2 = B^2 = C^2 = (ABC)^2 = 1 \rangle,$$

correponds to the triangular tessellation of the 2-dimensional euclidean plane, as illustrated in Fig. 2.1(b). Plenty of other examples can be found in the classic book by Coxeter and Moser [C-M].

We are now ready for precise definitions concerning cellular automata.

2.3 Local Rules and Global Maps

A cellular automaton is a model of a parallel computer. Roughly, the idea is to interconnect a number of simple processors into a network in a *uniform* fashion and let the input to a given processor come from other processors to which it has been interconnected. Initially, each processor is given an initial amount of data. At successive (discrete) time steps, each processor looks up data in its neighbors and decides according to its program how to modify its own data. All processors are supposed to operate synchronously, and in parallel. They are also supposed to be as simple (for instance, usually binary states are allowed) and uniform as possible (all are copies of a single processor, i.e., use the same program). The interconnection pattern of the network is to be as *homogeneous* as possible and the fsm's are to have two-way communication with each other as well. Common networks include the euclidean lattices of integer points. The input is encoded to the machine as a pattern of states initially assigned to the machines and the output is similarly retrieved after a certain number of cycles or steps of the network. Thus, as just defined, this is a closed, autonomous model that does not exchange information with its environment in the course of its computation (in contrast with *systolic arrays*). These models have been considered under many names, including *uniform arrays*, *tessellation structures*, *homogeneous structures*, *mosaic automata*, ⋯.

Cellular automata have held a lasting appeal through decades for a number of reasons. Notable among them is the simplicity of their *finitary* specification. The richness of their behavior without informationally exorbitant hypotheses (e.g., that a finite amount of space must hold an infinite amount of information) seems to fit more tightly with our intuition about physical space. Their homogeneity bears promise of a simple hardware implementation. The resulting models of physical phenomena they provide shed a ray of hope for 'programmable matter' that might not even call for construction, but rather could be simply picked in the wild in nature, if we only knew where and what for to look. And last, but not least, the assumptions of simple components and symmetric interconnections hide hopes that the arising computing systems may prove susceptible to some type of tractable analysis that will shed light on the capabilities, nature and scope of fine-grained parallel computation.

2.3.1 Cellular Spaces

A general model of the type described above requires that the abstract graph indicating the sites where the processors are located and their interconnections be a Cayley graph (see Sect. 2.2). Since different networks of processors can all have the same interconnection network, it will be convenient to distinguish between a cellular automaton and the underlying space on which it is defined. Thus we will call the pair (Γ, Q) consisting of a Cayley graph Γ (including its fundamental neighborhood X) together with a set of states Q a *cellular space* containing a *quiescent state* denoted 0. The *immediate neighbors* of a given cell i (the *center cell*) will be denoted $i + j$ ($j \in X$), despite the fact that the corresponding group G may not be commutative (although in many cases it will be). Consistently, the identity of the group G will be referred to as 0.

Remark. The origin of the cellular space and the quiescent state are both denoted 0. There will be few occasions for confusion, and the context will quickly resolve the ambiguity.

In the following chapters we frequently encounter yet another finite set of vertices N that is usually (but may not be) equal to X. In that case, given a center cell i of Γ, the vertices $i + j$ ($j \in N$) are its *nearest neighbors*, and the set $i + N := \{i + j : j \in N\}$ is the neighborhood of i. Thus, assigning a fixed numbering to the nodes in N will induce a numbering (consistent under translation) to the neighbors of every node. We will assume that this numbering has been assigned once and for all and so (slightly abusing notation) we may refer to the cells in N as $1, 2, \ldots, d$ instead of $i_1, i_2, \ldots, i_d$. The center cell will be numbered 0 as well. The notation may sound a bit ambiguous at first, but it is convenient and the context usually resolves the ambiguity at any rate. Often the center cell needs to be considered together with its neighbors, so throughout we will use $N_\bullet$ to denote the neigborhood N expanded to include the center cell.

Remark. The notation is set up so that $i_1, \ldots, i_n$ can be used for the components of i when the underlying cellular space is a cartesian product of some sort (for instance, in the frequent case of euclidean or hypercube universes).

Some neighborhoods are used very often and have proper names. The *full neighborhood* X consisting of the $2d$ cells directly linked to the origin is the *Von Neumann neighborhood*. In commutative spaces with d generators, the *Moore neighborhood* consists of all the $3^d - 1$ cells closest to the origin in the ordinary visual sense. It can be shown that, at least as far as overall possibilities, all neighborhoods containing the full neighborhood X are equivalent (see Problem 13).

In order to describe the action of cellular automata, we will need some more notation. A *configuration* (or *total state* or *instantaneous description*) of the cellular space is a mapping

$$x : \Gamma \to Q$$

that assigns to each $i \in \Gamma$ a state x_i, the current state of site i. The symbol **C** denotes the set of all configurations, never the set of complex numbers (which will not be needed herein). The *support* of x, denoted $\underline{x}$, is the set of vertices of Γ given by

$$\underline{x} := \{i \in \Gamma : x_i \neq 0\}.$$

A configuration is *finite* if its support is a finite set. The all-quiescent configuration O is given by $O_i := 0$. The set of all finite configurations will be denoted $\mathbf{C}_0$. When displaying a configuration, the position of the state of the origin of the cellular space, if it is important, will be indicated by an underline or a subindex zero as in $\cdots x_{-1} \underline{x}_0 x_1 \cdots$. We will also use the term *array* in a technical sense, to describe the restriction of a configuration to a finite *connected* subset of its domain (generally the support of a finite configuration).

Cellular spaces are the 'physical spaces' where cellular automata live. Now, we need a description of the 'laws of motion' that govern how they run in cellular space.

2.3.2 Local Rules

Cellular automata are defined by local uniform processes in a cellular space. More precisely, they are conceived to operate as follows. A copy of a common fsm M (called the *computing* or *processing element*) occupies each vertex of Γ, which is then called a *cell* –or also a *site*. A cell in the quiescent state is said to be *unexcited* or *inactive*. *Synchronously*, each copy of M looks up its input in the states $x_1, \ldots, x_d$ of its neighboring cells $i + j$ $(j \in N)$ and its own state x_0, and then changes its state according to its local dynamics δ in (2.1). The cellular automaton thus performs its calculation by repeating this atomic move a (finite) number of times.

For future reference, we record these definitions precisely.

Definition 2.4 *A cellular space is a pair (Γ, Q) consisting of a regular graph Γ and a finite-state set Q containing a special state 0 (the quiescent state). A cellular automaton is a triple $\langle \Gamma, N, M \rangle$ consisting of a cellular space (Γ, Q), a finite set N of vertices of Γ and a finite-state machine M with input alphabet $Q^d := Q \times \ldots \times Q$ (d times) and local transition function*

$$\delta : \ Q \times Q^d \ \rightarrow \ Q \qquad\qquad (2.1)$$
$$(x_i, x_{i+i_1}, \ldots, x_{i+i_d}) \ \mapsto \ \delta(x_i x_{i+i_1} \ldots x_{i+i_d})$$

where $d := |N|$ is the number of cells in the neighborhood N.

The local dynamics δ can be actually given in several equivalent ways. Most common is a *transition table* analogous to a truth-table, with rows describing the state of the nighborhood and a next-state column indicating the next-state of the sites. An alternative way is identify a rule by a number defined as follows. For a fixed cellular space and neighborhood ordering (usually lexicographic, interpreting the neighborhoods as the m-ary expressions of integers $0, 1, 2, \ldots$), one can interpret the next-states as the digits of a number in m-ary, which is referred to as the *Wolfram number* of the rule. A third description as *block rules*, or *partitioned cellular automata* is fairly rare. It requires partitioning the cellular space into at least two (but finitely many) sets of blocks, each of finitely many types. The rule is actually specified by listing the way each block is updated. Updates with only one partition block quickly become stable or periodic, so it is necessary to provide a schedule (like even/odd times) to switch between partitions in order to allow information to spread locally from block to block –see Problem 14. For local rules on the 1D euclidean line, a fourth description is possible as the *de Bruijn digraph*. The vertices of this digraph are all m-ary strings of length $2r$, where r is the distance of the farthest cell in a

N	111	110	101	100	011	010	001	000
$\delta(N)$	1	0	0	1	0	1	1	0

(a) Decimal $150 = 10010110_2$ (binary)

$$000, 011, 101, 110 \mapsto 0 \quad 001, 010, 100, 111 \mapsto 1$$

(b) Neighborhood lists for next-states 0 and 1

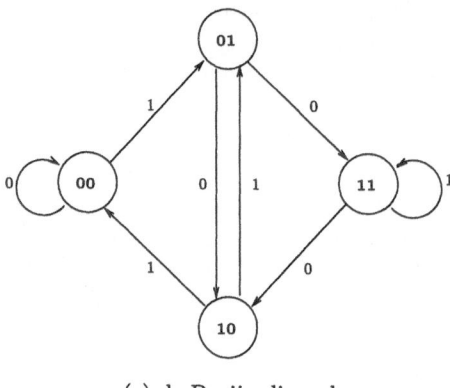

(c) de Bruijn digraph

Fig. 2.2. Three presentations of elementary rule 150

neighborhood from the center cell. There is an arc from node $p\omega$ to ωq, where p, q are cell states labeled $\delta(p\omega q)$. Figure 2.2 shows the various presentations for 1D rule 150.

An automaton in which $Q := \mathbf{B}$ (the states are binary boolean values) is called *elementary* as with cellular spaces. Sometimes, especially on the 1D euclidean cellular space, there is also the requirement that N be the Moore neighborhood. (When no state set or neighborhhod is specified, binary states and nearest-neighbor rules are assumed.)

Local rules can also be regarded as plain words over an alphabet the size of the state set, just by interpreting the Wolfram number as a word on the state alphabet. The digits correspond to next states, arranged according to the given ordering of the neighborhoods. One can measure the difference between two local rules by defining a *Hamming distance* between them. This distance can be defined in at least two ways. Either count the number of next-states where they differ for identical neighborhoods,

$$|\delta, \rho| := \sum_{\delta_l \neq \rho_l} 1\,,$$

or give these differences some weight, say the distance $\| *, * \|$ between them modulo m (on a circle),

$$|\delta, \rho| := \sum_{\delta_l \neq \rho_l} \|\delta_l - \rho_l\|\,.$$

It is very easy to verify that these formulas indeed define metrics in local rule space. (See Problem 12.)

2.3.3 Global Maps and Dynamical Systems

A cellular automaton can be viewed as a dynamical system of discrete type. Similarly to the situation with fsm's (but more generally), the action of the local function of a cellular automaton on a given configuration transforms, in one iteration, every one of its pixels, into a new set of pixels, i.e., a new configuration. Thus the local dynamics induces a *global dynamics*

$$T: \mathbf{C} \;\rightarrow\; \mathbf{C} \tag{2.2}$$
$$T(x)_i \;:=\; \delta(x_i, x_{i+N}) = \delta(x_i x_{i_1} \ldots x_{i_d})\,,$$

where $x_{i_1}, \ldots, x_{i_d}$ are the states of the cells in the neighborhood of i, and x_i, the state of i. Moreover, the restriction of T to $\mathbf{C_0}$ maps into $\mathbf{C_0}$. This restriction will be denoted T_0. Thus each can be iterated as a dynamical system (under composition) any number of times. Their t-th iterations will be denoted T^t and T_0^t, respectively ($t \geq 0$).

The global dynamics of a cellular automaton can be visualized in the so-called *phase space* of T. It consists of a digraph where configurations x have shrunk to simple vertices connected by arcs $x \rightsquigarrow T(x)$. On a finite cellular space with n cells, the phase space has 2^n cells, but it is uncountable otherwise (i.e., it can be put in one-one correspondence with the set of all subsets of natural numbers). Because T is a function, each node (configuration) in phase space has outdegree 1.

As a dynamical system, the most basic question about a global map T is perhaps the effect of its repeated application in phase space to a random given configuration x. The theory of dynamical systems has standard concepts to trace their behavior. The *orbit(space–time)* of x is the sequence of configurations

$$\{T^t(x)\}_{t \geq 0} := x, T(x), T^2(x), T^3(x), \cdots$$

resulting from successive iteration of the global rule on x. In phase space, orbits can be identified with walks from an initial node x through the successor configurations $T(x), T^2(x), T^3(x), \ldots$. As the automaton is iterated over two configurations x, y, their orbits may merge whenever $T^t(x) = T^{t'}(y)$, or they may remain disjoint forever.

A configuration is *periodic* (or more precisely, *temporally periodic* or just *t-periodic*), if its orbit eventually returns to a previous configurations (i.e., the orbit only contains a finite number of distinct points). In this case, the orbit is split into a *transient part* and a *cyclic* part, the number of configuration in which is called the *period* of x. This is, of course, the number of iterations of T needed to repeat any configuration in the cycle. A configuration may, in general, remain eternally in its transient phase and be *aperiodic*. In phase space, periods of periodic configurations appear as cycles and transients appear as paths leading to cycles. Any other configuration y eventually merging with that of a given x of necessity enters the same cycle. They all make up a connected component of phase space, which, in fact, looks like a directed tree with arcs pointing toward a root that is actually a cycle (perhaps, a fixed point). The entire phase space is thus a *forest of unicyclic graphs*.

From a biologically oriented point of view, local rules can be regarded as genotypes (genetic codes) for evolution of individuals from certain initial conditions. The evolution gives rise to a phenotype, an individual continuously changing under the influence of the genetic code and the initial conditions.

With basic definitions in place, we finish this chapter with a sketch of the fundamental problems posed by cellular automata.

2.4 Fundamental Questions

As with fsm's, there are two fundamental problems in the study of cellular automata. The **forward problem** is the following: given a *local rule* that determines the local interaction of each cell with its neighbors (namely the dynamics δ), characterize the *global* effect of this rule on an arbitrary initial configuration x. This means (*a*) determine a very specific description of (a certain aspect of) the orbits of arbitrary configurations under T as defined in equation (2.2); and/or (*b*) identify their long-term (asymptotic) behavior. From an engineering point of view, it is the **inverse problem** that is of paramount importance. Given a (global) effect on configurations of the cellular space that defines a desirable transformation T of configurations, determine whether there exists (and if so, find) a local rule δ as in (2.1) whose induced global rule is precisely T. This local rule δ provides a very simple *parallel algorithm* (namely, a sequence of local decisions δ) for all the parallel processors thus connected to solve the problem of computing $T(x)$ from any configuration x in $\mathbf{C_0}$, or even perhaps $\mathbf{C}$.

Somehow, this problem is a discrete version of a classical problem in dynamical systems. A local rule δ is, in fact, some sort of *discrete* differential equation that determines the local interaction of a particle, say, with its few immediate neighbors (in contrast with uncountable many others that are infinitesimally close in real euclidean space). The initial configuration plays the role of the initial conditions. Solving the equation usually means determining the global transformation induced by the given equation. Except here, the cellular space cannot contain infinitely many neighbors and none of them can encode an infinite amount of information (as does, for instance, a real number). For further discussion of this analogy the reader can consult S. Wolfram's preface to the proceedings in [F-T-W]. The reader familiar with sequential computation cannot help noticing that this is the parallel version of the old synthesis problem for finite-state machines. But this time, regular languages, which prove to be an adequate tool to handle the synthesis of finite words, fall far short of being sufficient for the infinitary complexity of configurations and their tranformations by cellular automata, even if generalized to infinite strings and sequential machines by 'limiting processes' (see Problem 11).

The analogy only extends along the lines of dynamical systems. For arbitrary cellular spaces are, in a precise sense, identical to the well-known Cantor set (just as the notation indicates) and a global dynamics is just an iteration of this transformation. Thus, cellular automata transformations can be regarded as dynamical systems over a 'familiar' object, namely the Cantor set, only that 'smoothness' no longer means differentiability but just continuity. These topological aspects of cellular automata (maybe cryptic for the moment) will be a leading theme in some later chapters.

The kind of global or asymptotic behavior of T that may be of interest as either forward or inverse problems is very diverse. From the point of view of applications, the most interesting aspect is the type of information processing that cellular automata are capable of. A rigorous study of information processing capabilities requires a suitable formulation of the notion of 'information' or, at least, 'information representation'. These remain basic issues as cellular automata are generalized to neural networks, and they will be dealt in the following chapters. In general, problems studied so far include, among others:

(a) Whether T 'preserves information' contained in the initial conditions. This usually takes the form of whether it is *one-one, onto, and/or bijective*; but it can also be asked at a more general and sophisticated level, using concepts borrowed from measure and ergodic theory (traditionally used for systems on the continuum), such as 'entropy', 'Lyapunov exponents', and 'randomness'. The general problem is to understand and quantify the capabilities of cellular automata as mechanisms that create, store, transform, and/or destroy information.

(b) Whether T preserves some metric or *topological properties* of configuration space (for example, continuity). This question can be posed, more generally, as the search for invariants (or conservation laws) of cellular automaton evolution. It brings up the question of whether one can find a correspondence between cellular automata and (an appropriate family of) continuous systems, particularly as the number of sites and states approaches infinity.

(c) The behavior of T on some particular class of configurations (for example, finite, spatially periodic, recursive, random), particularly amenable to mathematical analysis.

(d) Whether T eventually develops some 'self-organization' in arbitrary configurations (existence and type of limiting configurations). The type of organization can be described by the structure of phase space, for instance, how fast the automaton contracts the space of all configurations (i.e., how fast they are excluded) as the system is iterated. From the opposite end, one may ask how random are the configurations generated by a given T.

(e) The existence of numerical functions (for example, the number of states in regular languages describing successive generations in the evolution of T; or energy functions assuring convergence to a stable state; or the average number of nonquiescent states at time t generated from a random initial configuration) that help determine the nature of the asymptotic behavior of arbitrary initial configurations. In physical terms, one may ask for the 'thermodynamical' behavior of cellular automata, i.e., the average overall behavior as a system with many components.

(f) Decision problems: the (non)existence of algorithms to decide whether or not an arbitrary given local rule induces a global rule with given properties (e.g., one-one, onto), etc. Computations performed by cellular automata seem to be, for the most part, computationally irreducible in the sense that their results cannot be obtained faster or at all by sequential devices (e.g., Turing machines).

(g) Classification problems: is it possible to classify all rules on a given cellular space (say 1D euclidean) into natural families according to some criterion (say the ultimate uniform fate of most configurations)?

(h) Decomposition theory: can we generate *all* rules from a few primitive rules by a number of 'simple' operations (as in Kleene's theorem)?

(i) Observability and implementation problems: how can an ideal system such as a cellular automaton with an infinite number of components be implemented or observed on actual hardware (including nature) with all the limitations of a physical construction (concomitant noise, faults, imperfections and/or measurement errors)?

(j) Computational problems: how can we design cellular automata for specific tasks. Well known examples are the firing squad synchronization problem, language recognition in parallel, and, more recently, numeric parallel computation.

However, the most important problems, for example (g)–(j), remain open. The richness of the behavior and interpretation of the variety of rules makes such a description very difficult.

Research on cellular automata can be broadly classified into two categories, *experimental* and *analytic*. Much of the research on cellular automata and neural networks in the last two decades has been of an experimental type. A number of software packages and devices have been written to simulate cellular automata. Particularly worth mentioning are Rudy Rucker's CALAB (the simplest and most accessible software running on IBM-compatible or Macintosh machines) [R], the CAM-6 and CAM-8 printed circuit boards (for IBM compatibles with color graphics) [Ca], and CellSim, a package for workstations developed by C. Langton and D. Hiebeler at Los Alamos. The architecture of the Connection machine CM-2 of Thinking Machines Co. is essentially that of a cellular automaton on a 2D torus [CM]. Even theoretical results support the prediction that experimental work will continue to be an important part of research in cellular automata.

However, the chief purpose of the following chapters is to survey some of the basic *analytical* results available on the solutions to these problems for both cellular automata and their generalizations, neural and random networks. No attempt has been made to make them comprehensive, although virtually every one of what is considered a major result has been included. Moreover, proofs of these results are only provided when they are relatively recent and the techniques are of general interest. The amount of recent work and literature in cellular automata has been exploding and collecting it into a comprehensive volume would be an exceedingly demanding, and perhaps premature, undertaking. In particular, experimental research on specific applications of the subject and heuristic aspects of it have been only implicitly considered. The reader may consult the bibliography listed below and in Chapter 12 for references to the many aspects of cellular automata not considered here.

2.5 Notation

The following notation (with or without subindices and/or superindices) will be used consistently throughout.

Set of vertices/nodes	V
Set of arcs	$\mathcal{A}$
Cardinality of a set S	$|S|$
A (di)graph	$G\,(D)$
State/activation sets	$Q, \mathbf{B}$ (boolean)
A finite ring (field)	$A\,(F)$
Length of a string ω	$|\omega|$
A machine (finite-state, Turing)	M, M'
The empty string, eigenvalues	λ
Integers	$l, m, n, p, q, r, s, \tau \ldots \in \mathbf{Z}$
Time	t
Regular (di)graphs/grids	$\Gamma, \mathbf{B}^d, \mathbf{Z}^d, \mathbf{F}_d$
Group generators	$A_1, A_2, \cdots$

Fundamental neighborhoods	X
Indeterminates	$X, X^{(i_1, i_2)}$
Cellular spaces	$\Gamma, \mathbf{Z}^d, \mathbf{B}^d, \mathbf{Z}_n^d, \mathbf{F}_d, \dots$
Sites, nodes in a cellular space	$i, j, k, (i_1, i_2), \cdots$
(Expanded) cell neighborhoods	$(N_\bullet), N, N_o$
Local rules	$\delta, \delta', \eta, \rho, \sigma, \dots$
Global dynamics	$T, T', \dots$
Iterates of a map $f, T, \dots$	$f^t, T^t, \dots$
The all-zero (-one) configuration	$O, \mathbf{0}, (\mathbf{1})$
Arrays, finite configurations	$a, b, c, \dots$
Configurations	$c, \dots, x, y, z, \dots$
Configuration space (Cantor set)	$\mathbf{C}, Q^V, \mathcal{C}, \dots$
Finite configuration space	$\mathbf{C}_0, Q_0^N, \dots$
Sets of real numbers	$\mathbf{R}, X_1, X_2, \dots$
d-Dimensional Euclidean space	$\mathbf{R}^d$
Set of integers (rationals)	$\mathbf{Z}(\mathbf{Q})$
Scalars, eigenvalues	$\lambda, \lambda_1, \dots$
Abstract spaces	$X, Y, \cdots$
Functions, maps	$f, g, h, \cdots$
Families of functions	$\mathcal{D}, \mathcal{E} \cdots$
Complexity classes	$\mathbf{P}, \mathbf{NP}, co\mathbf{NP}, \mathbf{PSPACE}, \mathbf{CA}, \mathbf{NN}, \dots$

2.6 Problems

Problems marked CaLab are best solved using the software package CALAB. Likewise for CAM. (See the historical notes after the problems). Problems marked *
may need to be looked up in the literature. [Brackets give hints.]

CAYLEY GRAPHS

1. Draw a portion of (until you identify a general pattern in the structure
   of) the Cayley graphs associated with the following groups:

$$
\begin{aligned}
\mathbf{B}^d &:= \langle A_1, \dots, A_d \,|\, \forall m, n(2A_m = 1, A_m + A_n = A_n + A_m) \rangle \\
\mathbf{Z}^d &:= \langle A_1, \dots, A_d \,|\, \forall m, n(A_m + A_n = A_n + A_m) \rangle \\
\mathbf{Z}_3 &:= \langle A \,|\, 3A \rangle \\
\mathbf{Z}_2 \times \mathbf{Z}_4 &:= \langle A, B \,|\, 2A, 4B, A + B = B + A \rangle \\
\mathbf{S}_3 &:= \langle A, B \,|\, 2A, 3B, A + B = B - A^{-1} \rangle \\
\mathbf{F}_d &:= \langle A_1, \dots, A_d \,| \qquad \rangle
\end{aligned}
$$

2. Draw the Cayley graph of the modular group presented by

$$
\langle A, B \,|\, 2A = 3B = 0 \rangle.
$$

(Recall that the operation in the group is not necessarily commutative.)

3. Draw the Cayley graph of the group presented by

$$\langle\, A, B, C \mid 2A = 2B = 2C = 0 \,\rangle.$$

4. Show that every graph can be regarded as a metric space with the distance between any two nodes given by the length of the shortest path connecting them. [The defining axioms of a metric space require that (a) $|i, i| \geq 0$ and it is 0 iff $i = j$; (b) the distance be *symmetric*: $|i, j| = |j, i|$; and (c) the *triangle inequality* hold: $\forall i, j, k$ $|i, k| \leq |i, j| + |j, k|$.]

5. Show that the grid in the euclidean cellular space $\mathbf{Z}^d$ has polynomial growth $O(n^d)$. [The *growth* of a graph is the sequence $\gamma_n := |B[0, n]|$ counting the number of nodes within distance n from a fixed origin 0.]

6. Show that the Cayley graphs $\mathbf{F}_d$ of free groups on d generators have exponential growth $O((2d - 1)^n)$.

REGULAR LANGUAGES

7. Describe the regular sets of strings accepted by the diagrams in Figs. 1.1 and 1.2, Chapter 1 (regular expressions preferred).

8. Show that regular languages are closed under complementation.

9. Show that regular languages are closed under union and intersection. [Use a cartesian product for the state set and select appropriate transitions and final states.]

10. Prove that the set of images $T_0(\mathbf{C}_0)$, i.e., the set of finite subwords of image configurations in $T(\mathbf{C})$ of a cellular automaton T, is a regular language. Likewise for a fixed number of iterates T_0^t of T_0.

11. Prove that one-way infinite subconfigurations in the image set of a cellular automaton form an ω-regular language. (A set of one-way infinite strings is ω-*regular* if there exists a fsm which enters a final state infinitely often upon reading initial finite segments iff the string is in the language.)

LOCAL RULES

12. Verify that Hamming distance in rule space is a metric. [See Problem 4.]

13. Show that every cellular automaton can be simulated by one with neighborhood of radius 1 on an arbitrary cellular space. [See Problem 4.]

14. Show that every partitioned cellular automaton is indeed a cellular automaton per Definition 2.4. [Attach a modular clock.]

15. Show that, up to symmetries, there are only 88 distinct rules in the set of 256 possible 1D elementary rules. (A symmetry of a rule $\delta(x_{i-1}x_i x_{i+1})$ is defined as complementation $\bar{\delta}(\bar{x}_{i-1}\bar{x}_i\bar{x}_{i+1})$ of values 0 and 1, a reflection $\delta(x_{i+1}x_i x_{i-1})$ of the cellular space about the origin, or a combination of the two.)

GLOBAL DYNAMICS

16. Determine the behavior of the global dynamics (per the problems defined in Sect. 2.3) of the elementary rule SUM_MOD2 on the graphs $\mathbf{Z}_n$. (CaLab).

17. Same as Problem 16 for $\mathbf{S}_3$. (CaLab).

18. Show that shifts are cellular automata. A *shift* S_k for a given location k is defined by $S_k(x) := x_{i-k}$.

19. Show that every cellular automaton commutes with arbitrary shifts, i.e., $T \circ S_k(x) = S_k \circ T(x)$, for every configuration x.

20. Prove that an elementary shift $\sigma := S_A$ is bijective. Prove that its inverse is also a cellular automaton. Likewise for any shift.

21. Give an example of a self-map of configuration space that commutes with shifts but is not induced by a local rule.

22. Show that the composition of two cellular automata is again a cellular automaton.

23. Show that if x is a spatially periodic configuration for a cellular automaton T, then $T(x)$ is also periodic (perhaps with a smaller period).

24. For 1D euclidean rules, show that a configuration can be folded into a biinfinite path of the de Bruijn digraph so that its image is obtained by reading off the labels of the arcs traversed.

MISCELLANEOUS

25. Find the SCCs of the de Bruijn presentation of rule 150. (The Strongly Connected Components SCCs of a digraph are the elements of the partition of the vertex set generated by the equivalence relation of being connected by a path.)

26. Show that de Bruijn graphs are Hamiltonian and strongly connected. [A digraph is *strongly connected* if it only contains one SCC. It is *Hamiltonian* if it is traversed by a cycle, i.e. a closed circular walk spanning all vertices.]

27. Find the SCCs of the cartesian product of the de Bruijn graph B_{150} and itself constructed so that two nodes $(s\omega, \omega u)$ and $(t\omega, \omega v)$ are adjacent iff $\delta(s\omega t) = \delta(u\omega v)$.

28. Show that the diagonal Δ of the square of the de Bruijn graph of a cellular automaton is isomorphic to the original de Bruijn graph of the rule.

29. Give an algorithm to eliminate all transient nodes in a digraph D. (A node is *transient* if it fails to be in a biinfinite walk.) [Observe that a node is nontransient iff it lies in an SCC containing at least one edge or it lies in a path from one SCC to another. Start by marking all nontrivial SCCs and successively mark their successors following a topological sorting of the SCCs. Delete unmarked nodes. Repeat with all the arcs in the digraph reversed. The survivors are the nontransient nodes.]

30. Show that regular languages are closed under homomorphism, inverse homomorphism and regular substitutions. (A *homomorphism* $h : \Sigma^* \to \Delta^*$ is a map that preserves catenation, i.e. $h(uv) = h(u)h(v)$, so it is entirely defined by its values at atomic symbols.) [Use regular expressions. See Hopcroft–Ullman [H-U, Sect. 3.2].]*

31. Show that an observer watching just the origin cell of a 1D cellular automaton will only see a context-sensitive language in the course of finite evolution of arbitrary initial configurations. (The evolution of a configuration x for t steps can be traced as a word $x_0 T(x)_0 \cdots T^{t-1}(x)_0$ over the state alphabet. The class **NSPACE**(n) consists of *context-sensitive* languages that can be recognized by a, perhaps nondeterministic, Turing machine using only a number of auxiliary cells bounded by a linear function of the length of the input word.) [The state T^{t-1} only depends on the array $x_{[-rt,rt]} := x_{-rt}x_{-rt+1} \cdots x_0 \cdots x_{-rt-1}x_{-rt}$, where r is the neighborhood radius. A Turing machine can guess one such candidate window in an input configuration and verify in linear space that its time evolution indeed leaves the given word as a trace at the origin.]

32. Same as Problem 31 for an arbitrary finite window of fixed finite width.

2.7 Notes

Cayley graphs have a long history. They were originally introduced by A. Cayley [C] over a century ago as a geometric tool in the description of finite groups, and later extended by M. Dehn to a topological object (the *Gruppenwild*) in order to include infinite groups. Theorem 2.3 was pointed out precisely long ago by Magnus–Karrass–Solitar [M-K-S]. Gentle introductions to the more group-theoretic aspects of Cayley graphs can be found in Grossman–Magnus [G-M], Coxeter–Moser [C-M], and/or Magnus–Karrass-Solitar [M-K-S], where the reader can find detailed presentations. In more recent years, the symmetry of Cayley graphs has become a fundamental property for interconnection design for distributed networks. Its potential to realize optimality of global communication times between processors, design cost, recursive scalability and ease of analysis has intrigued network designers for a long time. Many parallel machines are based on Cayley graph interconnection networks. The interested reader can consult any number of references on the subject for further details –see, for example, [C-N-R, L-J-D].

Cellular automata models were introduced by John von Neumann [VN] upon a suggestion of S. Ulam and have since been objects of fascination to many people. Early studies appeared in Codd [Co] and Burks [B]. They have been used since the early 1970s in parallel image processing on CLIP machines [P-D, D-F]. In the 1980s, Wolfram's studies on computation universality, classification and physical analogies [W2] struck the field with renewed interest, approaches, and a redefinition of fundamental problems [W1]. More recently, cellular automata have experienced an upsurge of experimental investigation with the emergence of powerful massively parallel computing devices (such as the connection machines [H]) made possible by VLSI technology. In particular, they have opened up entirely new possibilities in physics and engineering [W1], where they are being studied as 'imaginary' physical universes, an interaction beneficial both to physics and to computation. In this direction, two recent books are particularly useful: the Wolfram collection of reprints on physical, statistical and theoretical aspects of cellular automata [W2]; and the Califano–Margolus–Toffoli [T-M] tutorial for a printed circuit board that simulates up to 16-state 2D cellular automata on the cellular space $Z_{256} \times Z_{256}$. Proceedings of cellular automata conferences can be found in [F-T-W, D-G-T, JCS, Gut]. Many journals (e.g., *J. of Computer and System Sciences, SIAM J. Discrete Mathematics, Physica D, Int. J. of Theoretical Physics, J. Statistical Physics*) regularly publish on the subject. A new journal, *Complex Systems* [JCS], was originally conceived as a forum for publication of cellular automata like research and its applications. De Bruijn digraphs were first used by Wolfram to describe cellular automata in [W2]. They have been used since in several ways to solve questions about cellular automata, as described in later chapters. Recently, an electronic journal, *Complexity International: a hypermedia journal of complex systems research*, has become available on the internet and offers rapid dissemination of research papers in complex systems. The project *Tierra*, in fact, contains a subproject *CAM-brain* "aiming at growing/evolving an artificial brain, which contains thousands of interacting neural network modules, inside special hardware called Cellular Automata Machines." [Tie, DG].

Probabilistic and asynchronous models of cellular automata are also interesting. Probabilistic cellular automata appear naturally in the form of percolation models in physics [S-A]. (See Kesten [K] for mathematical aspects of percolation theory.) They may also play an important role in the design of cellular automata for discrete modeling at a very general level (see for instance, Rujàn [Ru], Gutowitz [Gut], Drescher [D]). Unfortunately, research in these areas is just awakening, and it will not be formally covered in this volume. However, the reader can find some pointers to a number of applications of cellular automata in the literature in the annotated bibliography to be found in Chapter 12.

References

[B]   A.E. Burks: Essays on cellular automata. U. of Illinois Press, Chicago, 1972

[C]   A. Cayley: On the theory of groups. American J. of Math. **50**(1878) 50–82

[Ca]   A. Califano, N. Margolus, T. Toffoli: CAM-6: a high-performance cellular automata machine (Users guide and hardware manual). MIT Lab for Computer Science, Cambridge, MA, 1987

[CM]   Connection machine models, CM-2 technical summary. Thinking Machines Co., Cambridge MA, 1989

[Co]   E.F. Codd: Cellular automata. Academic Press, New York, 1963

[C-N-R] M. Cosnard, M. Nivat, Y. Robert: Algorithmique paralléle. Masson, Paris, 1992.

[C-M]   H.S.M. Coxeter, W.O. Moser: Generators and relations for discrete groups. Springer-Verlag, New York, 1972

[DG]   H. de Garis: The "CAM-BRAIN" project, Parts I (fundamentals) and II (a billion neuron artifical brain). Preprints Brain Builder Group, Evolutionary Systems Department, Kansai Science City, Kyoto.

[D-G-T] J. Demongeot, E. Goles, M. Tchuente: Dynamical systems and cellular automata. Proc. Journeés de la Société Mathématique de France. Academic Press, New York, 1985

[D]   G.L. Drescher: Demystifying quantum mechanics. Complex Systems **5**:2 (1991), 207–237

[D-F]   M.J.B. Duff, T.J. Fountain: Cellular logic image processing. Academic Press, New York, 1986

[F-T-W] D. Farmer, T. Toffoli, S. Wolfram (editors): Cellular automata. Proc. of an interdisciplinary workshop at Los Alamos. North Holland, New York, 1984

[G-M]   I. Grossman, W. Magnus: Groups and their graphs. Random House, New York, 1964

[Gut]   H. Gutowitz (ed.): Cellular automata: theory and applications. Proc. 3rd. Int. Conf. Cellular Automata, Los Alamos, 1991. Physica D **45** (1990) 431–440 Also issued as a separate book by MIT Press, Cambridge MA, 1992

[H]   W.D. Hillis: The connection machine. MIT Press, Cambridge MA, 1985

[H-U]   J.E. Hopcroft, J.F. Ullman: Introduction to automata theory, languages and computation. Addison-Wesley, Reading MA, 1979

[K]   H. Kesten: Percolation theory for mathematicians. Birkhauser, Boston, 1982

[L-J-D] S. Lakshmivarahan, J-S Jwo, S.K.Dhall: Symmetry in interconnection networks based on Cayley graphs of permutation groups: a survey. Parallel Processing **19** (1993) 361–407

[M-K-S] W. Magnus, A. Karrass, D. Solitar: Combinatorial group theory. Dover, New York, 1975

[P-D]   K. Preston, Jr., M.J.B Duff: Modern cellular automata (theory and applications). Plenum Press, New York, 1984

[R]   R. Rucker: CALAB: Cellular automata laboratory. Autodesk Inc., Sausalito CA, 1989

[Ru]   P. Rujàn: Cellular automata and statistical mechanical models. J. of Statist. Physics **49**:1/2 (1987) 139–222

[S-A]   D. Stauffer and A. Aharony: Introduction to percolation theory. Taylor & Francis, Philadelphia, 1991

[Tie]   ALife Digest No. 117 (1994): On the Internet, January 31.

[T-M]   T. Toffoli & N. Margolus, Cellular automata machines (a new environment for modeling). MIT Press, Cambridge MA, 1987

[VN]   J. von Neumann: Theory of self-reproducing automata. U. of Illinois Press, Chicago, 1966

[W1]   S. Wolfram: Twenty problems in the theory of cellular automata. Physica Scripta **9** (1985) 170–183. Reprinted in [W2]

[W2]   S. Wolfram: Theory and applications of cellular automata. World Scientific, Singapore, 1986

[JCS]   S. Wolfram, editor: Complex systems, Center for Complex Systems Research, U. Of Illinois, Champaign, volumes starting **1** (1987)

3. Linear Cellular Automata

In the theory of classical systems, linear concepts play a fundamental role for at least two reasons: (a) they are usually about the only ones that admit a satisfactory mathematical analysis; and (b) nonlinear systems can usually be studied through linear approximations. At least part (a) has been true in the study of cellular automata. Emerging evidence suggests that (b) may bear some truth as well. In this chapter we introduce linear cellular automata and study their basic properties. We assume, as usual, that the nodes in the center cell's neighborhood N have been numbered in a fixed (but arbitrary) order $x_1 x_2 \ldots x_n$ and that $N_\bullet$ denotes N expanded to include the center cell.

3.1 Linear Rules

Linearity requires *scalars* and *vectors*. Therefore, one usually assumes that the state set Q is a finite field F (e.g., the integers $\mathbf{Z}_p$ modulo a prime p). Thus $\mathbf{C}$ becomes a (usually infinite dimensional) vector space, or *module*, over A. The standard basis of this vector space is given by the *pixel* configurations e^k given by

$$e_i^k := \begin{cases} 1 & \text{if } i = k \\ 0 & \text{otherwise} \end{cases} .$$

Remark. In principle Q can be, more generally, a finite ring with identity A, or even a module over a ring. These possibilities are explored later in the chapter.

With respect to the standard basis, every configuration $x \in \mathbf{C}$ can be expressed in the form

$$x = \sum_{i \in \underline{x}} x_i e^i ,$$

a fact which will be often used without explicit mention. The pixel e^0 will be denoted e. The basic idea of linearity is the validity of the *superposition principle*, as stated next.

Definition 3.1 *A global dynamics T is linear if, for every pair of configurations $x, y \in \mathbf{C}$,*

$$\forall \lambda \in A, \; T(\lambda x + y) = \lambda T(x) + T(y)$$

For example, it is easy to see that trivial automata given by $T(x) := O$ are linear. The *shift* $\sigma_k : \mathbf{C} \to \mathbf{C}$ given by

$$\sigma_k(x)_i := x_{i-k}$$

for a fixed element $k \in \Gamma$ (i.e., a shift by k) are also linear, as can be easily checked (see Problem 2.18). The effect of a shift is to translate the pixels of the input configuration x uniformly in the direction of k so that cell k assumes the state of cell 0.

Lemma 3.2 *If T is a linear global dynamics that (a) preserves O, i.e., $T(O) = O$; (b) commutes with shifts; and (c) $T(e)$ has finite support, then T is a cellular automaton. In this case, linearity is equivalent to the local condition*

$$\delta(x_0 x_1 \ldots x_n) = a_0 x_0 + a_1 x_1 + \cdots + a_n x_n \tag{3.1}$$

for some $a_0, a_1, \ldots, a_n \in A$.

A dynamics of type (3.1) will be denoted by $\langle a_0, \ldots, a_n; A \rangle$, or simply by $\langle a_0, a_1, \ldots, a_n; m \rangle$, if $A = \mathbf{Z}_m$. The scalar a_i is called the ith *weight* or *coefficient* of the cells in the neighborhood N.

Proof. First, we observe that $T(e^i)$ has finite support for every i by condition (c) and only differs from $T(e)$ by a shift. Let N be the support of $T(e)$ and put $a_i := T(e)_i$ for $i \in N$. Let δ be the local rule with neighborhood N given by

$$\delta(x_0 \ldots x_n) = \sum_{i \in N_\bullet} a_i x_i.$$

Since every configuration $x \in \mathbf{C}$ decomposes as $x = \sum_{j \in \underline{x}} x_j e^j$ and T is linear, it follows that

$$
\begin{aligned}
T(x)_i &= \sum_{j \in \underline{x}} T(x_j e^j)_i = \sum_{j \in \underline{x}} x_j T(e^j)_i \\
&= \sum_{j \in \underline{x}} x_j T(e)_{i-j} = \sum_{j \in \underline{T(e)}} x_{i-j} T(e)_j \\
&= \sum_{j \in \underline{T(e)}} x_{i-j} a_j,
\end{aligned}
$$

as claimed. Thus T is induced by the local dynamics δ defined above. $\square$

In the rest of this chapter we will assume that T satisfies the conditions of Lemma 3.2 so that T is a linear global dynamics. In this case, there is a more elegant way to express the linearity property by means of an operation that also plays an important role in image processing.

Definition 3.3 *Let a be a configuration with finite support and x an arbitrary configuration. The* convolution *$a * x$ of a and x is given by*

$$(a * x)_i = \sum_{j \in \underline{a}} a_j x_{i-j}\,.$$

Thus, the configuration a defined above by $a := T(e)$ has finite support and defines a global rule by convolution with a,

$$T(x) = a * x \tag{3.2}$$

identical to T. This configuration a will be called the *template* (or *kernel* or even *window*) of T. As we will see, it plays a role similar to that of the ordinary slope for linear dynamics on the real interval.

It is easy to check that the operation of *convolution* is an associative and commutative operation which distributes over addition, and which has e as neutral element. Hence pointwise addition and convolution make configurartion space **C** what mathematicians call a *commutative ring with identity*, where elements may look weird but can be manipulated just as though they were integers. Note that convolution is defined for arbitrary products whenever all but at most one of the factors have finite support. The sums are meaningless in this discrete setting if the supports of both factors are infinite. Even if the sums are still finite, algebraic properties are lost if two factors have infinite support. For instance, in a 1D elementary cellular space $(x * y) * z \neq x * (y * z)$ for $x = \ldots 000\underline{1}1000\ldots$, $y = \ldots 1\underline{0}000\ldots$, $z = \ldots 000\underline{*}111\ldots$, as one can easily check. (Recall that the position of the underline $\underline{*}$ indicates the origin of the cellular space and $*$ is here the wild character.)

Many properties of linear automata follow from this characterization. First, the iterates

$$T^t(x) = a^t * x, \text{for every } t \geq 0\,, \tag{3.3}$$

where a^t is the iteration of convolution of a with itself t times. This implies that if T has some property for a configuration a (for example $T^t(a) = O$ for some t that may depend on x), then T satisfies this property *uniformly* (in the example, T is nilpotent, that is, some power is the trivial dynamics). The reason is that the asymptotic properties of T depend only on a sequence of configurations $(a^t)_t$, which is precisely the orbit of pixel e. The following section gives more precise characterizations of global properties. Later sections deal with the solutions of most of the problems mentioned in Chapter 2 for linear automata.

3.2 Basic Properties

In this section we prove that many of the important questions for a global dynamics mentioned in Chapter 2 (such as, is T one-one? Surjective?) can be answered satisfactorily for linear dynamics. First, we treat the case $Q := \mathbf{Z}_m$ on a cellular space defined by an abelian group, such as $\mathbf{Z}^n$. Recall that equality in $\mathbf{Z}_m$ means congruence modulo m of the corresponding integers. We assume $m > 1$ throughout.

3.2.1 Global Injectivity and Surjectivity Modulo m

Given arbitrary integers $a_0, a_1, \ldots, a_d$, any integer m can always be decomposed into a product of integers

$$m = W_m P_{m,0} \ldots P_{m,d} Q_m$$

defined as follows:

W_m := product of powers of the prime factors of m that divide *all* the a_i's;

$P_{m,i}$:= product of powers of prime factors of m that *fail* to divide a_i, but divide all the a_j's $(j \neq i)$;

Q_m := product of powers of prime factors of m that do *not* divide at least two a_is .

Example 3.4 If $m := 2^8 3^4 5^6 7^8$, $a_1 := 2^7 5^6 7^4$, $a_2 := 2^2 3^2 5^3$, $a_3 := 3^7 5^2 7$, then

$$W_m = 5^6 ; \quad P_{m,1} = 3^4, \quad P_{m,2} = 7^8, \quad P_{m,3} = 2^8 ; \quad Q_m = 1 .$$

If no prime power satisfies the given condition, the product is 1 by definition. In particular, $W_m = 1$ means the a_i's and m are relatively prime. Note that the $P_{m,i}$'s, Q_m, and W_m are also pairwise coprime and their product is m.

Theorem 3.5 *A linear global dynamics $T = \langle a_0, a_1, \ldots, a_d; m \rangle$ is onto if and only if $W_m = 1$, i.e., all the coefficients and the modulo are relatively prime.*

Proof. If the coefficients of T are not coprime, they must be divisible by some prime integer p, and hence $a * x$ has all cells in states $(a * x)_i \equiv 0 \mod p$. Therefore, $T(x) \neq e$ for all x, so T is not onto.

For the converse, we must prove that $T(y) = a*y = e$ for some configuration y. If so, $T(y * x) = a * (y * x) = e * x = x$ for all others. Coprimality of the a_is implies that the equation $\sum a_j x_j = 1$ has a solution y_j. However this is not enough since the other sites in the image must be certainly 0. Rather, we will prove that T is one-one, and hence onto by Richardson's theorem (see Chapter 7) by induction on $|N|$, the size of the neighborhood N. Assume $T(x) = O$. In case $|N| = 1$,

$$(a * x)_i = a_0 x_i = 0 \ (mod \ m),$$

hence $x_i = 0$ since a_0 is coprime to m. In general, if $|N| > 1$, Let N' be the neighborhood $N' := \{i : a_i \neq 0\}$ (at least one such a_i exists since $W_m = 1$). Thus, if a' is the corresponding template of T', then

$$(a' * x)_i = \sum_{i \in N'} a_i x_i = 0.$$

By induction, $x_i = 0$ for all i, that is, $x = O$. $\square$

The better known linear dynamics are those of rules defined on euclidean cellular spaces. The next result characterizes one-one linear dynamics on such spaces. It requires the following technique to convert d-dimensional configurations supported in the first ortant (octant if $d = 3$) of $\mathbf{Z}^d$ into diophantine polynomials or power series in d indeterminates. For simplicity we only consider the case $d = 2$ (the general case is analogous). Let $\mathbf{Z}[X]$ be the ring of polynomials with integers coefficients in two indeterminates $X = (u, v)$. It is a well known fact that $\mathbf{Z}[X]$ is a unique factorization domain, i.e., there are certain objects called *irreducible* (polynomials) which generate by multiplication all the elements of $\mathbf{Z}[X]$ in essentially a unique way (up to trivial factorizations by units of the type $\frac{1}{u}u$). Given a pair of integers $i = (i_1, i_2)$, denote by X^i the monomial $u^{i_1}v^{i_2}$. Define the mapping

$$\begin{aligned} \wp : \mathbf{C} &\rightarrow \mathbf{Z}[u, v] \\ \wp(x) &= \sum x_i X^i. \end{aligned} \qquad (3.4)$$

Note that $\wp$ is a *ring homomorphism* from the ring $\mathbf{C}$ into $\mathbf{Z}[X]$, that is, for all $x, y \in \mathbf{C}$,

$$\wp(x + y) = \wp(x) + \wp(y) \quad \text{and} \quad \wp(x * y) = \wp(x)\wp(y).$$

In particular, $\wp(e) = 1$, $\wp(a * x) = \wp(a)\wp(x) = X$, $\wp(e_{(1,0)}) = u$ and $\wp(e_{(0,1)}) = v$. A similar construction produces a mapping $\tilde{\wp}$ that maps arbitrary (even infinite) configurations into the ring of formal power series $\mathbf{Z}[[X]]$ with integer coefficients.

Theorem 3.6 *The following conditions are equivalent for a linear dynamics T on a euclidean cellular space:*

1. *T is one-one.*

2. *$W_m = Q_m = 1$.*

3. *T_0 is onto.*

4. *T_0 is bijective.*

As we will see later in Chapter 7, (1) $\Rightarrow$ (3) is a general property of arbitrary local maps over euclidean universes, but the converse is not true in general. Rather, we will prove also that, in general, if T_0 is onto then T_0 is 1-1.

Proof. Recall that T commutes with any shift σ_k, i.e., $T(\sigma_k(x)) = \sigma_k(T(x))$ for all x. (1) $\Rightarrow$ (3) will be proved in Chapter 7 (without using Theorem 3.6 of course). To prove (3) $\Rightarrow$ (2) let $\wp$ be the map defined above and let a be the template of T. Since T_0 is onto, $T(b) = a * b = e$ for some $b \in C_0$. An application of $\wp$ yields

$$\wp(a * b) = \wp(a)\wp(b) = 1,$$

and hence any prime divisor p of m divides at most one a_i. Moreover $T(b * x) = a * (b * x) = x$, i.e., T is onto.

(2) $\Rightarrow$ (1) is proved by induction on m. The statement is clear for $m = 1$. If a prime p divides m, condition (2) implies that some divisor p of m divides $P_{m,i}$, i.e., p does not divides a_j ($j \neq i$) but it does divide a_i. Therefore $x_{i+k} = 0 \pmod{p}$, for all k since $\sum a_i x_{i+k} = 0 \pmod{m}$. It follows that $x_k = 0 \pmod{p}$ for all k. Now consider $x' := \frac{1}{p}x = 0$ over $\mathbf{Z}_{m/p}$. We still have $W_m' = Q_m' = 1$. By induction, $x' = 0 \pmod{m}$, whence $x = px' = 0 \pmod{m}$.

Obviously (4) $\Rightarrow$ (3). Conversely, if T_0 is onto, then $T(b) = a * b = e$ for some $b \in C_0$, so T has an inverse with template b. $\qquad\square$

Definition 3.7 *A configuration x is T-periodic (shift-periodic), respectively) if $T^t(x) = x$ ($T^t(x) = \sigma_k(x)$, respectively) for some $t \geq 1$, i.e., if some iteration of T (that may depend on x) just returns to (a shift of) x. The global dynamics T is T-periodic (respectively, shift-periodic) if it makes every configuration T-(shift-)periodic. A configuration x is (spatially) s-periodic if it is σ-periodic, i.e., if $\sigma_k(x) = x$ for some $k \in \Gamma$, $k \neq 0$. The configuration x is finitely periodic if $\sigma_k(x) = x$ for each of the coordinate directions k in the cellular space.*

In particular, if T is 1D euclidean, but not in general, finitely periodic means the same as periodic. When $Q = \mathbf{Z}_m$, x is congruent to y modulo m if $x_i = y_i \bmod m$ for each i. Note that the definition of s-periodic by itself does not imply the existence of a smallest integer $k > 0$ that makes every configuration T-periodic. Even if no power of T is a shift, it is still a priori possible that T should make a copy of the nonquiescent part of a configuration c elsewhere in the cellular space. In this case c is called a *self-reproducing* configuration.

In order to characterize s-periodic configurations with respect to a given dynamics and their possible periods, let us prove first a property of powers of a sum in $\mathbf{Z}[[X]]$.

Lemma 3.8 *If $p_1, \ldots, p_r$ are the distinct prime factors of an integer $m > 1$ and $m_0 = m/(p_1 \ldots p_r)$ then*

$$(f + p_1 \ldots p_r g)^{m_0} \equiv f^{m_0} \pmod{m}$$

for all $f, g \in \mathbf{Z}[[X]]$.

Proof. It is easy to show by induction on t that

$$(f + pg)^{p^{t-1}} = f^{p^{t-1}} \pmod{p^t}$$

for all such f, g using the binomial theorem. Using this property, it follows that, for each prime divisor p_l of m,

$$(f + p_1 \ldots p_r g)^{m_0} = ((f + p_1 \ldots p_r g)^{p_l^{r^l-1}})^{m_0/p_l^{r^l-1}}$$
$$\equiv f^{m_0} \pmod{p_l^{r_l}}.$$

Since the p_l's are relatively prime, the assertion follows. $\square$

Recall that the value of Euler's totient function ϕ at m is given by the number of integers between 1 and m which are coprime to m. A classical result of Fermat's asserts that when an integer n is coprime to m,

$$n^{\phi(m)} = 1 \pmod{m}.$$

Theorem 3.9 *The following properties are equivalent for a linear dynamics* $T = \langle a_0, a_1, \ldots, a_n; m \rangle$:

1. $T^t(e) = e^k$, *for some* k.

2. T *is shift-periodic.*

3. $T^{\phi(m)} = \sigma_k$ *for some* k.

4. $m = P_{m,i}$ *for some* $1 \leq i \leq n$.

Proof. Recall that we are assuming that $m > 1$. To prove (1) $\Leftrightarrow$ (2) note that if a is the template of T and $T^t(e) = \sigma_k(e) = e^k$, then

$$T^t(x) = T(\sum_{i \in \underline{x}} x_i e^i) = \sum_{i \in \underline{x}} x_i T(e^i)$$
$$= \sum_{i \in \underline{x}} x_i \sigma_i(T(e)) = \sum_{i \in \underline{x}} x_i \sigma_{i+k}(e)$$
$$= \sigma_k(\sum_{i \in \underline{x}} x_i \sigma_i(e)) = \sigma_k(\sum_{i \in \underline{x}} x_i e^i)$$
$$= \sigma_k(x),$$

since T commutes with shifts. Therefore the s-period of T is exactly the T-period of e.

Since $T^t(a_0 e) = a_0^t e^k$, this s-period equals the least t for which $a_i^t = 1$ (mod m) for all a_i, which is precisely $\phi(m)$. Obviously (3) $\Rightarrow$ (2). This proves the equivalence of conditions (2) and (3).

Observe that condition (4) implies that $T^{\phi(m)}$, and hence T, is one-one, so that, by Theorem 3.5, $W_m = Q_m = 1$, whence $m = P_m$. To prove the converse, it suffices to show that $T^{\phi(m)}(e) = e^k$ for some k. Using the definition of $P_{m,l}$ and the map $\tilde{\wp}$ constructed above, it follows that

$$\tilde{\wp}(a) = a_l u^{r_1} v^{r_2} + p_1 \ldots p_s h(u, v),$$

for some integers r_1, r_2 and $h \in \mathbf{Z}[[X]]$ with finite support, where a_i are the coefficients of the template a of T and $p_1, \ldots, p_s$ are the prime divisors of $P_{m,l}$.

Thus, if $m_0 := m/(p_1 \ldots p_r)$, then $\phi(m_0) = m_0(p_1 - 1) \ldots (p_r - 1)$, so that by Lemma 3.8,

$$\begin{aligned} \tilde{\wp}(T^{\phi(m)}(e)) &\equiv (a_i u^{r_1} v^{r_2})^{m_0(p_1-1)\ldots(p_s-1)} \mod m \\ &\equiv a_1^{\phi(m)} u^{\phi(m)r_1} v^{\phi(m)r_2} \mod m. \end{aligned}$$

Since the a_is and m are coprime the result now follows by Euler's theorem.   □

Finally, one can give a complete characterization of the dynamical behavior of linear automata on abelian (and, in particular, euclidean) cellular spaces over $\mathbf{Z}_m$. Factoring up the modulo m leads immediately to prime power moduli (see Problem 37). Now, in $\mathbf{Z}_{p^r}$, every element $v \in Q$ is either coprime to $m = p^r$ or it is divisible by p, and hence $v^{p^r} = 0$, i.e., it is nilpotent. It turns that that essentially the same property is inherited by linear cellular automata: they all are either nilpotent or surjective depending on the number of *unit* coefficients (those invertible modulo m) in the kernel of the local rule.

Theorem 3.10 *Every linear dynamics T with kernel a on an abelian cellular space Γ over $\mathbf{Z}_{p^r}$ satisfies exactly one of the following conditions:*

1. *T is nilpotent, when all coefficients are zero-divisors;*

2. *T is bijective and shift-periodic, when exactly one of the coefficients is a unit;*

3. *T is surjective but not injective, when at least two coefficients are units.*

Proof. (1) If all coefficients are divisible by p, the kernel $a = pc$ for some configuration $c \in \mathbf{C}_0$, so that $a^{p^r} = p^{p^r} c^{p^r} = 0$ and hence T is nilpotent.

(2, 3) Let a_U (respectively, a_Z) denote the finite configurations obtained from a by replacing zero-divisor (unit) coefficients a_i with 0, so that $a = a_U + a_Z$. Since binomial theorem holds in the commutative ring LCA(Γ) (see Problem 35), an argument similar to the proof of Lemma 3.8 shows that

$$(a_U + a_Z)^{p^{r-1}} = a_U^{p^{r-1}} \pmod{p^r}.$$

In case 2, the power $a_U^{p^{r-1}}$ is a pixel kernel, it yields is a power of the shift, and so is bijective. In either case, T is surjective since so is its p^{r-1}-th power by Theorem 3.5. In case 3, however, $T^{p^{r-1}}$ is not bijective since since it has $Q_m > 1$ by Theorem 3.6.   □

It is now straightforward to derive necessary and sufficient conditions for the dynamical behavior over composite moduli (see Problems 37–39).

Another important aspect of the dynamical behavior of cellular automata concerns the relationship between the length of the spatial and temporal periods of periodic configurations (see Problems 2.23). For euclidean automata these questions are in fact equivalent to similar questions over finite spaces and will be dealt with in Sect. 3.2.3.

3.2.2 Self-reproduction with Linear Automata

One of the early questions for cellular automata was whether some local dynamics would reproduce arbitrary initial finite configurations an arbitrary number of times in 1D euclidean space. It is known now that euclidean linear automata on finite fields can be constructed to produce at a specified time t any number of copies of an initial configuration x each multiplied by a specified scalar. This amounts to the construction of a template a such that

$$a^t = \lambda_0 a_0 e + \lambda_1 a_1 e^1 + \cdots + \lambda_n a_n e^n,$$

so that $T^t(x) = \lambda_1 x + \cdots + \lambda_n x$, as can be easily verified. Thus, if the pixels e^l are at a Manhattan distance on Γ of at least the diameter of x, these copies of x will appear as disjoint scaled copies of the original configuration in a cellular space of any dimension.

3.2.3 Linear Automata on Rings and Semigroups

The investigation of linear (let alone more general) automata over more general state sets such as (mathematical) rings and modules has been very limited so far, and understandably so, since, as can be seen from the results in this section, their properties depend to a great extent on the (yet unknown) structure of arbitrary state sets. We present without proof several results for this type of automata and refer the interested reader to the original sources or his ingenuity for their proofs (see the proof of Theorem 3.10 and Problem 27). Recall that a *unit* in a ring A is an invertible element, that is, an element that possesses an inverse for multiplication in A. Likewise, an element s is *nilpotent* if $s^t = 0$ in A for some integer $t > 0$.

Theorem 3.11 *If the coefficients a_i of a $(i \in \underline{a})$ are units in the ring A then*

1. *If $|\underline{a}| = 1$ then T is bijective.*

2. *T is injective if and only if $|\underline{a}| = 1$.*

3. *If $|\underline{a}| \geq 2$, then for every configuration x:*
   if $d = 1$: $|T^{-1}(x)| = |A|^{n_r - n_l} \geq |A|$, where $n_l = \min \underline{a}$, $, n_r := \max \underline{a}$;
   if $d \geq 2$: $|T^{-1}(x)| = \infty$.

4. *T is surjective.*   □

More is known if the state set is a commutative ring since their structure is better known. For example, there are only finitely many ideals (an ideal is an abelian subgroup that absorbs multiplication by arbitrary elements of A), say $J_1, \cdots, J_n$, and they give rise to subrings A which are local rings (i.e., they have a unique maximal ideal J_l) and direct summands of A:

$$A := A_1 \oplus \cdots \oplus A_n.$$

Moreover, every element of a local commutative ring is either a unit or is nilpotent and every nilpotent element is included in a maximal ideal J. Memberships in these ideals determine injectivity and surjectivity of the global dynamics as follows. For each coefficient a_i of the rule, let its *spectrum* $\zeta(a_i)$ be the set Ψ of indices l that contain it: $a_i \in J_l$. The spectrum of A is the union of the spectra of each direct summand.

Theorem 3.12 *If A is a commutative ring, then the global dynamics T of a linear rule is*

1. *surjective iff $\cup_{i \in N} \zeta(a_i) = \{1, \cdots, n\}$;*

2. *injective iff T is surjective and $\zeta(a_i) \neq \zeta(a_j)$ for $i \neq j$;*

3. *shift-periodic (i.e., $T^t = \sigma$ for some $t > 0$) iff some coefficient a_i has full spectrum $\sigma(a_i) = \{1, \cdots, n\}$ but all other a_j have empty spectrum $\sigma(a_j) = \Phi$ ($j \neq i$).*

Another generalization of the property of linearity gives rise to analyzable automata. Put a semigroup structure in the state set, i.e., define a single associative binary operation. XOR is a simple example. This operation among states can be extended componentwise to an operation in configuration space, which will be denoted $\bullet$ (or just catenation) for the remainder of this section. If there is an identiy element, it will play the role of the quiescent state. A map T is said to satisfy the *superposition principle* with respect to $\bullet$, if

$$\forall x, y, \quad T(x \bullet y) = T(x) \bullet T(y).$$

In the sequel, the $\bullet$ is usually omitted and the operation indicated by simple concatenation of the operands. The following straightforward lemma gives sufficient conditions for the principle to hold. The conditions are sufficient as well if the states form a monoid.

Proposition 3.13 *If $\bullet$ is commutative and associative on states, then T satisfies the superposition principle with respect to $\bullet$. If further $\bullet$ has a neutral element, then commutativity and associativity are also necessary conditions.*

Proof. The proof is illustrated for one-sided rules with a neighborhood of radius $r = 2$. In this case,

$$T(x)_i := x_{i-1} x_i$$

so that the principle is equivalent to

$$
\begin{aligned}
T(xy)_i &= (x_{i-1} y_{i-1})(x_i y_i) \\
&= (x_{i-1} x_i)(y_{i-1} y_i) \\
&= T(x)_i T(y)_i
\end{aligned}
$$

i.e., the principle is equivalent to the following condition on states,

$$\forall s, t, u, v \quad (st)(uv) = (su)(tv).$$

Hence commutativity and associativity are sufficient conditions. Vice versa, if superposition holds and ε is an identity element, then

$$(st)u = (st)(\varepsilon u) = (s\varepsilon)(tu) = s(tu).$$

i.e., associativity holds, as well as commutativity:

$$st = (\varepsilon s)(t\varepsilon) = (\varepsilon t)(s\varepsilon) = ts. \quad \square$$

Under a superposition principle one can obtain a fairly satisfactory description of the space–time diagrams for arbitrary initial configurations. The evolution of a single pixel under the linear rule is contained in a "cone of light" shaped as an angle with vertex at the pixel and sides the propagation of the activity fronts. In the case of XOR, this triangle is simply the reduction modulo 2 of *Pascal's triangle*, and we use the term by extension.

Theorem 3.14 *If T satisfies a superposition principle, all space–time diagrams are superpositions of Pascal triangles.*

Proof. For the sake of clarity, assume without loss of generality an elementary rule. Under the operation in the states set, a typical pixel ae has a finite period, say t. The states of the site at the origin are successively $a, a^2, a^5, a^{11} \ldots$. Eventually the exponents are reduced modulo t. The space–time thus is made up of a basic triangle up to the first repetition. This triangle occurs at the front of activity and tiles the borders of space–time. By superposition, other space–times are obtained from appropriate combinations. $\quad \square$

3.3 Global Dynamics via Fractals

In the previous sections we have studied the problem of predicting the behavior of a given linear T on specific configurations. The solution gives a "local" characterization of T by its template a. However, these results do not quite answer the question of the global asymptotic behavior of T on *all* configurations, and do not even make a simple answer to this problem possible. It would be interesting to have at least an estimate of the appearance of $T^t(x)$ for large values of t and a typical configuration x. A way to approach this problem is to compute $T^t(x)$ for a number of random configurations and large values of t, which may take very long and many trials.

The purpose of this section is to present a result which, theoretically, gives an ideal (although perhaps impractical) answer to this problem for linear dynamics. It says that a *single* object (a subset of euclidean space $\mathbf{R}^{d+1}$), hereby named

$$\lim T := \lim T^{2^t}(x)/2^t$$

exists which somehow summarizes the entire infinite run of T on x. Furthermore, $\lim T$ turns out to be independent of the initial configuration x, and hence is a *geometric invariant* of T. This implies that, in some sense, the information that comprises the entire past history (at least ignoring some powers) of an elementary euclidean cellular automaton T on *all* possible configurations can be summarized (losing some detail of course) into an object, which is conveniently called $\lim T$. Therefore, at least for elementary euclidean cellular automata, $T^t(x)$ can be estimated to a *large degree of accuracy* for large values of t, by the configuration

$$T^{2^t}(x) \approx 2^t \lim T ,$$

where the last product is obtained from L by pointwise multiplication by 2^t. The remainder of this section may be skipped without loss of continuity.

In this section we only consider binary configurations and euclidean cellular automata (although several of these results can be extended to more general state sets). A binary configuration will be regarded as a subset $\{i \in \Gamma : T(x)_i \neq 0\}$ of the host cellular space. Therefore they can be identified with their supports and regarded as sets of integer lattice points in $\mathbf{Z}^d$. A d-dimensional configuration can be embedded into $(d+1)$-dimensional configuration by adding 0 as a last time coordinate.

The following observations are easily proved for a linear dynamics T.

1. The powers of T are linear;

2. $i \in T(e)$ iff $2i \in T^2(e)$;

3. $i \in T(u)$ iff $2^t i \in T^{2^t}(e)$;

4. $T(x) = \sum_{j \in \underline{x}} T(e^j)$ and $T(x) \subseteq \cup_{j \in \Gamma} T(e^j)$.

Given a sequence of points x^t in $\mathbf{R}^d$, ordinary convergence of the sequence (x^t) to a point $x \in \mathbf{R}$ is denoted $x^t \to x$.

Definition 3.15 *The graph(space–time) of T is a $(d+1)$-dimensional global dynamics F given by*

$$F(y)_{(i,r)} = \begin{cases} 1 & \text{if } y_{(i,r-1)} = 1 \\ T(y_{(*,r-1)}) & \text{otherwise}, \end{cases}$$

where $y_{(*,r-1)}$ is the slice of the configuration y at the previous time $r - 1$.

The graph of T can be more easily visualized by thinking of the first d coordinates as the ordinary spatial coordinates of y and the last coordinate as time. F only updates quiescent cells to whatever state T would update them on the basis of the d-dimensional configuration at the previous time step.

If $X_i, X_2, \ldots$ is a sequence of subsets of $\mathbf{R}^d$, a sequence of points $(x^t) \subset \mathbf{R}^d$ is called a *choice sequence* from (X_t) if $\forall t\, (x^t \in X_t)$. A point x is called a *limit* (*cluster point*, respectively) of the sequence of subsets $X_1, X_2, \ldots$ if there exists a choice sequence $(x^t)_t$ from the X_ts (which has a subsequence, respectively) that converges to x. Let

$$\liminf X_t := \{x \in \mathbf{R}^d : x \text{ is a limit point of } X_t\},$$

that is, the set of all limits of choice sequences from (X_t). Let

$$\limsup X_t := \{x \in \mathbf{R}^d : x \text{ is a cluster point of } X_t\},$$

that is, the set of cluster points of choice sequences from the sets (X_t). Clearly,

$$\liminf X_t \subseteq \limsup X_t.$$

If they are equal, they are called the limit $limX_t$ of the sequence of sets X_t.

The graph F can be applied to a configuration $x \in \mathbf{C}_0$ as embedded in $\mathbf{Z}^{d+1}$. In this case,

$$(i, r) \in F^s(x) \Leftrightarrow i \in T^r(x) \ (0 \le s \le r).$$

In the rest of this section we will fix an elementary cellular automaton T and consider the sets $X_t(x) \subset \mathbf{R}^{d+1}$ defined for a configuration x (embedded in $\mathbf{Z}^{d+1}$) by

$$X_t(x) := \frac{1}{2^t} F^{2^t}(x).$$

Lemma 3.16 $\liminf$ *is invariant under translation:*

$$\liminf X_t(\sigma_k(x)) = \liminf X_t(x).$$

for all nonquiescent $x \in \mathbf{C}_0$. *Likewise for* $\limsup$.

Proof. This follows from the fact that T commutes with shifts, i.e., for all k's, $T(\sigma_k(x)) = \sigma_k(T(X))$, for every configuration x (see Problem 2.19). $\square$

Lemma 3.17 *For all nonquiescent finite configurations* $x \in \mathbf{C}_0$,

$$\liminf X_t(x) = \liminf X_t(e)$$

Likewise for $\limsup$.

Proof. We first prove that $\liminf X_t(e) \subseteq \liminf X_t(x)$, that is, every cluster point of a choice sequence from $(X_t(e)$ is the cluster point of a choice sequence from $X_t(x)$. By Lemma 3.16, assume $0 \in \underline{x}$. If $(i, r) = \lim \frac{1}{2^t}(i_t, r_t)$, for $i_t \in T^{r_t}(x)$, by the observations before Definition 3.15,

$$(i, r) = \liminf \frac{1}{2^\alpha 2^t}(2^\alpha i_t, 2^\alpha r_t).$$

We just need to prove that $2^\alpha i_t \in T^{2^\alpha r_t}(x)$ so that $(2^\alpha i_t, 2^\alpha r_t) \in F^{2^\alpha r_t}(x)$. This follows from the equality

$$T^{2^\alpha r_t}(x) = \bigcup_{j \in T^{2^\alpha r_t}(e)} \sigma_j(x),$$

which can be verified as follows:

$$
\begin{aligned}
T^{2^\alpha r_t}(x) &= T^{2^\alpha r_t}\left(\sum_{i \in \underline{x}} \sigma_i(e)\right) \\
&= \sum_{i \in \underline{x}} \sigma_i T^{2^\alpha r_t}(e) \\
&= \sum_{i \in \underline{x}} \sum_{j \in T^{2^\alpha r_t}(e)} \sigma_i \sigma_j(e) i \in \subseteq j \in_T^{2^\alpha r_t}(u) \\
&= \sum_{j \in T^{2^\alpha r_t}(e)} \sigma_j\left(\sum_{i \in \underline{x}} \sigma_i(e)\right) \\
&= \sum_{j \in T^{2^\alpha r_t}(e)} \sigma_j(e),
\end{aligned}
$$

whence the fact.

To prove that conversely, $\liminf X_t(x) \subseteq \liminf X_t(e)$, assume $(i_t, r_t) \to (i, r)$ for some point (i, r) in the left-hand side, so that $i_t \in T^{r_t}(x)$. Hence

$$i_t \in T^{r_t}(x) = \sum_{j \in \underline{x}} \sigma_j(T^{r_t}(u)),$$

again by the observations before definition 3.15. Therefore, $i_t \in \sigma_{j_t}(T^{r_t}(e))$ for some $j \in \underline{x}$, i.e., $i_t - j_t \in T^{r_t}(e)$. This means that $\frac{1}{2^t}(i_t - j_t, r_t) = (i, r) \in \liminf X_t(e)$, as claimed. The similar proof for $\limsup$ will be omitted. $\square$

Lemma 3.18 *The set $\liminf X_t(x)$ is independent of the choice of a configuration of finite support $x \neq O$, i.e.,*

$$\liminf X_t(x) = \liminf X_t(e) = \limsup X_t(e).$$

This is the set $\lim T$ referred to in the beginning of this section.

Proof. By the previous lemmas, it suffices to prove that $\liminf X_t(e) \subseteq \limsup X_t(e)$. If some subsequence $\frac{1}{2^{n_t}}(i_{n_t}, r_{n_t}) \longrightarrow (i, r)$ as $t \to \infty$, for some $(i_{n_t}, r_{n_t}) \in F^{r_{n_t}}(e)$, then $(i_{n_t}, r_{n_t}) \in F^{2^{r_{n_t}}}(e)$. In order to prove that $(i, r) \in \lim X_t(e)$ it suffices to verify that $\frac{1}{2^{n_t}}(i_{n_t}, r_{n_t}) \in \liminf X_t(e)$ since the latter is a *closed* set (i.e., it contains all the limits sequences of points in the set). Indeed, since $(2^t i_{n_t}, 2^t r_{n_t}) \in F^{2^t 2^{r_{n_t}}}(e)$, it follows that

$$\frac{1}{2^t 2^{r_{n_t}}}(2^t i_{n_t}, 2^t r_{n_t}) \to (i, t)$$

as $t \to \infty$. $\square$

It is interesting to further investigate the nature of the limit number sets produced by asymptotic behavior of linear cellular automata. It turns out that these sets may be very "thin" and "fractured" subsets of euclidean space, as we will see next. The following is intended as an informal discussion of the notion of *fractional dimension* and may be skipped if the reader is satisfied with this description of $\lim T$ or wants precise statements of fractal sets (see [H]).

Let X be an arbitray subset of $\mathbf{R}^{d+1}$ and $\rho \geq 0$. Given $\epsilon > 0$, let $m_\rho(X, \epsilon)$ be the "least" (infimum) total $(d+1)$-volume necessary to cover X by subsets of $\mathbf{R}^{d+1}$ of diameter less than ϵ. Next, let $m_\rho(X)$ be the "largest" (supremum) such possible $m_\rho(X, \epsilon)$, as ϵ runs over all reals $\epsilon > 0$. There is a threshold D between 0 and $d+1$ such that $m_\rho(X) = \infty$ if $\rho < D$ and $m_\rho(X) = 0$ if $\rho > D$. This threshold is called the *Hausdorff dimension*, or g-dimension (for geometric dimension) of X. The g-dimension is a generalization of the intuitive notion of dimension.

Due to the fact that many self-similar sets (obtained by iterated application of a construction, like removing the middle third of an interval) have a fractional dimension, these sets are called *fractals*. The middle-third Cantor set is an example of a fractal of dimension $\log_3 2 \approx 0.6309$. It turns out that the g-dimension of the limit sets of linear cellular automata may be fractional. For instance, the 1-dimensional global dynamics induced by

$$\delta(x_{-1}x_0x_1) := (x_{-1} + x_1) \pmod 2$$

has a limit set of global dimension $\log_2 3 \approx 1.5850$.

Thus these sets have, not surprisingly, a very complex structure. The calculation of the fractal dimension, in particular, knows no general approach. The simplest expression for the dimension is given by

$$\log \lambda / \log m \tag{3.5}$$

where λ is the largest eigenvalue of a certain matrix associated with a recurrence relation counting the number of 1s in the sequence of configurations $\{T^{2^t}(e)\}_{t \geq 0}$.

Expectedly, an odd power of T may have a different limit set since the limit set of a linear dynamics if obtained by sampling iterates of their graphs at powers of 2. In fact, the local rule

$$\delta(x_{-1}x_0x_1) := (x_{-1} + x_0 + x_1) \bmod 2$$

induces a global dynamics T whose third power has a different limit set. Nevertheless, the g-dimension of all powers of T is a numerical invariant of T.

Theorem 3.19 *If T is a linear dynamics, then for every integer $t \geq 0$,*

$$dim \lim T^t = dim \lim T.$$

Limit sets can be shown to exist for arbitrary linear automata with p^k elements with respect to the sampling time sequences p^t. Equality (3.5) remains true for a corresponding transition matrix. In fact, the dimension of a linear automata with $m = p^k$ states is identical to the dimension of an automaton with the same coefficients taken modulo p. More details can be seen in [Ta2].

3.4 The Role of Linear Rules

As mentioned in the beginning of this chapter, one must pose the question of whether linear rules may play the same central role that linear systems play with respect to continuous systems. Although the answer to this question is yet unclear, there are some encouraging beginnings. Part of the problem is that the global type of phenomena modeled by linear rules themselves is not entirely understood, not even in the elementary case in 1D euclidean space.

A first approach to the global behavior of a given rule is to attempt to define a linear operator that acts as some sort of derivative. The problem is that a local rule is more like a function of several *boolean* variables than a continuum, so a derivative as an instantaneous rate of change is not an appropriate notion, unless one is willing to settle for a boolean value and use the continuous case simply as a formal analog. In this case, a boolean variable x_j in a neighborhood N can only be varied by flipping its value. The flip may or may not change the value $\delta(x)_i = T(x)_i$ of the local rule at the center cell i, and it produces a boolean value 0 (no change) or 1 known as the partial *boolean derivative* $\partial \delta / \partial x_j$ of the rule. In terms of the XOR $\oplus$ and complement $\bar{x}_j$ operations, the boolean derivative is given by

$$\partial \delta_i / \partial x_j := \delta(x_i x_{j_1} \cdots \bar{x}_j \cdots x_{j_n}) \oplus \delta(x_i x_{j_1} \cdots x_j \cdots x_{j_n}). \qquad (3.6)$$

On a finite cellular space, these partial derivatives can be put together in a linear operator using the Jacobian matrix $[\partial \delta_i / \partial x_j]_{n \times n}$. A calculus can then be developed which is formally similar to the ordinary calculus, even for infinite cellular spaces. The main difficulty, however, is that the global approximation properties of the ordinary derivative are conspicuously lacking, and this seems to render the apparatus void of expected practical applications.

A second approach to the global behavior of a given rule consists of looking at the closest linear rule with respect to the Hamming distance. An easy count of all rules for a given state and neighborhood sizes shows that, asymptotically, there are far fewer linear rules. On the other hand, the subspace of linear rules is distributed uniformly in rule space, very much like the integer lattice is distributed in euclidean spaces, with an important difference. For a prime number of states, for example, there is only one possible distance $m^n - m^{n-1}$ between any pair of linear rules. For composite states, this distance can assume only a discrete spectrum as multiples of dm^{n-1}, where d is a divisor of m.

Moreover, perhaps surprisingly given their number and distribution, the average Hamming distance between linear rules is asymptotically identical to the distance between arbitrary rules. When the number of states is prime or a field's prime power, this distance agrees with the maximum distance of a random rule to the nearest linear rule. These facts may be taken as encouraging evidence that at least certain rules may admit a linear approximation that is robust in time and on a more global scale beyond just the range of local rule tables.

3.5 Problems

Problems marked * may need to be looked up in the literature.

LOCAL RULES

1. Show that the number of linear rules goes to 0 asymptotically compared to the total number of possible rules.

2. Show that the average distance between two random rules is given by $\theta(m^n - m^{n-1})$ [See Sect. 2.3.2 for the definition of Hamming distance.]

3. Show that the distance between two linear rules is $m^n - dm^{n-1}$, where d is the greatest common divisor of m and of all the differences of their coefficient pairs.

4. Show that if the number of states is prime, the average distance between two linear rules is $m^n - m^{n-1}$.

5. Give a polynomial-time algorithm to find the shortest distance of a given rule to a linear rule on a fixed neighborhood size.

6. Is there a subexponential time algorithm for computing the distance of a given rule to the closest linear rule for arbitrary neighborhoods? *

7. Show that a linear rule is its own Jacobian on any finite cellular space.

8. Establish the expected properties of the boolean derivative for constant functions and addition modulo 2.

9. Establish chain and product rules for boolean derivatives. *

10. Compute the boolean derivatives and Jacobians of the 88 distinct 1D elementary rules. (See Problem 2.15.)

GLOBAL BEHAVIOR

11. Prove that the elementary shift $\sigma := \sigma_1$ is linear and bijective. Prove that its inverse is also a linear cellular automaton. Likewise for any shift.

12. Give an example of a linear self-map of configuration space that is not a local rule. [Define it on a basis so it does not commute with shifts.]

13. What can be said about a cellular automaton T which commutes with all others? Does it have to be a shift? (See Problem 2.21.)

14. Show that the composition of two linear cellular automata is a linear cellular automaton.

15. Prove that a bijective linear cellular automaton has an inverse self-map on configuration space which is also linear. Conclude that the set RCA(Γ, m) of reversible (i.e., bijective) cellular automata on a given cellular space Γ on m symbols form a group under composition.

16. Prove that every rule in 1D euclidean space is the composition of a right-shift and a one-way right rule (i.e., one whose neighborhood set only contains cells to the right of the center cell). Likewise for left one-way rules.

17. Generalize the previous result about one-way rules to higher-dimensional euclidean spaces.

18. Show that all nodes in the phase space of a linear rule have the same degree (possibly infinite). In particular, each node has the same number of predecessors and the trees rooted at each periodic configuration on a finite cellular space are all isomorphic.

19. Prove that the all-ones configuration **1** is never a garden of Eden on an arbitrary finite binary cellular space. (A *garden of Eden* for an automaton T is a configuration a outside the range of T, i.e., one for which the equation $T(x) = a$ has no solution.)

20. Show that if a linear automaton has gardens of Eden, then it has mutually indistiguishable configurations. Is the converse true? (Two configurations are called *mutually indistinguishable* for T if $T(x) = \sigma T(y)$ for some shift σ.)

21. Show that the global dynamics induced by extending the group operation among states in the following local rule δ to all configurations pixelwise is linear. The cellular space is the Cayley graph presented by

$$\langle\, A, B, C \mid 2A = 2B = 2C = 0 \,\rangle.$$

on the state set the Klein group $Q := \{0, 1, 2, 3\}$, where the elements are added bitwise in their binary expansion without carry, e.g., $2 + 3 = 10 + 11 = 01 = 1$. The local rule $\delta(x_i x_{i+A} x_{i+B} x_{i+C})$ is given by ($*$ is a wild character)

$$0100, 0200, 0010, 0030, 0002 \to 1 \qquad 0300, 0200, 0001, *000 \to 0;$$

$$\begin{aligned}
\delta(0x_{i+A}x_{i+B}x_{i+C}) &= \delta(0x_{i+A}00) + \delta(00x_{i+B}0) + \delta(00x_{i+C}) \\
\delta(x_i x_{i+A}x_{i+B}x_{i+C}) &= \delta(x_i000) + \delta(0x_{i+A}x_{i+B}x_{i+C}).
\end{aligned}$$

22. Prove that the rule in Problem 21 has gardens of Eden, i.e., is not surjective.

23. Show that the support of the space–time of $T(x)_i := x_{i-1} + x_{i+1} \pmod 2$ is the ordinary Pascal's triangle of binomial coefficients reduced modulo 2.

PERIODIC POINTS

24. Show that if a single pixel is periodic of period d under a linear rule, then the periods of all other configurations are divisors of d.

25. Show that the sequence of states $\{T^t(x)_0\}_t$ of a linear rule satisfy a linear recurrence relation and hence can be computed without evaluating the entire evolution of an initial configuration x. Likewise at any other cell i instead of the origin.

26. Show that the site value at time t of a cell in a finite torus can be computed in time $O(\log t)$. (A brute-force algorithm requires time $O(t^2)$.) *

ALGEBRAIC–COMBINATORIC

27. Show that if the state set is a local commutative ring, a linear map T is surjective iff some coefficient a_i is a unit. [Decompose the polynomial representation as a sum of a polynomial with unit coefficients and another with nilpotent coefficients, analogously to the proof of Theorem 3.10.]

28. Show that the cyclic semigroup generated by a 2D linear rule of type (3.1) modulo m under the Kronecker product

$$\langle\, a_0, a_1, \ldots, a_n; m \,\rangle \diamond \langle\, b_0, a_1, \ldots, b_n; m \,\rangle := \langle\, a_0 b_0, a_1 b_1, \ldots, a_n b_n; m \,\rangle$$

is a group is and only if $\sum a_j$ is a unit and $a_j a_{j'} = 0$ for all $j \neq j'$. (Such a map is said to have a *group structure*.) [Use the 2D polynomial representation.]

29. Show that every injective map with a group structure over $\mathbf{Z}_m$ has a neighborhood N of size at most n and an inverse with a symmetric neighborhood $-N$. [See Problem 28.]

30. Show that every injective local map T has an iterate T^t with a group structure. [See Problems 28-29.]

31. Show that a 1D euclidean linear rule over a ring A can be given by multiplication by a fixed polynomial $k(z)$ in two indeterminates $1/z$ and z, if configurations are encoded as polynomials $x(z) = \sum_{i\in\mathbf{Z}}$ in $A[\frac{1}{z}, z]$ (with the coefficient of z^i holding the state x_i), i.e.,

$$T(x) := k(z)x(z).$$

(We will refer to $k(z)$ as the kernel of the map again.)

32. Show that, in the polynomial representation of Problem 31, the space–time of a 1D configuration $x(z)$ is given by a rational function

$$M(z,t) = \sum_{i\in\mathbf{Z},t>0} x_{i,t} z^i \tau^t$$

in the variables $\frac{1}{z}, z$ and an auxiliary indeterminate t that keeps track of time, so that $x_{(i,t)} = T^t(x)_i$.

33. Show that the rational function of Problem 32 is actually given by

$$M(z,t) = x(z)/[1 - k(\frac{1}{z}, z)]$$

in variables $\frac{1}{z}$, z and is, therefore, algebraic (satisfies a polynomial equation in the variables in sight, namely $p(x, k, y) := y(1 - xk) - x = 0$).

34. Show that, in the setup of Problem 32, linear rules with the following kernels generate the objects described from the given initial configurations, if allowed to have the states indicated:

   (i) $k(z) := (\frac{1}{z} + z)$; Pascal's triangle from z^0; integers.
   (ii) $k(z) := (\frac{1}{z} + z)$; Catalan numbers at cell 0 from z; integers.
   (iii) $k(z) := (\frac{1}{z} + z)$; paper folding sequence from z^{-1}; 5-tuples of binary states (for a binary automaton with 5-step memory). [Fold a strip of paper an infinite number of times in the same direction, then code the unfolded strip by a 0-1 sequence according to the up or down position of the folds.]

35. Show that the set of all linear rules $\mathrm{LCA}(\Gamma)$ on a given cellular space Γ is isomorphic with the space of finite configuration under the same operations. [The set of rules is endowed with operations $+, *$ of addition and convolution that make it a commutative ring in the algebraic sense.]

36. When the state set is a field with q elements, the space of linear rules over a cellular space Γ is a so-called *group algebra*. Show that the subring of units of this algebra (i.e., the finite configurations with an inverse under convolution) is isomorphic to $\mathrm{RCA}(\Gamma, q)$, the ring of reversible cellular automata with operations of addition and composition.

37. Show that if m and n are relatively prime, then the state sets $\mathbf{Z}_{m \times n} \cong \mathbf{Z}_m \times \mathbf{Z}_n$ are isomorphic rings. Generalize to a finite number of coprime factors. [An isomorphism $\varphi(q) := (X, Y)$ can be defined as the solution (X, Y) of the equation $nX + mY = q \pmod{mn}$ which is unique modulo m in the first component X and modulo n in the second component Y. Use induction on the number of factors.]

38. Prove that every linear cellular automaton L on an arbitrary cellular space Γ where the state set is a finite abelian group A, which decomposes as a product $A \cong \prod_{l=1}^{n} \mathbf{Z}_{p^{r_l}}$ of cyclic groups of prime power order, can be decomposed as a cartesian product of n cellular automata T_l over $\mathbf{Z}_{p^{r_l}}$ whose evolutions are independent componentwise. In particular (a) T is nilpotent iff every T_l is nilpotent; and (b) T^t reduces to a product of the corresponding powers of nonnilpotent factors. (Such a decomposition is always possible.) [See Problem 37.]

39. Generalize Problem 38 to the case where L is a linear self-map of configuration space (not necessarily induced by a local rule).

3.6 Notes

The term *linear* has been used in the cellular automata literature in a number of places to refer to cellular automata over the integer euclidean lattice **Z** of dimension 1. In that case the word *additive* has been used to indicate the idea of linearity used in this chapter. For various reasons that will become apparent in the remaining chapters, we prefer a usage according to the more common mathematical notion of linearity:

Linear local rules have been studied in many places. Early studies include Amoroso–Cooper[A-C] and Barto[B]. Early questions on self-reproduction with simple rules appear in Codd[C]. The treatment of linear rules in this chapter appears explicitly in Aso–Honda[A-H], where most of the results in Sects. 3.1 and 3.2, as well as the solution to Problems 35–38, appear. Theorem 3.11 is also due to Aso–Honda[A-H], who further conjecture that this result holds for general cellular spaces. Theorems 3.5 and 3.6 originally appeared in Ito–Osato–Nasu[I-O-N]. Most of the results in Sect. 3.2.1 are proved in [I-O-N]. The characterizations of injectivity and surjectivity over finite commutative rings is due to Sato [Sa1], where a proof as well as a solution to Problem 27 can be found. Problems 28–30 were motivated by results in his [Sa2]. Problem 19 comes from Sutner [Su1] and Takahashi [Ta1]. A solution to problem 26 appears in Robison[Rob]. S.J. Willson has studied geometric invariants of elementary automata. The (proof of the) results on limit sets and the calculations of the g-dimension in Sect. 3.3 come from Willson[Wi1], where references to other geometric invariants can be found. The generalization to arbitrary linear automata has been established by Takahashi [Ta1, Ta2] using extensions of the same techniques. A purely discrete alternative with similar results has been is given by Martin[M]. The idea of self-reproduction via linear rules in Sect. 3.2.2 is due to Barto[B]. More complex self-reproduction of entire populations of clones from a single individual requires nonlinear rules but has been proved possible on euclidean spaces with only 7 states by Langton[L] - see also Sect. 4.2.2.

An analog of the continuum derivative in the discrete has been long sought in the works of Akers [A], Thayse [T], and Robert [Ro]. The concept was specifically introduced for cellular automata by Vichniac [V], where the boolean derivatives of all 88 elementary 1D rules (up to symmetries) are calculated. Further attempts at relating the notion more explicitly to global behavior appear in the works of Bagnoli, Rechtman, and Ruffo [B, B-R-R1, B-R-R2]. Approximation of arbitrary rules by linear rules has been considered in Bartlett–Garzon[B-G], where details of the results mentioned in Sect. 3.4 can be found. The far more important problem of approximation of the *long-term* behavior of an automaton by linear rules has been addressed by Jen [Je3] for 1D euclidean spaces. These results will be considered in detail in Chapters 9.

Linear automata can have interesting practical applications. They easily generate fractal-like images on a computer screen. They can also be set up to compute recurrence relations. The examples in Problems 31–33 come from

Littow–Dumas[Li]. However, the problems of assigning meaning to and utilizing the beautiful patterns that they generate remain largely open.

Linear automata over other cellular spaces (for instance, cyclic groups) are interesting problems. For instance, boundary-value problems on euclidean 1D automata naturally give rise to automata over the cellular space $\Gamma = Z_n$, which is just a ring of cells. These automata have been studied by Martin–Odlyzko–Wolfram[M-O-W] by using a projection technique over modular dipolynomials (sums of polynomials in x and $1/x$) over a field (the state set) similar to the technique used in Sect. 3.2. Asymptotic behavior of the same kind of automata has also been studied by Jen [Je1, Je2] using linear recurrences over finite fields. Asymptotic behavior and bijectivity of linear automata over finite grids (on abelian group) have been studied by Guan–He [G-H] using circulant matrices. The rule in Problem 21 is due to Muller [Mu] and will provide a counterexample in Chapter 7 – see Problem 7.14. Probabilistic linear automata are considered by several authors, such as Gilman [G] and Bramson–Griffeath [B-G].

In view of how tractable linear automata are, one might try to generalize their results to other type of cellular machines with less uniform connectivity schemes, such as neural networks. This is the subject of later chapters. Little is known, however, about the validity of the results in this chapter if one insists on keeping the superposition principle without locality. A discussion of the superposition principle can be found in Reimen [R], where Theorem 3.13 originated.

References

[A]   Akers, as cited in [V].

[A-C]   S. Amoroso, G. Cooper: Tesselation structures for reproduction of arbitrary patterns. J. Comput. Syst. Science **5** (1971) 455–464

[A-H]   H. Aso, N. Honda: Dynamical Characteristics of linear cellular automaton. J. Comput. Syst. Science **30** (1985) 291–317

[B]   F. Bagnoli: Boolean derivatives and computation of cellular automata. Int. J. Mod. Physics C **3** (1992) 307

[B-R-R1] F. Bagnoli, R. Rechtman, S. Ruffo: Damage spreading and Lyapunov exponents in cellular automata. Phys. Lett. A **186** (1993)

[B-R-R2] F. Bagnoli, R. Rechtman, S. Ruffo: Maximal Lyapunov exponent for 1D boolean circuit automata. In: Cellular automata and cooperative systems. N. Boccara, E. Goles, S. Martinez, P. Picco (eds.). Kluwer, Dordrecht 1993, pp 19–28

[B-G]   R. Bartlett, M. Garzon: Distribution of linear rules in cellular automata rule space. Complex Systems **6**:1 (1992) 519–532

[B]   A.G. Barto: A note on pattern reproduction in tesselation structures. J. Comput. Syst. Science **16** (1978) 445–455

[B-G]   M. Bramson, D. Griffeath: Flux and fixation in cyclic particle systems. preprint

[C]   E.F. Codd: Cellular Automata. Academic Press, New York, 1968

[C-D-D]   P. Cordovil, R. Dilaõ, A. Noronha da Costa: Periodic orbits for additive cellular automata. Discr. Comput. Geom. **1**:3 (1986) 277–288

[G]   R. Gilman: Classes of linear automata. Ergodic Th. and Dynam. Syst. **7** (1987) 108–118

[G-H]   P. Guan, Y. He: Exact results for deterministic cellular automata with additive rules. J. Statistical Physics **43**:3/4 (1978) 445–455

[H]   W. Hurewicz and H. Wallman: Dimension theory. Princeton University Press, Princeton, 1948

[I-O-N]   M. Ito, N. Osato and M. Nasu: Linear cellular automata over Z_m, J. Comput. Syst. Science 27 (1983) 291–317

[Je1]   E. Jen: Cylindric cellular automata. Comm. Math. Physics **118** 1988) 569–590

[Je2]   E. Jen: Linear cellular automata and recurrence systems in finite fields. Comm. Math. Physics **119** (1988) 13–28

[Je3]   E. Jen: Exact solvability and quasi-periodicity of one-dimensional cellular automata. Nonlinearity **4** (1991) 251–276

[L]   C.G. Langton: Self reproduction in cellular automata. In: Proc. Los Alamos workshop on Cellular Automata. North-Holland, Amsterdam, 1983

[L-P]   W. Li, N. Packard: The structure of the elementary cellular automata rule space. Complex Systems **4**:3 (1990) 281–297

[Li]   B. Littow, Ph. Dumas: Additive cellular automata and algebraic series. Theoret. comput. science **119**:2 (1993) 345-354

[M]   B. Martin, Self-similar fractals can be generated by cellular automata. In: Cellular automata and cooperative systems. N. Boccara, E. Goles, S. Martinez, P. Picco (eds.). Kluwer, Dordrecht 1993, pp 463-471

[M-O-W]   O. Martin, A.M.Odlyzko, S. Wolfram: Algebraic properties of cellular automata. Comm. Math. Phys. **93** (1984) 219–258

[Mu]   E.D. Muller, unpublished classnotes, University of Illinois, Urbana.

[R]   N. Reimen, Superposable trellis automata. Dissertation, LITP, Université de Paris VI, 1993. In: I.M. Havel, V. Koubek (eds.), Lecture Notes in Computer Science, Vol. 629. Springer-Verlag, 1992, pp. 472–482

[Ro]   F. Robert: Discrete iterations: a metric study. Springer-Verlag, Berlin, 1986

[Rob]   A.D. Robison, Fast Computation of additive cellular automata. Complex Systems **1**:1 (1987) 211–216

[Sa1]   T. Sato, Decidability of some problems of linear cellular automata over finite commutative rings. Inform. Process. Lett. **46** (1993) 151–155

[Sa2]   T. Sato, Group structural linear cellular automata. J. Comput. Syst. Sci. **49**:1 (1994) 18–23

[Su1]   K. Sutner: Linear celular automata and the garden of Eden. Math Intelligencer **11**:2 (1989) 49–53

[Su2]   K. Sutner: σ-automata on Graphs. Complex Systems **2**:1 (1988) 1–28

[Ta1]   S. Takahashi: Fractal sets in linear cellular automata. In: Proc. 3rd Int. Conference on Cellular Automata, CNLS Los Alamos, *Physica D* **45** (1990) 36–48

[Ta2]   S. Takahashi: Self-similarity of linear cellular automata. J. Comput. Syst. Science **44**:1 (1992) 114–140

[T]   A. Thayse: Boolean calculus of differences. Springer-Verlag, Berlin, 1981

[V]   G. Vichniac: Boolean derivatives on cellular automata. Physica D **45** (1990) 65–74

[Wi1]   S.J. Willson: Cellular automata can generate fractals. Discr. Applied Math. **8** (1984) 91–99

[Wi2]   S.J. Willson: Computing fractal dimension for additive cellular automata. *Physica D* **24** (1987) 190–206

4. Semi-totalistic Automata

Aye on the shores of darkness there is light,
And precipices show untrodden green,
There is a budden morrow in midnight,
There is a triple sight in blindness keen.
                                          John Keats

In view of how tractable linear rules turn out to be, one is encouraged to investigate similar questions for more general rules. A natural next step is to make the next state of a center cell depend, not linearly on the full local distribution of neighboring cells, but rather on the their *density* and, possibly, its own state. For instance, under a majority rule for an elementary celullar automaton the center cell polls its neighbors for a state and goes with the majority (ties are broken arbitrarily by the center cell, for instance by keeping its current state). These rules are called semi-totalistic. In the particular case of elementary automata, it is necessary to reduce this total count to a binary value. The simplest way to achieve this reduction is to set up a minimal threshold value for the count to become 1.

Totalistic rules are no slight generalization, however, since they will be shown in this chapter to be already computation universal on euclidean spaces, even in dimension $d = 1$. In other words, they are capable of implementing arbitrary algorithms (simulating arbitrary Turing machines). In particular, their global behavior is algorithmically unpredictable because there is no general procedure to shortcut the computation performed by arbitrary Turing machines. The only way to predict their behavior is to take the whole tour of running them from the given initial configurations.

4.1 Semi-totalistic Rules

In the treatment of semi-totalistic rules it makes sense to consider arbitrary state sets with an additive structure. However, since renaming symbols (formally, going to an isomorphic activation set) will not really yield different rules, we can rename the m elements of a ring so it becomes a subset $\{0, 1, \ldots, m-1\}$ of the set of nonnegative integers $\mathbf{N}$. We will also assume that the quiescent state (which we assume is the neutral element for addition) has been renamed 0. Recall that $N_\bullet$ is the expanded neighborhood including the center cell.

Definition 4.1 *A local (global) dynamics δ is* totalistic *if there exists a function* $f : \mathbf{N} \to Q$ *such that* $f(0) = 0$ *and*

$$\delta(x_0 x_1 \dots x_d) = f(\sum_{j \in N} x_j).$$ (4.1)

The rule is semi-totalistic if for each state $x_0 \in Q$, there exists a function $f_{x_0} : \mathbf{N} \to \mathbf{N}$ such that

$$\delta(x_0 x_1 \dots x_d) = f_{x_0}(\sum_{j \in N} x_j),$$

that is, if each of the restrictions of δ to a fixed center cell state is a totalistic rule.

Thus, the next-state of a semi-totalistic rule may depend not only on the neighborhood sum of the center cell but also on the state of the center cell. If further, the next-state is independent of the state of the center state as well, the rule is called totalistic.

Semi-totalistic rules are, in a sense, about the mildest generalization of linear rules. Unlike linear rules, however, their global behavior is already enormously complex. For instance, the usual algorithmic problems posed by their local rules are unsolvable.

4.1.1 An Example: Conway's Game of LIFE

LIFE is a semi-totalistic elementary rule defined on a 2D euclidean space with a Moore neighborhood. It is almost the perfect example of perfectly planned social development. A cell is born (i.e., it switches states from 0 to 1), if exactly three of its eight neighbors (parents?) are alive in the previous generation. A cell dies either of overpopulation (it has more than 3 live neighbors) or of isolation (it has fewer than 2 active neighbors). It is very easy to write a computer program (or just use an electronic spreadsheet) to observe successive generations of LIFE on a computer screen. A great variety of patterns and behavior can be observed for small initial configurations such as pixels and other *seeds* (initial configurations). One can even construct pieces of softmatter that behave like electrons or logical gates, and combine them into soft von Neumann computers that compute by evolution of their rules. Thus, despite its apparent simplicity, LIFE is unpredictable. The reader is referred to [B-C-G, Chap. 25] for a thorough description of the most interesting features of LIFE, including a garden of Eden.

4.1.2 Nomenclature for Totalistic Rules

The local rule of a semi-totalistic cellular automaton can be displayed on a table each of whose rows are headed by the current states of a cell and each of whose columns are headed by net-input sums. The entry in the x_0-row and the s-column is the next-state of the center cell when it is in state x_0 and has total neighborhood sum s. For instance, LIFE is given in Table 4.1.

Table 4.1. The moves of LIFE

$x_i \backslash \sigma$	8	7	6	5	4	3	2	1	0
0	0	0	0	0	0	1	0	0	0
1	0	0	0	0	0	1	1	0	0

Table 4.2. The moves of rule 10

$x_i \backslash \sigma$	8	7	6	5	4	3	2	1	0
0,1	0	0	0	0	0	1	0	1	0

A totalistic rule is defined by a table with identical rows since the next state is independent of the center cell. In this case, the *Wolfram number* of the rule can be somewhat simplified since the only row can be seen as the the expansion in radix m (the number of states) of an integer. Vice versa, given a *fixed* cellular space and a *fixed* neighborhood N, a nonnegative integer n defines a unique totalistic rule on the corresponding cellular space with the given neighborhood whose defining row is the m-ary expansion of n, (possibly with padding 0s as highest significant values to complete the required $(m-1)|N|$ digits to base m). For instance, in the 1D elementary euclidean cellular space, rule 10 is the familiar SUM_MOD2 rule. In 2D elementary euclidean space with the same neighborhood though, rule 10 is given in Table 4.2, whereas the SUM_MOD2 rule has number 170.

4.2 Construction and Computation Universality

Computation universality refers to the ability of a device to implement any algorithm. The term goes back to von Neumann's investigation [vN] of the computation universality of a 2D euclidean cellular automaton, which was initially considered with the property of construction universality. Construction universality requires the existence of a configuration c^0 with the following two properties:

1. c^0 is self-reproducing, that is, if at time $t = 0$ the initial configuration is c^0 (or more precisely, the restriction to its support), then at some later time there will be two disjoint copies of (the nonquiescent part) of c^0 in the current configuration of the host universe;

2. Upon being given another configuration (or its description) x in its neighborhood, c^0 will proceed to build a disjoint copy of x.

Von Neumann initially constructed a rule with 29 states capable of simulating an arbitrary Turing machine. Several authors have refined the same ideas and reduced the number of states to 8. Their description of the rule, however, still requires over 100 printed pages. A summarized version of the constructions can be found in [L]. In this section we deal with simpler, although partial, solutions to some of these questions.

4.2.1 Computation Universality of LIFE

More recently, careful examination of the development of particular configuration in the game of LIFE has been used to give a "pragmatic" proof of universality. Specifically, using certain configurations called 'gliders', LIFE enthusiasts were able to assemble streams of electron-like particles running on soft-wires. When directing them into certain other configurations, they were also able to simulate logical gates ANDs, ORs, and NOTs. The rest is easy to guess: gates of this type can now be assembled into a soft implementation of any (finite) *physical* computers, which are supposed to be capable of implementing any algorithm. Therefore, the game of LIFE is a 2D computation universal semi-totalistic cellular automaton, in the sense that one can build within it a copy of every von Neumann machine that behaves as such under the evolution of the rule.

One can add one more important property to universal models of computation. A *reversible* cellular automaton BBM, called the Billiard Ball model of computation, is a *block* rule, that is, one in which blocks of cells are updated together (see Problem 2.14). The 2D euclidean space is partitioned as follows.

x_4	x_3	x_4	x_3
x_2	$\mathbf{x_1}$	$\mathbf{x_2}$	x_1
x_4	$\mathbf{x_3}$	$\mathbf{x_4}$	x_3
x_2	x_1	x_2	x_1

One partition consists of blocks shown in the foreground (centered, with entries in boldface) and the other consists of displaced blocks as shown in the background. On odd (even) iterations, the background (respectively, foreground) blocks are updated. Thus information is transmitted between blocks. The local rule is rotationally symmetric, and so given by the following 6 transitions up to a rotation:

$$0000 \mapsto 0000 \quad 0110 \mapsto 1001 \quad 0101 \mapsto 0101$$
$$1101 \mapsto 1101 \quad 1111 \mapsto 1111$$

The rule can be more compactly represented in polynomial form if the four binary inputs to the local rule are coded in hexadecimal, i.e., as integers in $\mathbf{Z}_{16}$. An input block $(x_1, ..., x_4)$ is mapped to an output block given by the 4-variate polynomial of the form

$$
\begin{aligned}
p_{16}(x_1, x_2, x_3, x_4) = \ & c_1 x_1 + c_2 x_2 + c_3 x_3 + c_4 x_4 + \\
& c_5 x_1 x_2 + c_6 x_1 x_3 + c_7 x_1 x_4 + c_8 x_2 x_3 + c_9 x_2 x_4 + c_{10} x_3 x_4 + \\
& c_{11} x_1 x_2 x_3 + c_{12} x_1 x_2 x_4 + c_{13} x_1 x_3 x_4 + c_{14} x_2 x_3 x_4 + \\
& c_{15} x_1 x_2 x_3 x_4
\end{aligned}
$$

An additional bit x_0 is required to indicate how to interpret (foreground or background) the input block: set $x_0 = 1$ for odd iterations (foreground) and $x_0 = 0$ for even iterations (background). Thus, a polynomial representation for the BBM is given by

$$\delta_{32}(x_0, x_1, x_2, x_3, x_4) = x_0 p_{16}(x_1, x_2, x_3, x_4) + (1 - x_0)p_{16}(x_4, x_3, x_2, x_1) + 16(1 - x_0) \pmod{32}$$

The last term $16(1 - x_0)$, serves to set $x_0 = 1$ on even iterations. Note that the coefficients c_l of p_{16} must now be taken modulo 32 and can be found by solving a system of equation using the local rule as follows. Plugging in the values of single pixels one obtains

$$c_1 = 0001_2 = 1 \quad c_2 = 0010_2 = 2 \quad c_3 = 0100_2 = 4 \quad c_4 = 1000_2 = 8$$

Plugging in values with two pixels on one obtains

$$c_1 + c_2 + c_5 = 1100_2 \quad \longrightarrow \quad c_5 = 11$$
$$c_1 + c_3 + c_6 = 1010_2 \quad \longrightarrow \quad c_6 = 5$$
$$c_1 + c_4 + c_7 = 0110_2 \quad \longrightarrow \quad c_7 = 29$$
$$c_2 + c_3 + c_8 = 1001_2 \quad \longrightarrow \quad c_8 = 3$$
$$c_2 + c_4 + c_9 = 0101_2 \quad \longrightarrow \quad c_9 = 27$$
$$c_3 + c_4 + c_{10} = 0101_2 \quad \longrightarrow \quad c_{10} = 23$$

Plugging in values with three pixels on one obtains

$$c_1 + c_2 + c_3 + c_5 + c_6 + c_8 + c_{11} = 1110_2 \quad \longrightarrow \quad c_{11} = 20$$
$$c_1 + c_2 + c_4 + c_5 + c_7 + c_9 + c_{12} = 1101_2 \quad \longrightarrow \quad c_{12} = 1$$
$$c_1 + c_3 + c_4 + c_6 + c_7 + c_{10} + c_{13} = 1011_2 \quad \longrightarrow \quad c_{13} = 13$$
$$c_2 + c_3 + c_4 + c_8 + c_9 + c_{10} + c_{14} = 0111_2 \quad \longrightarrow \quad c_{14} = 28$$

Finally, all 4 pixels on gives

$$\sum_{i=1}^{14} c_i + c_{15} = 1111_2 \quad \longrightarrow \quad c_{15} = 15$$

Thus, we have a polynomial representation of degree 5 for a computation universal *reversible* CA.

4.2.2 Constructibility and Self-reproduction

LIFE and BBM, however, are not construction universal, and the question remains whether computation universality is a necessary condition for construction universality. That this is not the case was already established in Chapter 3 since

```
2 2 2 2 2 2 2
2 1 7 0 1 4 0 1 4 2
2 0 2 2 2 2 2 0 2
2 7 2       2 1 2
2 1 2       2 1 2
2 0 2       2 1 2
2 7 2       2 1 2
2 1 2 2 2 2 2 2 1 2 2 2 2 2
2 0 7 1 0 7 1 0 7 1 1 1 1 1 2
  2 2 2 2 2 2 2 2 2 2 2 2 2
```

time = 0

```
2 2 2 2 2 2 2                          2 2
2 7 0 1 7 0 1 7 0 2                   2 1 1 2
2 1 2 2 2 2 2 2 1 2                     2 1 2
2 1 2       2 7 2                       2 1 2
2 1 2       2 0 2                       2 1 2
2 1 2       2 1 2                       2 7 2
2 1 2       2 7 2                       2 0 2
2 0 2 2 2 2 2 2 0 2 2 2 2 2 2 2 2 2 1 2
2 4 1 0 4 1 0 7 1 0 7 1 0 7 1 0 7 1 0 7 2
  2 2 2 2 2 2 2 2 2 2 2 2 2 2 2 2 2 2 2
```

time = 70

```
2 2 2 2 2 2 2        2 2 2 2 2 2 2
3 0 1 1 1 1 1 7 0 2    2 1 7 0 1 7 0 1 4 2
2 4 2 2 2 2 2 2 1 2    2 0 2 2 2 2 2 2 0 2
2 1 2       2 7 2    2 7 2       2 1 2
2 0 2       2 0 2    2 0 2       2 4 2
2 4 2       2 1 2    2 1 2       2 0 2
2 1 2       2 7 2    2           2 1 2
2 0 2 2 2 2 2 2 0 2  2 2 2 2 2 2 2 2 1 2
2 7 1 0 7 1 0 7 1 0 7 1 0 7 1 0 7 1 1 1 2
  2 2 2 2 2 2 2 2 2 2 2 2 2 2 2 2 2 2 2
```

time = 120

```
2 2 2 2 2 2 2        2 2 2 2 2 2 2
2 0 1 7 0 1 7 0 1 2    2 4 0 1 4 0 1 1 1 2
2 7 2 2 2 2 2 2 7 2    2 1 2 2 2 2 2 2 1 2
2 1 2       2 0 2    2 0 2       2 1 2
2 1 2       2 1 2    2 7 2       2 7 2
2 1 2       2 7 2    2 1 2       2 0 2
2 1 2       2 0 2    2 0 2       2 1 2
2 1 2 2 2 2 2 1 2    2 2 7 2 2 2 2 2 2 2 2 7 2
2 0 4 1 0 4 1 0 7 1 5 2 1 0 6 1 0 7 1 0 2
  2 2 2 2 2 2 2 2      2 2 2 2 2 2 2 2
```

time = 127

```
                                  1
                                2 1 2
                                2 7 2
                                2 0 2
                                2 1 2
2 2 2 2 2 2 2        2 2 2 2 2 2 2        2 2 2 2 2 2 2        2 2 2 2 2 2 2
2 1 7 0 1 7 0 1 7 2    3 0 1 4 0 1 1 1 2    2 1 1 1 7 0 1 7 0 2    2 1 7 0 1 4 0 1 4 2
2 0 2 2 2 2 2 2 0 2    2 4 2 2 2 2 2 2 1 2    2 1 2 2 2 2 2 2 1 2    2 0 2 2 2 2 2 2 0 2
2 7 2       2 1 2    2 1 2       2 7 2    2 1 2       2 7 2    2 7 2       2 1 2
2 1 2       2 7 2    2 0 2       2 0 2    2 0 2       2 0 2    2 1 2       2 1 2
2 1 2       2 0 2    2 7 2       2 1 2    2 4 2       2 1 2    2 0 2       2 1 2
2 1 2       2 1 2    2 1 2       2 7 2    2 1 2       2 7 2    2 7 2       2 1 2
2 1 2 2 2 2 2 2 7 2  2 0 2 2 2 2 2 2 0 2  2 0 2 2 2 2 2 2 0 2  2 1 2 2 2 2 2 2 1 2 2 2 2 2
2 1 0 4 1 0 4 1 0 5  2 7 1 0 6 1 0 7 1 2  2 4 1 0 7 1 0 7 1 2  2 0 7 1 0 7 1 0 7 1 1 1 1 1 2
  2 2 2 2 2 2 2 2      2 2 2 2 2 2 2 2      2 2 2 2 2 2 2 2      2 2 2 2 2 2 2 2 2 2 2 2 2
```

time = 128   time = 151

Fig. 4.1. The basic self-reproductive cycle in Langton's rule

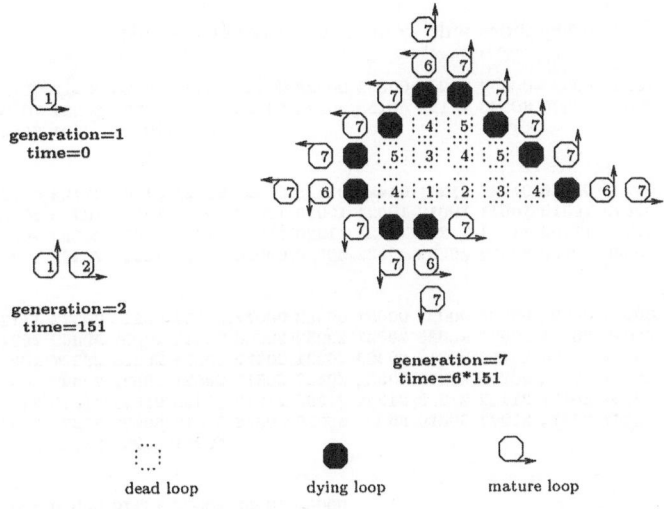

generation=1
time=0

generation=2
time=151

generation=7
time=6*151

dead loop   dying loop   mature loop

Fig. 4.2. Successive generations in the evolution of Langton's colony

some linear rules have the self-reproducing property, which may be interpreted as saying that in their spaces the all-quiescent configuration is construction universal, but they were seen to be fairly predictable (see, for example, Problems 3.25–26) and hence, not computation universal.

A perhaps more interesting example is Langton's rule, in which a specific configuration self-reproduces to generate an entire population of individuals with a finite life span in a 2D universe (a weak form of construction universality). The basic idea goes back to von Neumann's encoding of a genotype of the individual that can be either *interpreted* as a process that generates a new offspring, or just copied *uninterpreted* into it to finish the reproductive process. Communication is effected by 'signals' consisting of two-state packets 70 traveling at light speed down a wire of 1s sheathed by 2s. The local rule produces a copy of the signals when it hits a T-joint, as illustrated in Fig. 4.1 at time $t = 70$. If the wire is capped with a lid of 2s, six signals 07 extend the wire by six cells, while two signals 40–40 build a left-hand corner at the end of the wire. By folding the wire into an Adam loop and placing in it the appropriate signals (the genotype) 70-70-70-70-70-70-40-40, as shown at time $t = 0$ in Fig. 4.1, one obtains a loop that produces a side and corner at every run around until it's built another square attached to the original parent (time $t = 120$) by a sort of umbilical chord. The collision of signals at the joint creates a signal 5 that severes the chord back into the parent (time $t = 128$) and travels upward to regenerate its arm at 90° with respect to the original one (time $t = 128$-151). At the same time, the collision has generated in the offspring a signal 60 which travel down its side, pierces its corner, and builds a copy of the parent's arm, while also spawning a 70-signal that will regenerate a copy of the parent's geno-

Table 4.3. NEWS neighborhoods' transitions in Langton's rule

00000	00020	40122	00030	00050	40221	00022	00023	00032	00121	00220	00222	00230	
00250	10217	20052	20225	30122	40121	40152	40225	50025	50222	51222	60221	70121	$\longrightarrow$ 0
									70122	70152	70221	70227	

00070	01223	01224	01226	01227	01257	01422	01423	01424	01427	01652	01722	01725	
01726	01727	02572	10010	10011	10012	10021	10015	10060	10110	10111	10212	10221	$\longrightarrow$ 1
11121	11122	11152	11162	11221	11222	11223	11226	11252	11322	20070	20223	20525	
21162	30024	30040	30120	30215	40222	40322	60010	60020	61231	70222	70223	70252	

00010	00011	00012	00021	00026	00027	00031	00062	00072	00120	00223	00522	11125	
20010	20012	20020	20021	20022	20025	20027	20032	20040	20042	20051	20062	20072	
20120	20121	20122	20124	20127	20212	20220	20221	20222	20224	20226	20227	20230	
20324	20250	20251	20254	20255	20272	20321	20422	20512	20521	20522	20622	20627	$\longrightarrow$ 2
20721	20722	20724	20727	21122	21222	21242	21262	21272	21422	21522	21622	21722	
22244	22264	22267	22272	22277	30020	30026	50020	50032	50072	50220	50221	50227	
										50251	51221	51224	51227

00060	10262	20024	20270	30010	30021	$\longrightarrow$ 3

10042	10142	10224	10242	10246	11142	11224	11242	12234	12242	12342	50242	$\longrightarrow$ 4

00025	00221	01225	01752	12452	12572	20075	20527	50012	50022	61221	61222	70225	$\longrightarrow$ 5

10220	10226	20023	20312	20322	30070	40223	$\longrightarrow$ 6

10070	10072	10172	10223	10227	10272	10276	10524	11172	11227	11272	12245	12272	$\longrightarrow$ 7
										12372	12462	20015	70070

type. The net effect is that the initial loop has generated an identical copy to its right and has rotated itself 90° clockwise (see times $t = 0$ and $t = 151$). By the homogeneity of the cellular space, parent and child now proceed to live 'independent' lives, repeating the same job every 151 generations until they run out of space. When a parent does, it finally stabilizes into a periodic cycle (death?). The relevant evolutionary process of the entire colony can be appreciated in the sequence Fig. 4.2. All is due to the surprisingly few laws in the local rule on the 2D von Neumann neighborhood NEWS given by Table 4.3.

4.2.3 Provable Computation Universality

A formal proof of universality, however, must take the form of simulation of an arbitrary Turing machine, or an equivalent most powerful sequential computing device, with its potentially infinite memory. It is easy to simulate a Turing machine

$$M = \langle Q', \Sigma, \delta', q_0, F \rangle$$

by a cellular automaton on the 1D euclidean universe with the Moore neighborhood. The state set of the automaton is the set $\Sigma \times (Q' \cup \Sigma)$. Words over Σ are encoded as the corresponding configurations of finite support and given as input configurations to the automaton, except that the left-most symbols of the input x_0 word is encoded as (q_0, x_0). At any time during the evolution of the initial configuration, only *one* cell is in a state whose second component coming from Q' encodes the current position of the tape where the finite control of the original Turing machine would be located. The local rule δ just updates the state of a cell to reflect the actions of this head by δ' on its tape. Thus an initial configuration is quasiperiodic if and only if the Turing machine M halts on its input. This is a *real-time* simulation, that is, it only takes one step of the cellular automaton to simulate one step of the Turing machine.

One may now ask if *totalistic* 2D rules are also computation universal. Wolfram [Wo2] conjectured that even totalistic 1D euclidean automata are capable of simulating a universal Turing machine as well. This conjecture has been confirmed.

Theorem 4.2 *Every Turing machine with m tape symbols and n states can be simulated in twice real-time by a 1D totalistic cellular automaton on the Moore neighborhood with $O(m^2 n)$ states.*

Based on known constructions of universal Turing machines with few states and tape symbols [S, K-O], one can further obtain computation universal 1D cellular automata with nearly 10,000 states. This result poses the question of whether the number of states is optimal. One can decrease the number to 967 by considering semi-totalistic rules.

Theorem 4.3 *Every Turing machine with m tape symbols and n states can be simulated in twice real-time by a 1D semi-totalistic cellular automaton on the Moore neighborhood with $O(m^3 n)$ states.*

This result raises the question of the most efficient simulation by a totalistic automaton. Improvements are certainly possible.

Theorem 4.4 *For every universal cellular automaton with n states there exists a computation universal totalistic 1D cellular automaton with $4n$ states.*

Using the same cellular automata, one can reduce the number of states to 14. Clearly, there is a tradeoff between the number of states, the slow-down factor and the size of the neighborhood of the cellular automaton. Gordon [Go] conjectures a bound $\Omega(m^3 n)$ on the number of states required by a 1D totalistic computation universal cellular automaton in real time. The minimun number of states required by computation universality may be much smaller, given the complex nature of automata with 2 and 3 states. For instance, universal 1D cellular automata exist for radius $r = 1$ with 7 states, and $r = 2$ and 4 states

[L-N]. Banks [B] has shown that 3 states suffice for 2D automata on finite configurations only, and the minimum 2 works if periodic infinite configurations are allowed in coding inputs.

Naturally, all the results in this section remain valid for higher dimensional automata, since every 1D cellular automaton can be obviously embedded in higher dimensional spaces.

Proof. (of Theorem 4.4). The idea is to code configurations of a given cellular automaton by configurations in an enlarged cellular space so that the center cell can recognize the states of the various cells in its neighborhood despite the fact that it only sees the total sum of their states. One can do this efficiently by considering the states in Q' of the given cellular automaton δ' as integers in base $\rho := n + 1$ and the additional states obtained by multiplying them by 10, 100, and 1000 (in base ρ). Over the state set

$$Q := \{1, 10, 100, 1000\} \times Q',$$

a subblock of length 4 such as $x_{-2}x_{-1}x_0x_1$ in a configuration x over Q' can be encoded as

$$100x_{-2} \; 10x_{-1} \; x_0 \; 1000x_1$$

while the rest of x is coded likewise in blocks of length 4. A totalistic automaton δ can now recognize its relative position by the missing digit in base ρ of the total neighborhood sums (in ρ-ary) $0x_{-2}x_{-1}x_0$, $x_10x_{-1}x_0$, $x_1x_20x_0$ and $x_1x_2x_30$ at its disposal. Thus it is clear that the totalistic map given by

$$\begin{aligned}
f(0x_2x_{-1}x_0) &:= 10\delta'(x_{-2}x_{-1}x_0) \\
f(x_10x_{-1}x_0) &:= \delta'(x_{-1}x_0x_1) \\
f(x_1x_20x_0) &:= 1000\delta'(x_0x_1x_2) \\
f(x_1x_2x_30) &:= 100\delta'(x_1x_2x_3)
\end{aligned}$$

will provide a simulation of δ' under the given encoding.   □

4.3 Restricted Totalistic Rules

Along the same lines, one can ask what other properties can be assumed on a totalistic cellular automaton without taking away computation universality. A desirable property may be spatial symmetry of the local rule, i.e.,

$$\delta(x_ix_{i+N}) = \delta(x_ix_{i-N}),$$

which is particularly interesting because it extends to the global dynamics. For example, in euclidean dimension 1 with the von Neumann neighborhood this property amounts to the identity

$$\delta(x_{-1}x_0x_1) = \delta(x_1x_0x_{-1}).$$

Theorem 4.5 *Every 1D cellular space on the von Neumann neighborhood can be simulated in real-time by a 1D symmetric cellular space with the same neighborhood.*

Another possibility is to restrict the ability of the center cells to detect information from some of the cells in the neighborhood. It is fairly easy to see that it suffices to have at least one cell in a set of independent directions and their opposites (e.g., NEWS in the euclidean plane; see Problem 2.13). It is not immediately obvious that the same can be done with a yet smaller neighborhood. This is the case, in fact, in euclidean dimension 1.

Theorem 4.6 *A 1D cellular automaton with n states can be simulated at half-speed by a one-way 1D automaton with at most $n^2 + n$ states.*

The simulation produces a cellular automaton capable of simulating any other 1D cellular automaton (even on infinite configurations), i.e., this automaton is *universal* in the class of all 1D cellular automata.

There is another property that plays a basic role in computational considerations, namely "backward determinism", or more precisely, reversibility. Usually computation is seen as a one-way, forgetful, time irreversible process. It's impossible to resolve the ambiguity in the operands knowing only the total sum. Yet this property is not incompatible with universality or even one-wayness.

Theorem 4.7 *There exists a computation universal one-way 1D cellular automaton.*

Proof. Recall that computation universal Turing machines themselves can be made reversible, even with one tape and 2 states. The automaton can actually be contructed as such a Turing machine. The construction requires a so-called *partitioned cellular automaton* (pca for short in this proof). Each state $s \in Q$ consists of three components s^l, s^c, s^r referred to as the left l-state, center c-state and right r-state, respectively. The local rule is given a next state $\delta(x_{-1}^r x_0^c x_1^l)$ that depends on the nearest neighbors as indicated.

The key property of δ to establish is that it is reversible iff the global map is reversible. It will be seen in Chapter 7 that reversibility is equivalent to injectivity. So assume $T(x) = T(y)$ for $x \neq y$, say with $x_k \neq y_k$, i.e., x_k and y_k differ in at least one of the of three components. Possibly after a shift, assume $k = 0$. Since $T(x)_k = T(y)_k$, also

$$\delta(x_{-1}^r x_0^c x_1^l) = \delta(y_{-1}^r y_0^c y_1^l), \tag{4.2}$$

so δ cannot be injective either. For the converse, assume that T is reversible but δ is not. Let x_0, y_0 be two states collapsed by δ. One can easily extend these to two configurations identical in the remainder of the space so they would contradict the injectivity of T. □

This topic of universality for classes of cellular automata (not just algorithms) will be resumed later after neural networks have been introduced.

4.4 Threshold Automata

As a particular case of a totalistic cellular automaton, and inspired by an obvious biological analogy, one can consider rules that consist of a two-step transition. In the first step, the states of the neighboring cells are multiplied by real numbers acting as weights and the products added to obtain a weighted sum, called *net-input*. In the second step, a special type of totalistic rule, a threshold function f, is applied to the net-input. The rules thus obtained can be called *linear-threshold* local rules.

Definition 4.8 *A cellular automaton is* linear-threshold *if the function f in equation (4.1) is of the form*

$$\delta(x_0 x_1 \ldots x_d) = \begin{cases} 1, & \text{if } A \cdot x \geq \theta; \\ 0, & \text{otherwise} \end{cases} \tag{4.3}$$

where A is a $1 \times d$ vector of real numbers, x is the binary vector of neighboring states $(x_0 x_1 \ldots x_d)$, $A \cdot x$ is the usual dot product of A and x, and θ is a (threshold) real vector.

These automata will be of particular interest in relation to neural networks and they will be examined carefully in Chapter 7.

It is abundantly clear from all results in this chapter that most algorithmic questions about totalistic rules are expected to be unsolvable. Algorithmic problems about rules in general will be examined in detail in the next chapter. Here we only mention some results for finite spaces. On a finite boolean space $\mathbf{B}^d$, the global dynamics of a linear-threshold rule becomes periodic on every input configuration by the pigeon-hole principle, that is, every configuration is T-quasiperiodic. However, it is interesting to note that this period is at most 2 in case the weights in A are symmetric, i.e., the configuration eventually becomes stable or oscillating with period 2 when the weight on every arc ij equals the weight on the arc ji. The proof depends on the following lemma on the behavior of an elementary threshold automaton on the hypercube $\mathbf{B}^d$, the proof of which can be found in [G-M], along with many other results on finite automata and networks (see Problem 6).

Lemma 4.9 *If $T : \mathbf{B}^d \to \mathbf{B}^d$ is a symmetric global dynamics on $\mathbf{B}^d$ then every configuration has period at most 2.*

This result implies that if the elementary global dynamics of a local rule is nonexpansive, i.e., if there exists an absolute constant κ such that

$$\kappa \leq |\underline{x}|$$

for all x and $t \geq 0$, then T eventually becomes cyclic with period 2. In particular,

Corollary 4.10 *If T is an elementary symmetric linear-threshold rule on a finite cellular space, then for all x, there exists $s \geq 0$ such that*

$$T^{s+2}(x) = T^s(x).$$

4.5 Problems

1. Count the number of totalistic elementary cellular automaton rules on the 2D euclidean, hexagonal, and triangular cellular spaces. Show they are asymptotically a negligible fraction of the total number of rules.

2. Show that every 1D elementary cellular automaton can be represented by a polynomial of the form

$$\begin{aligned}
\delta(x_{-1}x_0x_1) =\ & a_0 + a_1x_{-1} + a_2x_0 + a_3x_1 + \\
& a_4x_{-1}x_0 + a_5x_{-1}x_1 + a_6x_0x_1 + \\
& a_7x_{-1}x_0x_1 \quad (\text{mod } 2)
\end{aligned}$$

3. Show that the polynomial representation of Problem 2 is unique. [Count distinct polynomials: they give rise to different rules.]

4. Show that LIFE admits the mod 2 polynomial representation of degree 8 given by

$$\delta(\mathbf{x}) = \sum_{|A|=3} X_A + x_0 \sum_{|A|=4}^{6} X_A + (1-x_0)\sum_{|A|=7} X_A \quad (\text{mod } 2),$$

Here $\mathbf{x}$ denotes (an enumeration of) the Moore neighborhood, and X_A denotes the monomial $\prod_{i \in A} x_i$ for subsets A of $\mathbf{x}$. (This polynomial is, in fact, unique.)

5. Show that LIFE is not a quadric rule over $\mathbf{Z}_m$, for any m. [A quadric rule δ expresses the next state as a homogeneous polynomial of degree 2, or equivalently, a bilinear form $x_{i+N}Bx'_{i+N}$, in the variables of the neighborhood vector x_N and its transpose x'_{i+N}.]

6. Prove Theorem 4.9 by showing that the following *bonding energy* function between a configuration x and its successor $y := T(x)$ given by

$$E(x,y) = -x \cdot (Wy - \theta) + y \cdot \theta$$

decreases with time on the orbit of every initial configuration (W is the symmetric weight matrix, θ the threshold vector and "$\cdot$" stands for the ordinary dot product in $\mathbf{R}^d$). (Energy functions are a useful technique in the analysis of global rules.)

7. Simulate a semi-totalistic rule with activation function

$$f_a(\sigma) = a\lfloor sin\sigma \rfloor$$

on various initial configurations. Can you figure out any systematic global behavior on arbitrary configurations? [CaLab]

8. Make up and test a "good" 1D euclidean life. [CaLab]

4.6 Notes

The game of LIFE was invented by J. Conway in the early 1970s and was popularized by an article in M. Gardner's column *Mathematical Recreations* in *Scientific American* [G]. The study of the behavior of LIFE on a number of configurations played a catalytic role in 1970s in the beginnings of the field – see [B-C-G, G]. A more detailed summary of the game can be found in [B-C-G, Vol.2, Chap. 25]. Automata similar to LIFE in other universes also exhibit interesting properties. For instance there is Golay's life on a hexagonal tesselation which has its own gliders and so forth – see Preston–Duff [P-D]. Bays [Ba1, Ba2, Ba3] has investigated meaningful LIFEs 4555, 5766 and 5655 for 3D euclidean spaces.

The question of computation universality was originally posed and solved jointly with construction universality by von Neumann [vN] using 29 states, and later simplified by Banks [B] and Codd [C] using 8 states. Construction universality has not been examined as carefully. The simple solution presented here is due to Langton [L], but it only clones structures of a very special type, so it is not as complete a solution as required by von Neumann (although, in defense, one may argue that a universal constructor for every configuration is somewhat of an overkill). Fredkin's Billiard Ball Model of computation is presented in [M] and elaborated in [T-M]. The polynomial representation is due to Bartlett [Bar]. It is one of the two standard ways to establish that a given type of automaton is complex enough to defy analysis by algorithms for decision problems. By contrast, the other way, analysis of formal universality was initiated by Turing himself [T]. Theorems 4.2 and 4.3 are due to Gordon [Go], while Theorems 4.4 and 4.6 are due to Albert–Culik [A-C]. Theorem 4.5 appears in Kobuchi [K] and Szwerinksi [Sw]. However, one may point out that the more interesting question of whether cellular spaces on more complex architectures can be used to *speed up* the execution of standard Turing machine programs has been left untouched.

Linear-threshold automata play an important role in image processing, where, in practice, most operations are combinations of linear and linear-threshold rules – see Preston–Duff [P-D] for more detail. Other local rules can be tailored for pattern recognition which are not necessarily totalistic – see Jen [J]. A classification of totalistic rules in 3D euclidean binary spaces is given in Bays [Ba4].

Life in 1D euclidean spaces is a puzzling idea. Some results along LIFE lines appear in Dewdney [D]. Rules on finite 1D spaces where the interaction between two cells is a decreasing function of their distance in the cellular space are considered in Goles–Tchuente [G-T]. Since rules on finite spaces are not the main subject of this book, the reader is referred to Goles–Martinez [G-M] for an exposition of these results, particularly with a view of applications and other analytical tools (modeling of physical phenomena, energy functions).

References

[A-C]   J. Albert and K. Culik II: A simple universal cellular automaton and its one-way and totalistic version. Complex Systems **1** (1987) 1–16

[B]   E.R. Banks: Information processing and transmission in cellular automata. Doctoral dissertation, MIT, 1971

[Bar]   R. Bartlett: Discrete computation in the continuum. Doctoral dissertation, Department of Mathematical Sciences, The University of Memphis, 1994

[Ba1]   C. Bays: Candidates for the game of Life in three dimensions. J. Complex Systems **1** (1987) 373–400

[Ba2]   C. Bays: The discovery of a new glider for the game of three-dimensional life. J. Complex Systems **4**:6 (1987) 373–400

[Ba3]   C. Bays: A new game of three-dimensional life. J. Complex Systems **5**:1(1991) 15–18

[Ba4]   C. Bays: Classification of semitotalistic cellular automata in three dimensions. Complex Systems **2**:2 (1988) 235–254

[B-C-G] E.R. Berlekamp, J.H. Conway, R.K. Guy: Winning ways for your mathematical plays, vol. 2. Academic Press, New York, 1982

[K-O]   H. K. Buening, T. Ottman: Kleine Universale mehrdimensionale Turing-maschinen. Elektron. Inf. und Kybernetik **13** (1977) 179–201

[C]   E.F. Codd: Cellular Automata. Academic Press, New York, 1968

[D]   A.K. Dewdney: Computer recreations. Scientific American **252**:5 (1985) 18–24

[G]   M. Gardner: The fantastic combinations of John Conway's new game of 'life'. Scientific American (Oct. 1970) 100–123

[G-M]   E. Goles, S. Martinez: Neural networks: theory and applications, Kluwer, Amsterdam, 1990

[G-O]   E. Goles, J. Olivos: Comportement périodique des fonctions à seuil binaires et applications. Discr. Appl. Math. **3** (1981) 93–105

[G-T]   E. Goles, M. Tchuente: Iterative behavior of one-dimensional threshold automata. Discr. Appl. Math. **8** (1984) 319–322

[Go]   D. Gordon: On the computational power of totalistic cellular automata. Math. Syst. Theory **20**:1 (1987) 43–52

[J]   E. Jen: Invariant strings and pattern recognizing properties of one-dimensional cellular automata. J. Stat. Physics **43**:1-2 (1986) 243–265

[K]   Y. Kobuchi: A note on symmetrical cellular spaces. Inform. Process. Lett. **25**:6 (1987) 413–415

[L]   C.G. Langton: Self-reproduction in cellular automata. In: Proc. Los Alamos workshop on cellular automata. North-Holland, Amsterdam, 1983, pp. 135–144

[L]   C. Lee: Synthesis of a cellular computer. In: Applied automata theory. Julius T. Ton (ed.): Academic Press, 1968, pp. 217–234

[L-N]   K. Lindgreen, M.G. Nordhal: Universal computation in simple one-dimensional cellular automata. Complex Systems 4:3 (1990) 299–318

[M]   N. Margolus: Physics-like models of computation. Physica D **10** (1984) 83–95

[P-D]   K. Preston, Jr, M.J.B. Duff: Modern cellular automata. Plenum Press, New York, 1984

[S]   A.R. Smith III: Simple Computation-universal cellular spaces. J. Assoc. Comput. Mach. **18** (1971) 339–353

[Sw]   H. Szwerinski: Symmetrical one-dimensional cellular spaces. Inform. and Control **67**:1-3 (1985) 167–172

[T-M]   T. Toffoli & N. Margolus: Cellular automata machines (a new environment for modeling). MIT Press, Cambridge MA, 1987

[T]   A.M. Turing: On computable numbers, with an application to the Entschei-dungsproblem. Proc. London Math. Soc. **42** (1936) 230–265. A correction, *ibid* **43** (1936) 544–546

[vN]   J. von Neumann, A. W. Burks (ed.): Theory of self-reproducing automata. University of Illinois Press, Urbana, 1966

[Wo1]   S. Wolfram: Statistical mechanics of cellular automata. Reviews of Modern Physics **55**:3(1983) 601–644. Reprinted in [Wo3]

[Wo2]   S. Wolfram: Universality and complexity in cellular automata. In: Proc. of a conference on Cellular Automata in Los Alamos, North-Holland, Amsterdam, 1983. Reprinted in [Wo3]

[Wo3]   S. Wolfram: Theory and Applications of cellular automata. World Scientific, Singapore, 1986

5. Decision Problems

She whom I love is hard to catch and conquer,
Hard, but O the Glory of winning were she won!
George Meredith

One of the puzzling aspects about local rules of cellular automata is that despite their behavioral complexity, they are describable by finite tables that require only finitely many symbols and can be fed as input to a conventional computer. It is therefore natural to ask whether one could, at least in principle, write a program for a von Neumann computer that would answer questions, as some sort of pre-processing or short-cut, about the effect in the large of local rule inputs. For instance one may ask, is the global effect of the input local rule injective? Is it surjective? If reversible, what is the inverse that will undo the change caused by the rule? And so forth.

Given the complexity exhibited by global rules of very simple types, one would naturally doubt that such programs exist. If they do not, one may still ask, how restricted must a subclass of local rules be to allow algorithmic solutions to these decision problems? We know that linear rules are very much predictable and that, in particular, most of these questions admit an algorithmic solution. On the other hand, in case a problem admits an algorithmic solution, the immediate question arises as to how difficult it is to implement the solution in practice. These considerations give rise to complexity questions that are also of interest from the point of view of prediction of global behavior.

The purpose of this chapter is to take a systematic look at the algorithmic (un)solvability of decision questions for cellular automata, particularly on euclidean universes. Tractability and complexity issues for solvable problems are also explored. The reader is assumed to be familiar with the basic facts of classical computability, as in, for example, Hopcroft–Ullman [H-U]), and complexity, as in, for example, Garey–Johnson [G-J], Bovet–Crescenzi [B-C] or Papadimitriou [P].

5.1 Algorithmic and Dynetic Problems

Classical computability theory is concerned with solving algorithmic problems, i.e., problems that could, in principle, be solved using a computer. For example, one cannot hope to determine whether or not an arbitrary real number is rational using a fancy piece of machinery. The reason is quite simple: an arbitrary real number cannot be actually *given* even to a Turing machine since, regardless of the choice of representation, some reals will always require infinitely many digits. On the other hand, one may ask whether or not a rational number is an integer, since every rational number is expressible as a finite string of symbols over the decimal alphabet. In this case, an *instance* of the algorithmic problem is a rational number given as a finite string.

In general, an *algorithmic problem* consists of a possibly infinite list of words (called *instances* of the problem) over a *fixed, finite* alphabet representing a number of questions, usually with a common theme. The algorithmic problem is usually presented by a typical instance followed by the question being asked about that string. For example, the integrality problem for rationals is the entire list of finite strings over the symbols $\{0, 1, \bar{0}, \bar{1}\}$ describing a rational expansion (the bar is used to express digits in a periodic part) specified as follows.

INTEGRALITY FOR RATIONALS
INSTANCE: a binary string ω over $\{0, 1, \bar{1}\}$
OUTPUT: YES, if ω represents an integer; NO, otherwise.

A more relevant example in this chapter is the problem of injectivity of a local rule over euclidean universes presented as

INJECTIVITY
INSTANCE: the transition table of a rule δ
QUESTION: is the induced global dynamics T injective?

Other problems can be obtained as subproblems by restricting the set of questions. For example, solving 2D INJECTIVITY only requires answering the questions for rules δ interpreted on a 2D euclidean cellular space.

A third example is the PARENT SEARCH problem, where one is not allowed to vary the rule, but rather a given configuration y. The problem consists in finding a predecessor (or parent) that gives rise to y under the fixed rule.

PARENT SEARCH(δ)
INSTANCE: a finite configuration y
QUESTION: does $T_\delta(x) = y$ have a solution for x?

In order to perform a systematic study of the computational difficulty of determining these properties of a cellular automaton, there are two requirements. First, a definition of what it means to solve algorithmic problems. Second, a generalization of the concept in order to include *pictorial* information not encodable as finite strings of symbols and readily available in the form of configurations on an infinite cellular space.

Intuitively, an algorithmic problem is a problem that might, in principle, be solved by a conventional computer with a fixed program but unbounded storage and running time. An algorithmic problem is *solvable* if there exists a (single!) Turing machine M (i.e., a sequential program) that satisfies the following conditions upon input any instance ω of the problem:

- M computes for a finite amount of time (that depends on the input ω), but *always* eventually halts;

- M leaves on the tape the *correct* answer to the question posed for instance ω by the problem it is solving;

- M never changes its transition function in the course of its computation over any particular instance ω.

Thus, Turing computability implicitly assumes that all kinds of information can be encoded as finite *strings of symbols* over a fixed character set. An ever increasing body of evidence, however, points to human use of pictorial representations rather than strings, even in the use of ordinary language. Although finite pictures may be, to some degree, encodable as strings, processing the semantic content of images turns simple tasks into expensive string processes. Sequential processing of images via a single processor only compounds the problem. What is needed is a representation of pictorial information that preserves local relations among image elements (pixels) while still requiring only finitely many symbols, and parallel processing of all pixels by simple, possibly local rules. The same kind of representation also captures very naturally other computational problems.

Cellular spaces afford a generalization of the concept of algorithmic problem for pictorial information processing. First, images and pictures are directly encodable as configurations in a cellular space without recoding into a string language. Secondly, and more importantly, *infinite* amounts of information can, in principle, be given in finite time to a network, and likewise, in a reply encoded as a configuration, a network can return an infinite amount of information as well. Third, due to massive parallelism, an (infinite) network can perform in a single step work that Turing machines will never accomplish in a lifetime of computation. (The issue of physical realizability of the hardware required to implement these processes will be discussed in a general framework in the next chapter.)

Thus, the leading question is now what problems can be solvable using fine-grained models of computation such as cellular automata. With configuration as the given encoding, one may assume that some cells are singled out as *input* nodes while others are looked up as *output* nodes. Initially, suitable states (activation levels) are clamped to the input cells, and the automaton is then *repeatedly updated*. After a certain period of time, the net stabilizes and the output is then "collected" from the output nodes. From an abstract point of view, one just has a set of initial configurations and a mapping (functional or relational) that associates other configurations to them. Thus we can formulate the most general type of problem that could, in principle, be solved using distributed representations as follows.

Definition 5.1 *Let* **C** *be configuration space and* $\mathcal{I}, \mathcal{J} \subseteq \mathbf{C}$ *arbitrary but fixed subsets of* **C**. *A (functional)* dynetic problem $\mathcal{P}$ *is a function*

$$\mathcal{P} : \mathcal{I} \to \mathcal{J}.$$

The members of $\mathcal{I}$ are called instances *or* inputs *of the problem $\mathcal{P}$ and the corresponding member of J is called a* solution *or* output *of the given instance. The symbol $\mathcal{P}$ can also stand for a relation $\mathcal{P} \subseteq \mathcal{I} \times J$ (more than one output is allowed for a given input).*

The concept of solution of a dynetic problem fullfills the motivation of the previous definition.

Definition 5.2 *A* solution *of a dynetic problem is a global dynamics $T : \mathbf{C} \to \mathbf{C}$ of some cellular automaton that, upon iteration, stabilizes on each instance of the problem at the correct output, i.e., it satisfies the following condition:*

> *for each $x \in \mathcal{I}$, there exists a positive integer t (which may depend on x) such that*
> $$T^{t+1}(x) = T^t(x) = \mathcal{P}(x).$$

If $\mathcal{P}$ has a solution it is said to be solvable. *In case $\mathcal{P}$ is a relation, $\mathcal{P}(x)$ can be any element in J so that $(x, T(x)) \in \mathcal{P}$. A dynetic* decision problem *$\mathcal{P} : \mathcal{I} \to J$ is one with only two possible answers (YES/NO configurations) in its range J. A dynetic decision problem is* weakly solvable *if there exists a neural network $\mathcal{N}$ that stabilizes on an input $x \in \mathbf{C}$ (although not always with the same or the correct answer) if and only if $\mathcal{P}(x) =$YES.*

Example 5.3 The STABILITY PROBLEM for neural nets consists of a set $\mathcal{I}$ of encodings $\langle \mathcal{N}, x \rangle$ of a network and its input, where

$$\mathcal{P}(\langle \mathcal{N}, x \rangle) := (\mathcal{N} \text{ stabilizes on input } x),$$

where the boolean answer can be represented by some configurations x_{YES} or x_{NO}.

In particular, when $\mathcal{I}$ and J consist of configurations of finite support over a 1D euclidean space and T is induced by a Turing machine, one recovers as a particular case the notion of an algorithmic problem. This is the case with which this chapter will be mainly concerned. The general case of a dynetic problem will be explored after cellular automata are generalized to include nonhomogeneous models such as neural networks and automata networks in the next chapter.

We first consider algorithmic problems in the simplest type of euclidean universes.

5.2 1D Euclidean Automata

One of the simplest questions to be asked about a fixed dynamics T is whether a given finite configuration y has a (finite) predecessor. The brute force solution to PARENT SEARCH is to enumerate all possible configuration and begin applying T until y is obtained. This procedure, however, requires a lot of work, in fact, too much for a machine. More seriously, it is not always going to halt at all (let alone with the correct answer). What is needed is some analysis that will effectively reduce the number of possible candidates that have to be searched. In the process, a larger problem will also be considered.

UNIFORM PARENT SEARCH

INSTANCE: a 1D transition table δ and a finite configuration y
QUESTION: Does y have a predecessor under the global dynamics T_δ?

Theorem 5.4 *Let $y \in \mathbf{C}_0$. The minimum size of any predecessor $x \in T^{-1}(y)$ satisfies*

$$|\underline{y}| - |N| + 1 \leq |\underline{x}| \leq 2|\underline{y}|(|N| - 1)$$

where $|N|$ is size of the neighborhood of T. Thus, UNIFORM PARENT SEARCH *is decidable on 1D euclidean universes.*

Proof. Information propagates only to neighboring cells in one step, so the lower bound is clear. Likewise, the support of y must be contained in the union of neighborhoods of nonquiescent cells of such an x. The worst case occurs when nonquiescent cells in y are spread so that their neigborhoods are disjoint and cells within distance $|N| - 1$ on each side may affect them.   □

Two other important problems in euclidean dimension one concern injectivity and surjectivity. Their decidability can be proven in several ways. The following proof makes use of the notion of de Bruijn graphs introduced in Chapter 2.

Theorem 5.5 1D INJECTIVITY *is recursively solvable in quadratic time for 1D euclidean cellular automata.*

Proof. Recall that for 1D automata, configurations become labels of biinfinite walks on the deBruijn presentation B_δ of the given rule δ. Let B_δ^2 be the cartesian product digraph with vertices pairs of vertices of B_δ and and arc between vertices (sw, uw') and $(wt, w'v)$ whenever $\delta(swt) = \delta(uw'v)$. Let $\bar{B}^2$ be the digraph resulting from eliminating its nontransient points (which can be done in a number of steps $\theta(n)$ using topological search – see Problem 2.29 in Chapter 2). Pairs of configurations x, y become single biinfinite walks on $\bar{B}^2$. Different states $x_i \neq y_i$ take the walk out of the diagonal Δ, the induced subgraph consisting of the vertices (sw, sw). Thus the existence of $x \neq y$ with the same image implies that $\Delta \neq \bar{B}^2$. Conversely, if Δ does not fill up $\bar{B}^2$, there is a biinfinite path going outside Δ, which can be unfolded into two different configurations with the same image (or label).   □

Theorem 5.6 1D SURJECTIVITY *is recursively solvable in quadratic time for 1D euclidean cellular automata.*

Proof. Use the fact (to be proved in Chapter 6) that a global rule T is onto iff its restriction T_0 to configurations of finite support is injective. In the construction of the previous proof, the latter is equivalent to the condition that the connected component Δ fills up the connected component of $\bar{B}_\delta^2$ that contains it, i.e., $\Delta = C_\Delta$. Indeed, two different finite configurations can be seen as infinite configurations, and as before, identical images give rise to an infinite path in C_Δ that goes outside Δ. Conversely, if C_Δ, by the strong connectivity of C_Δ one can choose an infinite path with only finitely many nodes outside Δ to obtain a witness against injectivity.   □

5.3 2D Euclidean Automata

Higher, even in one more dimension, the algorithmic properties are in sharp contrast with those for 1D automata in very restricted cases.

PIXEL's PARENT SEARCH
  INSTANCE: a local rule δ
  QUESTION: does the equation $T(x) = e$ have a solution x in $\mathbf{C}$?

Theorem 5.7 2D PIXEL's PARENT SEARCH *is recursively unsolvable, i.e., there is no algorithm to decide whether a pixel e has a predecessor (ancestor).*

Proof. Let $M = \langle Q', \Sigma, \Delta', q_0, q_h \rangle$ be an arbitrary Turing machine with halting state q_h. Consider the following instance of this decision problem on cellular automaton with neighborhood $N := \{(-1,0),(0,1),(1,0)\}$ and state set $Q = (Q \times \Sigma) \cup \Sigma \cup \mathbf{B}$. The transition δ is defined as follows:

The only *predecessor of 1:* $\begin{array}{ccc} & (q_0, b) & \\ 0 & 1 & 0 \end{array}$

Successive IDs (instantaneous descriptions) of M's tape can be stored in a 2D configuration in the first quadrant so that time increases in the positive direction of the y-axis. The remaining transitions of δ can be made to reflect the transitions of the Turing machine so as to ensure that Moore neighborhoods for δ (in the first quadrant) only contain chunks of consecutive rows that are legal successive IDs of M. If so, δ maps them to 0, but to 1 otherwise. Thus, if M never halts on blank tape, one can construct a predecessor for the pixel e, and conversely.   □

This result immediately implies that PARENT SEARCH (to which it is immediately reducible) is also unsolvable.

The other problems INJECTIVITY (equivalently, by a result of Richardson from Chapter 6, REVERSIBILITY) and SURJECTIVITY fare no better.

Theorem 5.8 INJECTIVITY *is recursively unsolvable for 2D euclidean rules.*

Theorem 5.9 SURJECTIVITY *is recursively unsolvable for 2D euclidean rules.*

The proofs of these results are given in the following subsections.

5.3.1 Reversibility is Unsolvable

The proof of unsolvability of INJECTIVITY makes heavy use of *tilings* of the plane. The decision problem of whether a set of tiles can tessellate the plane is known to be unsolvable. If we imagine we are given square tiles of various colors from a finite color set P (the palette), one may ask whether it is possible to tile up arbitrarily large squares of the plane using copies (of course, as many as desired) of the given tiles in such a way that they 'match' at the edges. The matching property is specified in advance in the form of a relation $R \subseteq P^d$ that tells him when tiles in neighboring cells are admissible. The neighborhood may be the ordinary Moore neighborhood or the neighborhood of a cellular automaton. This property can be equivalently stated as follows.

Definition 5.10 *Let N be a neighborhood of the 2D euclidean plane $\mathbf{Z}^2$ with d elements $i_1, \ldots, i_d$. Let P be a finite color set and $R \subseteq P^d$ a relation among the colors in P. The triple $\langle N, P, R \rangle$ is called a tiling system. A configuration $x : \mathbf{Z}^2 \to P$ assigning a tile color to each cell is a valid (or proper) tiling (or simply a tiling) of the plane if*

$$\forall i \in \mathbf{Z}^2 \ (x_{i+i_1}, \ldots, x_{i+i_d}) \in R.$$

One has the corresponding algorithmic problem.

TILING
  INSTANCE: a tiling system t
  QUESTION: does t afford a valid tiling of the euclidean plane?

It is a well known result that one cannot decide *algorithmically* if a set of tiles can afford a valid tiling, a fact that will be used shortly.

Theorem 5.11 2D TILING *is recursively unsolvable.*

The strategy of the proof for injectivity is really simple at this point. One proves that 2D TILING becomes a particular case of INJECTIVITY after a suitable reduction. Thus, if INJECTIVITY were solvable, so would be 2D TILING.

The reduction requires a digital version of the idea of a *plane filling* curve. Imagine each tile is associated a *passage*, i.e., a semaphor pair indicating a unique way into a tile and a unique way out of the tile, such as (n, se). There is also a starting passage. If an infinite path of neighboring tiles have matching passages, one can force a snaking infinite path through the plane.

Definition 5.12 *A tiling system* $t := \langle N, P, R \rangle$ *with a set of passages has the* plane filling property *if*

1. *it affords a valid tiling of the plane; and*

2. *On every tiling of the plane afforded by* t, *valid or not, the passages define a path which either contains a nonmatching tile or which visits all tiles of arbitrarily large squares.*

The reduction is accomplished in the next result.

Theorem 5.13 *If a tiling* t_0 *has the plane filling property, then for every tiling system* t *one can effectively construct a cellular automaton* M_t *that is not injective iff* t *affords a valid tiling.*

Proof. The state set of M_t consists of pairs of tiles, one from t_0, one from t, and a bit (0 or 1) attached to each passage of the t_0-component. M_t changes only these bits to indicate tiling errors in the Moore neighborhood in either of the components: if there is an error, the bit is not changed. Otherwise, each passage of the t_0-component adds the bit in the next passage on the path to its own current bit modulo 2. Thus, if t affords a valid tiling of the plane, one can construct configurations x_0, x_1 such that $T(x_0) = T(x_1)$ as follows. Both x_0 and x_1 have in the first two components the legal values of a valid tiling of their tiling systems. In x_0, all bits attached to passages are 0 while in x_1 they are 1. Since the tilings are correct everywhere, in both x_0 and x_1 all bits are changed to 0, as required. Hence T is not injective.

Conversely, if $T(x_0) = T(x_1)$ but $x_0 \neq x_1$, the tile components in their states agree in one time step but their passage bits differ at some position i. The bits in the next passage in the path must also differ (so they agree in the next time step). This argument can be repeated along a plane filling path of a tiling afforded by t_0 to obtain a path through arbitrarily large squares, which implies that the whole plane can be tiled with t. ☐

Therefore all is left to do is exhibit one plane filling tiling. One can be obtained from a refinement of Robinson's tilings in the original proof of undecidability, but the details are somewhat tedious and the construction will be omitted.

Proposition 5.14 *There exists a plane filling tiling of the plane with the Moore neighborhood.*

5.3.2 Surjectivity is Unsolvable

The same technique will not quite work to establish the undecidability of surjectivity. It is necessary to add a constraint to the tilings under consideration.

Assume now that there is a *blank* color in the palette P, denoted b, satisfying the property

$$(b, b, \ldots, b) \in R.$$

A tiling like this is called a *finite tiling* system. Now, every tiling system affords a tiling of the plane, but it may be *trivial*, consisting only of blank tiles. What is interesting now is the problem of whether nontrivial tilings are afforded by the system.

2D FINITE TILING
INSTANCE: a finite tiling system t

QUESTION: does t afford a nontrivial valid tiling of the euclidean plane?

It is not hard to reduce the HALTING PROBLEM for Turing machines on blank tapes to 2D FINITE TILING in a way analogous to the proof of Theorem 5.7.

Theorem 5.15 2D FINITE TILING *is unsolvable.*

One can now establish the unsolvability of surjectivity via Richardson's Theorem (see Chapter 6) by reducing the 2D FINITE TILING to injectivity of the restricted local rule T_0 on finite configurations.

Theorem 5.16 2D FINITE TILING *can be effectively reduced to* 2D INJECTIVITY *on finite configurations.*

A proof can be given by modifying the plane filling tiling of Sect. 5.3.1 to adjust for the trivial tiling, and it will be omitted.

5.4 Noneuclidean Automata

The results of the previous section raise the question of whether the positive results on decidability in dimension one can be maintained while enlarging the class of cellular automata to some other class of cellular spaces than higher dimensional euclidean. A generalization of this type has, in fact, been proven possible for a different type of parameter, namely the number of isomorphism classes of neighborhoods of infinity of the cellular space.

For example, in the 1D case, suppressing neighborhoods of the origin in the integer grid gives rise to digraphs with two connected components, regardless of the size of the suppressed neighborhood, and there are only finitely many (in fact, one) isomorphism types of such components. Likewise for an infinite binary tree. This section presents the proof that this ingredient is sufficient to guarantee algorithms for injectivity and surjectivity in these so-called 'context-free' spaces.

The motivation and the proof techniques are of a different nature than those occurring previously. The proofs consist in reducing the statements to sentences in a fragment of second-order logic of the cellular space, whose validity can be ultimately reduced to statements in a similar logic for the infinite binary tree. A deep and powerful result of Rabin on the decidability of this logic can then be applied. The reader unfamiliar with logical calculi may skip this section without loss of continuity.

The logical calculus in question is called the *monadic second-order logic* (MSL). Like any other formal calculus, it consists of words over a logical alphabet built from certain atomic formulas through boolean combinations and universal/existential quantification. The logic is monadic because quantification is additionally allowed over subsets of the universe of discourse, but under the important restriction that the only predicates allowed as atomic formulas are *one-place predicates*. These predicates originate in the notion of neighbor in a digraph Γ which is *finitely generated*, in the sense that it is locally finite, connected, of uniformly bounded degree, and whose edges are labeled from a finite alphabet. Cayley graphs are obvious examples.

A finitely generated graph is *context-free* if there are only finitely many isomorphism types in the punctured graphs obtained by successively deleting all vertices within a distance n from a fixed origin. An infinite binary tree is a prime example, with only one isomorphism type. Monadic second-order terms τ are either a constant (for the origin), set variables $x, y, z, \cdots$, vertex variables $i, j, k, \cdots$, or applications to these of unary predicates σ_A (one for each edge label A, signifying passage to a neighboring vertex of Γ). The *atomic formulas* are of the type $\tau \subseteq \tau'$. General formulas in the monadic second-order theory MST(Γ) of the cellular space Γ are obtained from the atomic ones in the usual way (boolean combinations, quantifications and particularizations of free variables in well-formed subformulas). A detailed example is in order.

Example 5.17 The properties of injectivity and surjectivity of a cellular automaton on Γ are expressible as well-formed formulas in MST(Γ). We illustrate with the elementary 1D euclidean cellular space, where $N := \{-1, 1\}$, and the rule δ that maps $\delta(101) = 1$ and to 0 elsewhere. We need to say in this restricted language that for every disjoint cover y^0, y^1 of Γ (describing a configuration y and its partition into pixels in state 0 and 1, respectively), there is another disjoint cover x^0, x^1 such that for every vertex i, $i \in y^l$ if and only if the pixels in the neighborhood of i are at the right values so that δ maps them to y_i, for each $l = 0, 1$. Now, it is easy to come up with a formula $dc(x, y)$ that expresses the property "x,y is a disjoint cover of Γ" (see Problem 9). Thus the following MST-logical formula expresses the surjectivity of this rule:

$$\forall y^0, y^1 \left[dc(y^0, y^1) \Rightarrow \exists x^0, x^1 \left(dc(x^0, x^1) \& \forall j [j \in y^1 \Leftrightarrow iA^{-1} \in x^1 \& i \in x^0 \& iA \in x^1]) \right) \right].$$

For each additional 1-successor in other rules a disjunction need to be added inside the last conjunct to guarantee the appropriate transitions. Additional states will require a longer 'dc' predicate and triple or more quantifier type formulas. Finally, note that every neighborhood N is describable by a sentence in the same calculus, so a uniform formula for surjectivity can be found for arbitrary cellular automata on Γ. Injectivity can be expressed in a similar fashion.

The main part of the proof of the following theorem ultimately goes back to Rabin's Theorem, whose proof is beyond the scope of this volume. (A simplified, more accessible proof of Rabin's Theorem can be found in [S-M-S].)

Theorem 5.18 INJECTIVITY *and* SURJECTIVITY *are uniformly solvable on context-free cellular spaces.*

Proof. It follows immediately from the expressability of the property in MST(Γ) and the solvability of the latter, a powerful result recently established by Muller–Schupp [M-S]. □

5.5 Complexity Questions

That an algorithmic problem has proven solvable has to be interpreted soberly from a practical point of view. A Turing machine guaranteed to halt on any instance will do so in an amount of time that will, in all likelihood, depend on the size of the instance. It may take, in fact, zillions of atomic moves of the machine. Even when executed at a fast speed (say billions of operations per second), the factual running time may turn out to be centuries or even millennia for very small input sizes (see Table 1.1 in Chapter 1). Certainly nobody can be patient enough to wait that long.

Thus it becomes necessary to outline the boundaries that separate problems solvable but *unfeasibly* so from problems that not only are solvable but, in fact, admit a fast *efficient* solution. When confronted with a complexity question of this kind, one only has two possibilities. Either find an efficient algorithm to solve the problem or prove that such an algorithm does not exist. Computational complexity is the area in computability concerned with questions of this kind. An algorithm is considered efficient when the number of instructions executed to solve an instance is polynomial in the size of the given instance. As discussed in Chapter 1, the class of problems admitting a polynomial time solution is denoted **P**. It is an important open problem whether this class coincides with the apparently larger class **NP** of problems whose purported solutions can only be *verified* in polynomial time, or with the apparently larger class of problems *co*NP whose purported solutions can be *refuted* in polynomial time. (See [B-C] or [P] for background definitions and results on computational complexity).

In the next two results, cellular automata have been extended to cellular spaces over arbitrary finite graphs while the local rule itself δ has been restricted to a fixed rule (e.g., SUM_MOD2). One can argue that the local rule is still uniform since the states are still elements that can be operated according to a given table.

Theorem 5.19 1D PARENT SEARCH(δ) *can be* **NP**-*complete even for some fixed rule* δ *on 3 states.*

One might think that the basic difficulty resides in the extent of the predecessor. However, this is not so. Consider

1D BOUNDED PARENT SEARCH(δ)

INSTANCE: a configuration y and a bound m
QUESTION: does $T(x) = y$ have a solution with $|x| \leq m$?

Theorem 5.20 1D BOUNDED PARENT SEARCH(δ) *is* **NP**-*complete even if restricted to some elementary rules δ.*

Theorem 5.21 REVERSIBILITY *is solvable in linear time for local rules on trees.*

Proof. Analogous to the proof of Theorem 5.18 (see Problems 2(b), 10, 11).   □

The complexity of the occurrence of subconfigurations has also been investigated. Here one is interested, not in obtaining a full configuration in a successor, but in obtaining an array as a subconfiguration of a successor. (Recall that an *array* is defined as the restriction of a configuration to a finite connected subset of pixels.)

Theorem 5.22 *The following problems are* **NP**-*complete for a suitable choice of 1D rule δ:*

1. ANCESTER(δ)
     INSTANCE: *an array a of length t*
     QUESTION: *does a occur in some $T^t(x)$, for some x?*

2. SUBCONFIGURATION RECURRENCE(δ)
     INSTANCE: *an array a*
     QUESTION: *will a reoccur in t time steps?*

3. TEMPORAL SEQUENCE(δ)
     INSTANCE: *a sequence of states $q_1, \ldots, q_m$ of a given cell i*
     QUESTION: *does the systems of equations*
         $$T_\delta^t(x)_i = q_t \quad t = 1, \cdots, m$$
     have a solution x?

Naturally, these problems remain at least as hard for higher euclidean dimensions.

Another type of complexity questions concern the structure of the global behavior of cellular automata. In particular, the *isomorphism problem* of determining whether two local rules define essentially the same dynamics has been investigated. Two cellular automata are said to be *isomorphic* if their phase spaces (i.e., their dynamics digraphs) are isomorphic digraphs. This isomorphism can be as labeled digraphs (the stronger notion) or as abstract (unlabeled) digraphs. The corresponding algorithmic problems are

WIT (SIT) Weak (Strong) Isomorphism Testing
INSTANCE: Two local rules δ_1, δ_2
QUESTION: Are T_{δ_1} and T_{δ_2} (strongly) isomorphic?

It is necessary to observe that at first sight, even for *finite* cellular automata on n cells, these problems would seem to have a higher complexity since there are 2^n configurations and so, in principle, $(2^n)!$ possible permutations to check in an instance of WIT. Nevertheless, SIT remain relatively low in complexity, although on the top of its class. SIT seems to provide a contrast to previous results on **NP**-completeness.

Theorem 5.23

1. SIT *is coNP-complete.*

2. WIT *is **NP**-hard.*

3. *Both* SIT *and* WIT *are solvable in linear space* $O(n)$ *when restricted to finite threshold automata on* n *cells.*

It is worth pointing out that *co***NP**-complete problems seem to be rare for complexity questions about cellular automata. Since SIT and WIT also formalize in different ways the classification problem for local dynamical systems, the proofs of these results will be given in Chapter 8.

Another type of decision problem concerns different objects related to the asymptotic behavior of global dynamics on infinite euclidean spaces such as limit sets (configurations that repeat infinitely often), and nonwandering sets (almost recurrent configurations to which the evolution returns arbitrarily closely). Naturally, these problems are unsolvable, although in dimension one they might be solvable. Since their proper definitions require some topological concepts, and since the concepts apply as well to neural networks, they will be dealt with after these concepts are put in place in the following two chapters.

5.6 Problems

Problems marked * may need to be looked up in the literature.

SOLVABILITY AND UNSOLVABILITY

1. Show that INJECTIVITY and SURJECTIVITY are uniformly solvable on finite cellular spaces.

2. Consider an automaton SUM_MOD2 on an arbitrary finite cellular space (where the digraph G is not necessaily regular), i.e., the next state of each

site is the sum modulo 2 of the states of its neighbors. Let T_G be the corresponding global dynamics.

(a) Show that in case $G := C_n$ is a cycle on n vertices, T_G is surjective iff $n \neq 0 \bmod 3$. [Hint: T_G is linear. Use the adjacency matrix of G to express $T_G(x)$ in matrix form.]

(b) Show that in case $G := P_n$ is a path on n vertices,

$$T_G \text{ is surjective iff } n \equiv 1 \bmod 3$$

(c) Show that in case G is a regular graph of odd degree d and d is odd, then T_G has a garden of Eden.

(d) Determine necessary and sufficient conditions for an arbitrary tree G to induce a reversible T_G.

(e) Determine necessary and sufficient conditions for an arbitrary finite graph G to induce a reversible T_G.

3. Design an algorithm that finds inverse maps of a linear rule on a euclidean space (see Chapter 2).

4. Show that INJECTIVITY, SURJECTIVITY and SHIFT-PERIODICITY are decidable for linear rules where the state set is a finite commutative ring. Are they uniformly decidable in polynomial time? [See Theorem 3.11.]

5. Show that the algorithmic problem of whether a given pair of local rules are inverses of each other is solvable.

6. Show that the set of 2D injective (equivalently, bijective) cellular automata is recursively enumerable, but its complement is not.

7. Show that there is no Turing computable function $\beta : \mathbf{N} \to \mathbf{N}$ such that every 2D injective local rule on a neighborhood of radius r has an inverse on a neighborhood of radius $\beta(r)$.

8. Show that 2D INJECTIVITY remains unsolvable when restricted to local rules on the von Neumann neighborhood. [Simulate the action of rules on the Moore neighborhood by a rule on the von Neumann neighborhood that proceeds in two phases: first collect the input from a Moore neighborhood, then find the appropriate next state for the given rule – see Problem 2.13.]

9. Write out a well-formed formula in MST(Γ) for the binary predicate dc of *disjoint cover* on the 1D euclidean binary cellular space Γ.

10. Write out a monary predicate in MST(Γ) for a neighborhood of radius 2 on the 1D euclidean binary cellular space Γ.

11. Show that the Cayley graph of a free group is context-free.

Complexity

12. Show that if a 1D cellular automaton of radius r on m states is invertible, then it has an inverse of radius at most $m^r(m^{r-1} - 1) + r - 1$ (and hence can be found algorithmically – see Problem 5). *

13. Show that an invertible 1D cellular automaton with a neighborhood of size 2 over a prime number of states p has an inverse with neigborhood size at most $p - 1$. *

14. The problems in Theorems 5.19 and 5.20 remain in **NP** for arbitrary local rules on regular graphs. Do they remain **NP**-complete for regular graphs?

15. Show that SIT is solvable in space $O(n)$ when restricted to finite cellular spaces with n sites. [An algorithm can simply generate all configurations in lexicographic ordering and verify identical transitions in linear time.]

16. Show that WIT is still solvable in space $O(n)$ when restricted to finite cellular spaces with n sites.

17. Are subconfiguration problems solvable for linear rules? If so, efficiently?

18. How hard are subconfiguration problems on finite spaces?

5.7 Notes

The issue of "complication" (or, in modern terms, computational complexity) of the global behavior of cellular automata can be traced back to von Neumann [vN]. In the period of development of Turing computability in the 1950s and 1960s, questions about cellular automata were consequently explored. Theorem 5.4 is due to Aladyev [A]. Moore and Myhill proved that injectivity implies surjectivity, as will be seen in the next chapter. The corresponding decision problems for 1D automata were proved solvable early on by Amoroso and Cooper in [A-P] by combinatorial methods. Culik [Cu] uses the fact that a one-dimensional cellular automata can be regarded as a special type of mapping (generalized sequential mappings), previously studied in formal language theory, to prove decidability of 1D INJECTIVITY and 1D SURJECTIVITY. Head [H] uses de Bruijn graphs in order to reduce them to ambiguity problems for finite automata. Sutner [Su3] finally provided the efficient algorithms given in Sect. 5.5. The technique of Sect. 5.2, now reducing to infinite paths on time dependent digraphs, can be further used to show that if a 1D euclidean rule has a predecessor for a recursive configuration, then it must also have a recursive predecessor, although one may not be, in general, effectively constructible – see Sutner [Su3]. Sutner's proof can be further used to prove that 1D PARENT SEARCH restricted to spatially periodic inputs is, in fact, complete in the class **NSPACE**$(log\,n)$ with respect to deterministic logspace reductions [Su4].

Early weaker results on undecidability appear in [Ya1, Ya2]. The unsolvability of the 2D TILING is due to Berger [B]. Finite tilings are introduced in Kari's elegant proofs in [K] to establish the unsolvability of 2D SURJECTIVITY. A simplified proof of the same result appears in Durand [Du1]. Durand [Du2] further shows that the restriction of 2D INJECTIVITY to finite rectangular arrays of size at most $n \times n$ is $co\mathbf{NP}$-complete, where n is the size of the description of the local rule of the instance automaton. The negation of the results in Theorems 5.8, 5.7 were falsely cojectured before [K] several times.

The solvability results about cellular automata on context-free graphs appears in Muller–Schupp [M-S], where other interesting combinatorial problems unsolvable for euclidean problems are also proven solvable on context-free graphs (membership and inclusion for reachability sets of vector addition systems). Little is known about the status of these problems on other infinite cellular spaces. Noncontext-free cellular spaces are known to have decidable monadic theories (although it has been conjectured by Muller–Schupp [M-S] that all context-free graphs can be obtained from context-free graphs (which, in turn, are essentially phase-spaces of the dynamics of push-down automata) by some sort of very restricted 'recursive pruning' and it is a priori possible that one can decide the corresponding formulas by more particular algorithms). Thus, characterizing those cellular spaces on which SURJECTIVITY or REVERSIBILITY are algorithmically solvable appears to be a challenging open problem.

Problem 2 is based on results of Andrasfai [An] and Sutner [Su1]. Other relations between injectivity, surjectivity and graph-theoretical concepts are discussed in Nasu [N]. The significance of reversibility from the point of view of computation is discussed in Toffoli [T].

The results on hardness of subconfiguration problems are discussed by Green [G] and Sutner [Su2]. Theorem 5.23 follows from results in Garzon–Zhang [G-Z] together with results in Chapter 7. The rule SUM_MOD2 has been studied by Sutner [Su2, Su1]. Orponen [O] explores the relationship between discrete symmetric (Hopfield) neural networks and sequential complexity classes such as P/POLY.

References

[A]   V.Z. Aladyev: Mathematical theory of nonhomogeneous structures and their applications. Central Statistics Board, Alinn-Valgus, Estonia, 1980

[A-P]   A. Amoroso, Y.N. Patt: Decision properties for surjectivity and injectivity of parallel maps for tesselation structures. J. Comput. Syst. Science 6 (1972) 448–464

[An]   B. Andrasfai: Cellular Automata in Trees. In: Finite and Infinite Sets. Math. Soc. János Bolyai, Eger, Hungary, 1981

[B]   R. Berger: The undecidability of the domino problem. Memoirs of the American Mathematical Society, Providence RI, 1966

[B-C]   D.P. Bovet, P. Crescenzi: Introduction to the theory of complexity. Prentice-Hall, Hertfordshire, 1994

[Cu]   K. Culik II: On invertible cellular automata. Complex Systems **1**:6 (1987) 1035–1044

[C-L]   G. Chartrand, L. Lesniak: Graphs and digraphs, 2nd edition. Wadsworth and Brooks/Cole, Monterey CA, 1986

[Du1]   B. Durand: Undecidability of the surjectivity problem for 2D cellular automata: a simplified proof. In: Fundamentals of Computation Theory, Z. Esik (ed.). Lecture Notes in Computer Science, Vol. 710. Springer-Verlag, Berlin, 1993. Full version to appear in J. Comput. Sys. Sci.

[Du2]   B. Durand: Inversion of 2D cellular automata: some complexity results. Theoret. Comput. Sci. **134**:2 (1994) 387–401

[G-Z]   M. Garzon, M. Zhang: Classifying neural networks. In: Proc. IEEE Southeast Conf. New Orleans, (1990) **II** 571–576

[G-J]   M.R Garey and D.S. Johnson: Computers and intractabillity: A guide to the theory of NP-completeness. W.H. Freeman, San Francisco, 1978

[G]   F. Green: **NP**–complete problems in cellular automata. Complex Systems **1**:3 (1987) 453–474

[H]   T. Head: One-dimensional cellular automata: injectivity from ambiguity. Complex Systems **3** (1989) 343–348

[H-U]   J.E. Hopcroft, J.F. Ullman: Introduction to automata theory, languages and computation. Addison-Wesley, Reading MA, 1979

[K]   J. Kari: Decision problems concerning cellular automata. Ph.D. Thesis, Department of Mathematics, University of Turku, Finland, 1990

[M-S]   D.E. Muller and P.E. Schupp: The theory of ends, pushdown automata, and second-order logic. Theoret. Comput. Sci. **37** (1985) 51–75

[N]   M. Nasu: Local maps inducing surjective global maps of one-dimensional tesselation automata. Math. Syst. Theory **11**:4 (1978) 327–351

[O]   P. Orponen: On the computational power of discrete Hopfield nets. Preprint, Dept. of Computer Science, University of Helskinki, 1993

[P]   C. Papadimitriou: Computational complexity. Addison-Wesley, Reading MA, 1994

[S-M-S]   A. Saoudi, D.E. Muller and P.E. Schupp: Recognizable infinite trees and their complexity. In: Foundations of Software technology and theoretical computer science (Bangalore, 1990). Lecture Notes in Computer Science, Vol. 472. Springer-Verlag, Berlin, 1990, pp. 91–103

[S]   T. Sato: Decidability of some problems of linear cellular automata over finite commutative rings. Inform. Process. Lett. **46** (1993) 151–155

[Su1]   K. Sutner: σ–automata on Graphs. Complex Systems **2**:1(1988) 1–28

[Su2]   K. Sutner: Additive automata on graphs. Complex Systems **2**:6 (1988) 641–661

[Su3]   K. Sutner: De Bruijn graphs and linear cellular automata. Complex Systems **5**:1 (1991) 19–30

[Su4]   K. Sutner: The complexity of finite cellular automata. Forthcoming

[T]   T. Toffoli: Computation and construction universality of reversible cellular automata. J. Comput. Syst. Sci. **15** (1977) 213–231

[vN]   J. von Neumann: Theory of self-reproducing automata. University of Illinois Press, Chicago IL, 1966

[W]   S. Wolfram: Theory and applications of cellular automata. World Scientific, Singapore, 1986

[Ya1]   T. Yaku: The constructibility of a configuration in a cellular automaton. J. Comput. Syst. Sci. **7** (1973) 481–496

[Ya2]   T. Yaku: Surjectivity of nondeterministic parallel maps induced by nondeterministic cellular automata. J. Comput. Syst. Sci. **12** (1976) 1–5

6. Neural and Random Boolean Networks

If a man will begin with certainties, he shall end in doubts;
but if he will be content to begin with doubts, he shall end with
certainties.

Francis Bacon

One of the fundamental features of cellular automata is the homogeneity of the underlying cellular space. It is natural to ask whether this is a fundamental restriction or just a convenient assumption. In order to gain some insight into this question, it is necessary to relax our restriction on the homogeneity of the space and allow more general interconnections between the sites of the space. This chapter discusses the nature of the resulting generalizations and compares the power of the resulting models vis-a-vis Turing machines and cellular automata.

The main difficulty about generalizations is to find the right type. At the heart of the matter is, on the one hand, the identification of core properties to be retained, and on the other, of features to be expanded. These issues are discussed in the first section. In order to facilitate the exposition, we follow a slightly different approach than for previous chapters. The following section presents justification and precise definitions of the models. Section 6.3 contains a summary of the results in the chapter. Their proofs appear in Sect. 6.4, and they may be skipped if the reader is not interested in technical details.

6.1 Types of Generalizations

Recent times have seen the coming of age, to various degrees of success, of at least three types of massively parallel computing devices: neural networks, cellular automata, and, to a lesser extent, automata networks. The variety of models utilized in each area is almost as great as the type of applications that they have seen. The activations (states) in use have been both continuous and discrete, the connections weighted by discrete and analog values, the time in the evolution taken to be discrete or continuous, and the local rule of evolution itself has ranged from differential (including stochastic) equations to boolean look-up tables. The importance of these models in a variety of fields ranging from cognitive science to applied mathematics is hardly questionable.

A particular case of this generalization, *artificial neural nets*, has recently been the focus of much attention. These models have also been called *connectionist models*, or *massively parallel computers* (there seems to be some convergence toward the term artificial neural networks, or simply neural nets). Neural nets differ from cellular automata in three important aspects. First, neural net cells are located at vertices of an arbitrary (possibly infinite) graph with arbitrary communication links. Consequently, the local transition function of a neural net is *not* homogeneous as it varies from site to site. Second, they are of a very restricted type (a slight perturbation of a linear cellular automaton rule – see Chapter 3). Third, the states (now called activation values) of the network are assumed to be real numbers with their ordinary properties for addition, multiplication, and differentiation. Much of the interest in the model, beyond their obvious apparent analogy with biological brains, stems from the possibility of applying classical optimization techniques (derivatives, gradient descent, etc.) in the design of architectures to perform significant tasks for which the classical discrete methods in computer science have proved short-ranged.

On the other hand, there is the fundamental issue of how *realistic* the generalizations should be. It would be clearly undesirable to develop a theory based on a model far removed from actual physical construction and utilization. One of the virtues of the classical sequential model is that it is ideal enough to be independent of physical approximations, while also close enough to offer results directly impinging on the power of physical machines in operation. Thus, an acceptable theory for massively parallel models must have as its fundamental goal the analysis of what can be *ultimately computed by arbitrary fine-grained models under realistic conditions*. These models include the widest variety of networks, from those designed or trained by any means, to random networks (with random connections between cells), and even to infinite networks. In particular, the admittedly important question of learnability of solutions, which does not impinge on computability per se, should not be relevant in first instance. Neither is the issue of their constructibility since the latter is independent of the effectiveness of the solutions as recursive procedures. Therefore, the results herein presented are relevant to the more theoretic aspect of sheer capability of neural networks rather than to their use in the analysis of, for example, learning algorithms.

This overarching goal imposes a number of restrictions on the models to consider. First of all, the restriction of *realistic* is to be interpreted along the guidelines of classical computability. This implies that realistic models can only consist of a finite number of pieces which assume a number of states that can actually be *observed* and be under the *control of a user* of the model as a computing tool. In particular, from this perspective, real numbers are not observable to unlimited precision, or, even if they were so stored in the form of activations or connection weights, technological constraints prevent *us* from measuring them with sufficient accuracy to really allow their effective use as inputs by artificial analog neurons in transition. Nonetheless, these considerations do not preclude their use (to bounded precision) as sets of activations or weights. (Learning considerations aside, one can further argue that networks with analog weights and activations that compute boolean-valued functions are provably input-output equivalent to networks whose activation and weights are integers with precision up to $O(n \log n)$ bits.)

Second, the analysis of *ultimate* capabilities imposes a seemingly contradictory kind of assumption. On the one hand, every physical network can only be assembled from finitely many components, notwithstanding the fact that it may consist of infinitely many particles arranged in a continuous fashion (as classical continuous modeling would assume). On the other, technology constantly reminds us that there is every reason to believe that networks of arbitrarily large size can, at least in principle, be constructed, and that therefore the number of pieces (say neurons) that compose a network cannot be bounded by any finite constant. Moreover, it would not be surprising if, in the course of the analysis, one were forced to make the simplifying assumption that the number of cells is so large as to be infinite for all practical purposes. Therefore, an analysis of ultimate capabilities forces the consideration of arbitrarily large networks. From the point of view of analysis, one makes the equivalent but more convenient assumption of a (countably) infinite number of neurons. However, in accordance with the discussion in the previous paragraph, the number of states that the various nodes may assume will be uniformly bounded, in fact coming from a common finite set. Additionally, these assumptions may provide a better understanding of the behavior of networks as the number of neurons scales up under realistic conditions, and thereby provide a better understanding of neural networks as computational devices.

Third, there is the question of how the inputs are to be given to the network. Our overriding assumption is that we are modeling truly massively parallel processing. Therefore, even in pipeline processing for example, inputs are entered *all at once* in high bandwidth, from one parallel processor to another. This assumption would seem to exclude a large chunk of what goes on today under neural network research. Upon reflection, however, the current largely sequential implementations of neural networks are not the expected scenario for the ultimate fate of massively parallel computing, in fact or intention. Truly parallel computers will probably communicate among themselves in a highly parallel fashion, with little (if any) entered by humans in the form of temporal sequences of symbols (to be assembled eventually into some sort of distributed representations). Even in tasks that appear to be inherently sequential (such as speech production or recognition), it is at least conceivable that at the very semantic inner level, language is really understood and produced in parallel, while sequentiality is forced solely at the physical level of material production or perception. Vision and other senses appear to be parallel, at least at the level of perception.

Under these considerations, we choose as an appropriate generalization cellular networks which are discrete, synchronous and deterministic. They consist of an infinite (countable) number of simple processing cells modeled by activation functions of a boolean type, and which operate synchronously in parallel. Two chief questions will be explored in the following sections. First, how does the computational capability of a neural net or random net compare to that of a cellular automaton? Second, how does this power compare with classical computational devices such as Turing machines? The following chapters will revolve about the problem of characterizing their computational power.

6.2 Other Parallel Models

It is *necessary* to have precise formal definitions of parallel models of computation in order to be able to make meaningful and precise statements about their computational power. While the definition of cellular automata is basically standard, there are many varieties of neural networks, and virtually no commonly accepted definition of random and automata networks.

We assume that each cell i is capable of storing some information represented by states coming from a *finite* set Q with a special *quiescent* activation level (or state) denoted 0. This set of states may be a subset Q_i peculiar to a given cell (as in neural or random networks) or uniform for all cells (as in cellular automata). At any rate, it contains at least one other nonquiescent state. It may also have a restricted structure (e.g., an additive-mltiplicative for neural networks) or be arbitrary (cellular automata or random networks).

Massively parallel processors presuppose an underlying hardware structured as a network of very simple processors. Formally, this means that random networks are built on a *digraph* consisting of *vertices* V (where the cells or nodes of the network are to be located), and *arcs* (directed edges representing the links of the network). This (di)graph has a very regular, homogeneous structure in cellular automata, but can be more irregular in neural and random networks. In both, each cell can communicate with other cells at adjacent nodes and perform relatively simple local computations. In order for the network to be physically realizable, each must have only finitely many incoming and outgoing arcs, i.e., the graph must be *locally finite*. In order to have full computational ability, we will simply assume a *countably* infinite number of cells. In the most general type of random network no restriction, other than finiteness conditions (boundedly many states and locally finite graphs), is imposed on the processors at the nodes of the network or on their communication digraph. Thus all state sets can be taken to be subsets of a single finite set, and each vertex to have finite in- and out-degree.

Since networks with distinct processors can have the same interconnection network and states, we again distinguish between a random network and the underlying space on which it is defined. The notion of *cellular space* can be extended to a pair $\langle D, \{Q_i\}_i \rangle$ consisting of a countably infinite, locally-finite, directed graph D (which may be a symmetric digraph, i.e., just a graph in the case of a cellular automaton) and a family of finite sets Q_i (each containing a common *quiescent* state 0), which are the possible activation levels (or states) of each cell i. Associated with every cellular space there is a *configuration space* $\prod_i Q_i$ (or simply **C**) consisting of all *configurations* (or *total states*, or *state vectors*) of the space, i.e., choice functions $x : V \rightarrow \bigcup_i Q_i$ that associate a state (level of activation) $x_i \in Q_i$ with each vertex i of the digraph D.

Both neural networks and cellular automata are particular cases of the more general type of random network considered here.

Definition 6.1 *A* random network *(also referred to as an* automata network*)* *is a pair* $\langle D, \{M_i\} \rangle$ *consisting of a cellular space* $(D, \{Q_i\})$ *and an associated family of finite-state machines* M_i *(only finitely many of which are distinct) with input alphabet* $\Sigma_i := Q_{i_1} \times \ldots \times Q_{i_d}$, *and local transition functions*

$$\delta_i : Q_i \times \Sigma_i \rightarrow Q_i$$
$$(x_i, x_{i_1}, \ldots, x_{i_{d_i}}) \mapsto \delta_i(x_i, x_{i_1}, \ldots, x_{i_{d_i}}),$$

where $i_1, \ldots, i_{d_i}$ *are the vertices with an arc into vertex* i *in* D.

A random network operates locally as follows. A copy of a finite-state machine M_i (called the *computing* or *processing element*) occupies each vertex i of D, which is then called a *cell* – or also a *site* or *node*. *Synchronously*, each copy M_i looks up its input in the states $x_{i_1}, \ldots, x_{i_{d_i}}$ of its *neighbor* cells and its own state x_i, and then changes its state according to a prespecified local dynamics δ_i. The random network thus performs its calculation by repeating this atomic move any (possibly very large) number of times.

As with cellular automata, the global evolution of a random network is best viewed as a discrete dynamical system. Given a configuration at a certain time t the local action of the random network transforms the current configuration of the cellular space into a new configuration according to a global dynamics $T : \mathbf{C} \rightarrow \mathbf{C}$ defined by

$$T(x)_i = \delta_i(x_i, x_{i_1}, \ldots, x_{i_{d_i}}). \tag{6.1}$$

Moreover, since the underlying digraph is locally finite, the restriction of T to $\mathbf{C}_0$ (configurations with only finitely many nonquiescent cells) maps into $\mathbf{C}_0$. The tth iteration of T will be denoted T^t.

This general definition captures the classical models of parallel computation as well as most synchronous discrete models. They clearly include cellular automata, since Cayley graphs are locally finite. They also include the discrete type of neural network considered here, as shown next.

Neural networks differ from cellular automata primarily in two aspects: (a) the underlying network is no longer homogeneous but an arbitrary digraph; and (b) the transition functions first compute weighted sums of inputs from neighboring cells (which makes the cell unable to detect where specific inputs come from). Thus we assume that the state set is finite and has an additive-multiplicative structure (what mathematicians call a finite ring A with unity), although some of the results of this chapter carry over to arbitrary rings, including the real numbers.

Each arc from vertex j to vertex i is labeled with a weight w_{ij}. Cells sum their weighted inputs and apply an *activation function* $f_i : A \rightarrow A$. Each f_i satisfies $f_i(0) = 0$ (in order to avoid spontaneous generation of activation). The weighted sum of the inputs to each cell from its neighbors is given by a function called net_i. Thinking of a neural network as a discrete dynamical system, at any time $t + 1$ (t a nonnegative integer), the net-input at i is given by

$$net_i(t+1) \;=\; \sum_j w_{ij}a_j(t), \tag{6.2}$$

where the sum is taken over all cells j supporting links into i, and

$$a_j(t) \;=\; f_j(net_j(t)) \tag{6.3}$$

is the activation of the cell j at time t.

Definition 6.2 *A (discrete) neural network is a triple $\mathcal{N} = \langle A, D, \{f_i\} \rangle$ consisting of a set with an additive-multiplicative structure A, a countable (finite or infinite), locally-finite, arc-weighted, digraph D, and a family of activation functions f_i, one for each vertex i in D. The local dynamics of $\mathcal{N}$ is defined by equations (6.2) and (6.3) above.*

The *global dynamics* $T : \mathbf{C} \to \mathbf{C}$ of the neural network $\mathcal{N}$ is defined as follows. For any configuration $x \in \mathbf{C}$, let

$$T(x)_i \;:=\; f_i(net_i(x)) \tag{6.4}$$

Thus the global dynamics of neural networks are also self-maps of configuration space. (Alternatively, one could interpret a configuration x as a net-input vector and obtain a closely related global dynamics.) This chapter deals with a large class of neural networks that encompasses virtually all discrete networks of practical interest, the class of networks of finite bandwidth.

Definition 6.3 *A labeling of a digraph D is a one-one mapping $\ell : V(D) \to \mathbf{N}$ that assigns a positive integer to each vertex of D. The bandwidth of a labeling ℓ is the maximum of the differences $|\ell(i) - \ell(j)|$ for every pair of adjacent vertices i, j, or ∞ if there is no maximum. The bandwidth of a (di)graph is the minimum bandwidth of all of its labelings. The class $\mathbf{NN_0}$ ($\mathbf{CA_0}, \mathbf{RN_0}$, respectively) is the subclass of $\mathbf{NN}$ ($\mathbf{CA}, \mathbf{RN}$) consisting of global dynamics defined on digraphs of finite bandwidth.*

Note that a digraph of finite bandwidth in a digraph must have bounded degree. For example, the 1D euclidean space has bandwidth 1 but 2D euclidean space has infinite bandwidth.

6.3 Summary of Results

Since Turing machines can be simulated by 1D cellular automata, the class of Turing computable (partial recursive) functions $\mathbf{TM}$ is included in $\mathbf{CA}$. This inclusion is proper, in general, for the class $\mathbf{CA}$. For instance, the prototypical recursively unsolvable algorithmic problem, the HALTING PROBLEM is solvable by a simple neural network (see Theorem 6.5 below). In a different vein, it is not too hard to prove that a 1D cellular automaton can take as input a real number (as an *infinite* binary expansion) and stabilize if and only if it is an integer. In other words, the problem of membership in the set of integers is weakly solvable, in the sense of Def. 5.2. A Turing machine cannot solve this problem since the integer 1 could be given as 0.9999... and hence the machine cannot even finish reading its input in finite time. This problem can also be solved by a neural network, as shown next.

Theorem 6.4 *Every self-map* $T : \mathbf{C} \to \mathbf{C}$ *realizable on a cellular automaton can be implemented by some neural network, i.e.,* $\mathbf{TM} \subset \mathbf{CA} \subseteq \mathbf{NN} \subseteq \mathbf{RN}$

This result raises the question of whether these classes of abstract dynamics $\mathbf{CA} \subseteq \mathbf{NN} \subseteq \mathbf{RN}$ are one and the same, i.e., whether every random network dynamics can be realized on a cellular automaton and/or a neural network. In order to establish an objective criterion, we compare them on the basis of the dynetic problems that they can solve on the assumptions discussed at the beginning of this chapter. In particular, there is the critical difference with classical computability that the input may, in principle, be an infinite configuration.

The following result makes patent the distinction between the notions of algorithmic problems and dynetic problems, introduced in Chapter 4. One will recall that the HALTING PROBLEM is one of the fundamental problems that are unsolvable by Turing machines.

Theorem 6.5 *The dynetic problem of membership in any countable set is neurally solvable. In particular, the* HALTING PROBLEM *for Turing machines is solvable as a dynetic problem by an infinite neural network of bandwidth 2.*

Despite its appearance, Theorem 6.5 is *not* a counterexample to the Church–Turing thesis since the thesis concerns only algorithmic problems solvable by *finitary* means, whereas the network involved in the proof of the theorem above requires the use of an infinite object, namely the underlying digraph. However, the theorem does establish that neural computability is essentially different from both classical Turing computability and from computability by *oracle* machines as well. Turing machines with a fixed oracle can recognize only countably many countable languages. Infinite neural networks, on the other hand, not only can solve a membership problem without oracles, but can also solve problems with uncountably many possible inputs (e.g., the integrality problem mentioned above).

Nevertheless, the usual diagonalization argument used to prove the Turing unsolvability of the HALTING PROBLEM still yields the existence of unsolvable problem in the stronger dynetic sense. Theorem 6.10 below provides a specific (and important) unsolvable dynetic problem.

The following is a converse of Theorem 6.4 for finite neural networks.

Theorem 6.6 *Every solution of a dynetic problem on a finite neural network of* n *cells can be implemented on a cellular automaton with* $O(n^3)$ *cells.*

We show below that this converse is true even for infinite networks of finite bandwidth.

The implementation in the proof of Theorem 6.6 is uniform in the sense that two dynetic problems on the same network are implemented by cellular automata defined on the same cellular space. This raises the question of whether the uniformity of this implementation can be extended to local dynamics, i.e., whether there exist universal neural networks.

The extent of this question requires some discussion. For one might argue that *infinite* neural networks acting on finite initial configurations are essentially equivalent to Turing machines. If so, the existence of a universal neural network would be tantamount to the existence of a universal Turing machine, a well-known result in classical computability. Although it is true that neural networks (even those of finite bandwidth acting on finite initial configurations) are at least as powerful as Turing machines, the converse is not true by Theorem 6.5. A Turing machine can only be given the *finite* amount of information contained in its transition table and its finite initial input. The *possibility of storing an essentially infinite amount of information (compared to what is encodable, for example, in a periodic decimal expansion representable by a finite string) in the pattern of interconnections among the nodes of an infinite neural network is a fundamental advantage indeed.*

Therefore the question of the existence of a universal neural network is *not* a consequence of the existence of a universal Turing machine. However, in analogy with the Turing notion of a (sequentially) universal machine, it is natural both from a theoretical and a practical point of view to ask whether there exist dynetic problems $\mathcal{P} : \mathcal{I} \to \mathcal{J}$ which are universal in the class of all dynetic problems. Such a universal network would be able to simulate, modulo a reduction, the action of every neural network $\mathcal{N}$ on any input x, given some encoding of both $\mathcal{N}$ and x. In the proof of the following theorem we provide the blueprint of a specific neural network that accomplishes precisely that for all networks of finite bandwidth, both finite and infinite.

Theorem 6.7 *There exists a universal neural network (over an extended activation set) for neural networks of finite bandwidth over any given activation set.*

This universal network $\mathcal{U}''$ has a row of cells numbered $1, 2, 3, \cdots$ and auxiliary cells some of whose states represent weights on the links of its input $\mathcal{N}$. The network $\mathcal{N}$ is encoded simply by arbitrarily numbering its nodes $1, 2, \cdots$ and clamping their initial activations as well as the weights of $\mathcal{N}$ onto the corresponding cells of $\mathcal{U}''$. Thus $\mathcal{U}''$ is a *programmable* neural network.

Theorem 6.8 *Every dynetic problem solvable on a neural network of finite bandwidth is also solvable on a 3-dimensional Euclidean cellular automaton.*

Theorem 6.9 *All three models, neural networks, cellular automata and random networks, are computationally equivalent when restricted to digraphs of finite bandwidth, i.e.,*

$$CA_0 = NN_0 = RN_0.$$

As pointed out before, the stability problem of a certain relatively simple class of neural networks is Turing undecidable. In view of the neural solvability of the HALTING PROBLEM, of which it is the Turing machine analog, a more germane question is whether it is undecidable *by neural networks* as well.

A *dynetic decision problem* $\mathcal{P} : \mathcal{I} \to \mathcal{J}$ is one with only two possible answers (YES/NO) in its range $\mathcal{J}$. For example, the STABILITY PROBLEM for neural networks consists of a set $\mathcal{I}$ of encodings $\langle \mathcal{N}, x \rangle$ of a network and its input, where

$$\mathcal{P}(\langle \mathcal{N}, x \rangle) := (\mathcal{N} \text{ stabilizes on input } x).$$

Recall that a dynetic decision problem is *weakly solvable* if there exists a neural network $\mathcal{N}$ that stabilizes on an input $x \in \mathbf{C}$ if and only if $\mathcal{P}(x) =$YES.

Theorem 6.10 *The dynetic problem of coSTABILITY for neural networks of finite bandwidth is not weakly solvable (i.e., no neural network of finite bandwidth over an activation set A takes as input some encoding of an arbitrary network over the set A and stabilizes if and only if its input does not).*

Weak unsolvability is a stronger notion of unsolvability than the one implied by Def. 5.2. Since neural solvability of a problem clearly implies weak solvability of the problem and its complement it follows that

Corollary 6.11 *The dynetic problem of STABILITY for neural networks of finite bandwidth is neurally unsolvable (i.e., no neural network over an activation set A as input some encoding of an arbitrary network over A and decides whether it eventually stabilizes on its input).*

The notions of *solvable* and *weakly solvable* by (infinite) neural networks can be thought of as natural analogs of the classical notions of recursive and recursively enumerable sets. But unlike the situation for recursively enumerable sets, problems which are not weakly solvable can be made so by a *finite* extension of the activation set of values (here playing the role of the tape alphabet of a Turing machine).

Theorem 6.12 *The dynetic problem of STABILITY for neural networks of finite bandwidth over an activation set A is weakly solvable by a neural network over some extended finite ring A'.*

The previous theorem indicates that this solution cannot be of finite bandwidth.

Corollary 6.13 STABILITY *for neural networks of finite bandwidth is weakly solvable by a 3D euclidean cellular automaton.*

6.4 Proofs

The purpose of this section is to provide missing proofs of the results stated in Sect. 6.3. Theorems 6.4–6.6 are proved in Sect. 6.4.1. Section 6.4.2 gives a high-level description of the universal neural network of Theorem 6.7. Section 6.4.3 indicates that the universal neural network can be implemented by a cellular automaton, which will thus become simultaneously a universal neural network and a universal cellular automaton, both for networks of finite bandwidth. This establishes the first equality in Theorem 6.9. The other equality is proved in Sect. 6.4.4. Finally, the results of stability are discussed in Sect. 6.4.5.

6.4.1 A Hierarchy

Proof (of Theorem 6.4). Let $Q := \{0, 1, \cdots, m-1\}$ be the states of a cellular automaton M with neighborhood of size d, where 0 is the quiescent state, and let $\delta :, Q \times Q^d \to Q$ be its local dynamics. Construct a neural network $\mathcal{N}$ over the activation set $A := Z_{m^d}$ of integers modulo m^d whose underlying digraph is the lattice of M as follows. Assign a fixed numbering to the neighbors of a vertex i of M's cellular space which is invariant under translation. Let the weight w_{ij} simply be $w_{ij} := m^j$ (independently of i) and define the activation function f_i by $f_i(r) := \delta(q_{d-1}, ..., q_0)$, where the q_j's are the digits of the m-ary expansion of r so that $r = \sum_j q_j m^j$. Thus $\mathcal{N}$ induces a global dynamics according to equations (6.2) and (6.3), i.e., if $a_j(t) := q_j$, the next activation level of cell i is given by

$$
\begin{aligned}
a_i(t+1) &= f_i(\sum_j m^j q_j) = \delta(q_{n-1}, \dots, q_0) \\
&= f_j(net_i(t)) = f_i(\sum_j w_{ij} a_j(t))
\end{aligned}
$$

and hence the new activation level of a node corresponds precisely to the new state of the cell in the cellular automaton M.

The other inclusion is obvious since a neural network cell is a finite state machine with local transition

$$
\delta_i(x_i, x_{i_1}, \dots, x_{i_d}) := f_i(\sum_j w_{i_j} x_{i_j}) \square
$$

Proof (of Theorem 6.5). For the sake of clarity, the proof below is given for the HALTING PROBLEM. A similar argument can be used to show that the membership problem for any countable set is neurally solvable. The existence of the appropriate network – but certainly not a recursive construction!– can be obtained by creating one node for each instance of the problem and connecting nodes of the same type together, It is then a matter of propagating the information to the left-most node, which will indicate which case it is. Thus the digraph "knows" in advance which instances are not halting and which are. All its dynamics will do is single out in which of the two groups the original input was clamped.

Precisely, the HALTING PROBLEM is formulated as a dynetic decision problem as follows. An instance $\langle M, x \rangle$ consisting of a Turing machine M and its input x is encoded as a configuration

$$000\cdots \quad \cdots \quad 0100\cdots \tag{6.5}$$

(to be explained below). The proof requires the construction of a network $\mathcal{N}$ and two configurations x_{yes} and x_{no} such that $\mathcal{N}$ stabilizes on every input at either x_{yes} or x_{no}, according as whether M halts on x or not.

Through a recursive Gödel numbering, each instance of the HALTING PROBLEM can be regarded as a positive integer. The cells of the network $\mathcal{N}$ are in one-to-one correspondence with the instances of the HALTING PROBLEM and are arranged in increasing order of instance. The underlying digraph consists of two disjoint one-way infinite paths of cells, called the *halting* and *nonhalting* cells, respectively, according as the corresponding instance halts or not. Each halting cell is connected to the previous *halting* cell while each nonhalting cell, except the first, is connected to the previous *nonhalting* cell. Thus the digraph consists of two disjoint connected components. The activation set will be $\mathbf{Z}_2$, the integers modulo 2. All weights will be 1. Since the (non)halting cells can be labeled (for the purpose of computing its bandwidth) with (odd) even numbers in consecutive order, this digraph has bandwidth 2.

The identity function serves as the activation functions of both halting and nonhalting cells. The left-most two cells are exceptional in that they are *persistent* (they have self-links to feed back their own activation), while all the others are not.

With these activation functions, the network operates as follows. A given instance $\langle M, x \rangle$ is encoded and clamped as a configuration (6.5) consisting of a 1 in the appropriate position corresponding to its encoding, and the remaining cells are quiescent. The connections and activation functions of the halting cells are set up in such a way that the 1 in the input cell will begin traveling left until it reaches the leftmost cell, at which point it stabilizes at 1. All halting instances always stabilize at the configuration $x_{yes} := 1000\cdots$, while the nonhalting instances do at $x_{no} := 0100\cdots$. $\qquad\square$

If you are feeling uneasy about the possibility of constructing or training such a network, you might want to read the discussion in the Notes at the end of this chapter about *computability* vs *constructability*. The key points is the theoretical solvability of the problem rather than the constructibilty of its solution. This statement of solvability is no different than classical statements of solvability in the absence of specific algorithms. Also, finding an initial chunk of the neural network that solves the HALTING PROBLEM is probably more likely than finding the forbidden subgraphs required by, say, the Seymour-Robertson Theorem. On the one hand, one would not expect random connections between an infinite number of cells to turn out to be just as described. On the other, connections of such type are *possible* (if not *likely* to be made by a determined character) for *some* numbering of the instances of the HALTING PROBLEM on a line.

This proof of Theorem 6.6 requires the following two lemmas. The first lemma deals with the occurrence of arbitrary graphs as induced subgraphs of Cayley graphs. Recall that a vertex subset $V \subseteq G$ of a Cayley graph $\Gamma(G, X)$ of a group G defines an *induced* subgraph whose edges are those edges of Γ connecting two vertices of V.

Lemma 6.14 *A graph D with n vertices is an induced subgraph of some Cayley graph of any (abelian) group with at least $(2 + \sqrt{3})n^3$ $(2.92n^3)$ elements.* $\quad\square$

The second lemma is a partial converse to Theorem 6.4 for identical underlying digraphs. Note that any Cayley graph $\Gamma(G, X)$ of a finitely generated group has bounded in-degree (in fact, at most $d := |X|$, the number of generators of the cellular space.)

Lemma 6.15 *Any neural network whose underlying directed graph is a subdigraph of a Cayley graph can be implemented by a cellular automaton over the same Cayley graph.*

Proof. Let $\mathcal{N}$ be a neural network, A be its set of weights and activations, and $\{f_i\}$ be its set of activations functions. Assume that its underlying digraph D is a subdigraph of a Cayley graph $\Gamma := \Gamma(G, X)$ with a (symmetrized, i.e., replacing arcs by undirected edges) finite neighboring set X with $d := |X|$ elements. Without loss of generality assume that every edge of Γ connecting two vertices in D actually appears in D as two opposite arcs, perhaps with zero weights, in $\mathcal{N}$.

Since A is finite there are only finitely many *distinct* activation functions at all the nodes. Since D is embedded in Γ, there are only boundedly many (at most d) inputs to each node of $\mathcal{N}$. Therefore, the nodes of $\mathcal{N}$ compute at most finitely many, say n, distinct functions

$$\delta_k : A \times A^d \to A \ \ (0 \le k < n) \, ,$$

including the all-zero function, which is assumed below to be δ_0. To every node i there is associated a node function δ_{k_i} (that depends on the activation function of the cell and the incoming weights). What is required is a *uniform* local dynamics of a cellular automaton $\mathcal{M}$ on $\Gamma(G, X)$ whose global dynamics (when restricted to the original network) is identical to the global dynamics of the given network $\mathcal{N}$.

Define a finite-state machine M with state set $Q := A \times \mathbf{Z}_n$ and transition function given by

$$\delta : Q \times Q^d \ \rightarrow \ Q$$
$$\delta[(a_i, k_i), \{(a_j, k_{i_j})\}_j] \ := \ (\delta_{k_i}(a_i, \{a_{i.j}\}_j), k_i).$$

where the index j runs over $X \cup X^{-1}$. Thus, M always maintains the initial value in the last coordinate of the ith cell state, and activates the corresponding dynamics δ_{k_i} in order to assign a new state to cell i, the first coordinate of which is the activation it would have as a cell in $\mathcal{N}$.

The simulation of $\mathcal{N}$ takes place as follows. The initial pattern of activation a in the various cells i of $\mathcal{N}$ is encoded as the initial configuration x with state $x_i := (a_i, k_i)$ if $i \in D$ and $x_i := (0, 0)$ otherwise. Thereafter, the evolution of $\mathcal{N}$ can be observed through the evolution of M by restricting attention to vertices in D and projecting their states into their first components. $\square$

Proof (of Theorem 6.6). To complete the proof of Theorem 6.6, let D be the underlying communications digraph of a neural network on n cells and D' the symmetrized graph. Apply Lemma 6.15 to the Cayley graph of a (abelian) group G containing D' as an induced subgraph obtained by Lemma 6.14. $\square$

6.4.2 A Universal Neural Network

The proofs of the rest of the results of this chapter depend heavily on the existence of a neural network capable of simulating any other one. This section provides a nontechnical high-level description of such a universal neural network.

Recall that a universal Turing machine is one which, when given a formal description of another Turing machine M and its initial input string w, will simulate the steps of M when processing w. Any common digital computer (von Neumann machine) can be thought of as an approximation of a universal Turing machine. When fed a program (a formal description of a Turing machine) and data (its initial input string), it implements the operations of the given program on the given data. Universal Turing machines are important, among other reasons, because they make possible diagonal arguments to prove the existence of Turing unsolvable problems.

Similarly, a universal neural network would receive as its initial configuration a formal description $\langle \mathcal{N}, x \rangle$ of a neural network of finite bandwidth and its initial configuration x, and would proceed to simulate the behavior of $\mathcal{N}$ when started on configuration x.

Given an activation set A with unity, a universal neural network $\mathcal{U}''$ for networks over A operates over an extended set A'. The network $\mathcal{U}''$ has *fixed* underlying digraph, weights, and activation functions for its nodes, all independent of the input networks that it is to simulate. The formal description $\langle \mathcal{N}, x \rangle$ of a (possibly infinite) input network of finite bandwidth can be given by its weights and activation functions laid out in a 2D grid, where the description

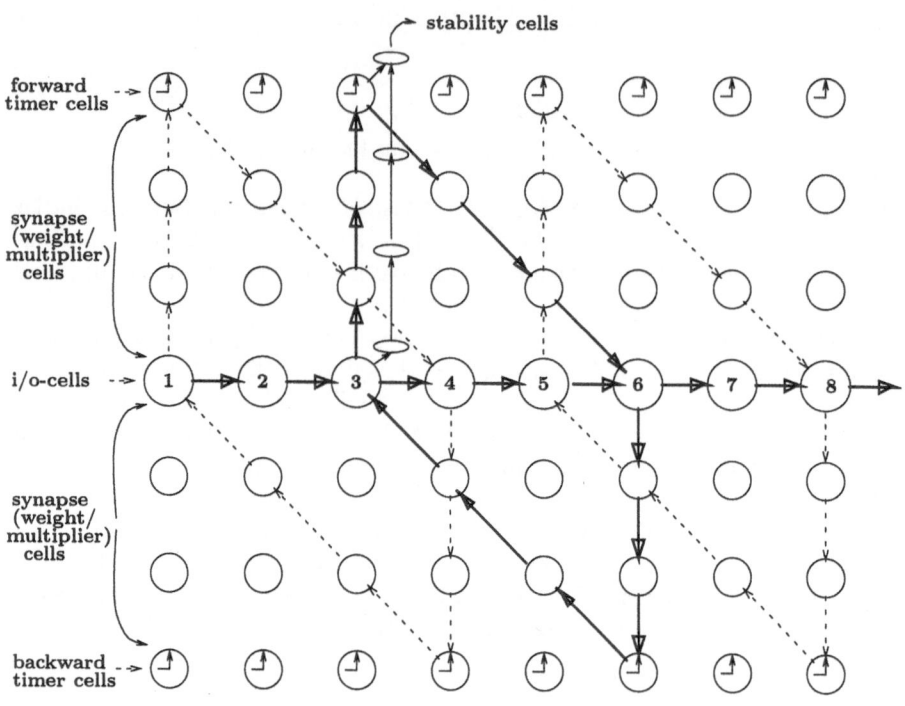

Fig. 6.1. Layout and basic operation of a universal neural network

$\langle \mathcal{N}, x \rangle$ is clamped as a pattern of activation. The cells of $\mathcal{U}''$ on the positive x-axis are designated as *io-cells*. The cells of $\mathcal{N}$ are labeled with natural numbers so as to correspond with the input cells of $\mathcal{U}''$ (although this can be done arbitrarily, a labeling of smaller bandwidth leads to a faster simulation.) The remaining cells above the x-axis are used as synapses cells that are orchestrated to collect the net input into each io-cell i from cells j with $j < i$ and effect its transition according to the characteristic transfer function stored at the cell. To that end, each synapse cell is endowed with appropriate adders, multipliers and timers (in the third dimension) that handle the weights and accumulated net input, as shown in Fig. 6.1. Similar cells below the x-axis effect the dual purpose from cells i to j.

The simulation of an input network $\mathcal{N}$ on input x proceeds as follows. The initial configuration of $\langle \mathcal{N}, x \rangle$ is clamped on $\mathcal{U}''$ at the io-cells together with their activation functions. The synaptic weights w_{ij} of $\mathcal{N}$ are loaded at the cell at position (i, j). Some additional activations are loaded at the height corresponding to the network's bandwith in order to ensure appropriate timing and the collection of net inputs. Thereafter, the operation of $\mathcal{N}$ can be observed in the levels of activation of the io-cells of $\mathcal{U}''$. Subsequent outputs are the activations of the io-cells as a result of their operation as part of network $\mathcal{U}''$. An external observer who can only see the io-cells of $\mathcal{U}''$ will witness precisely the operation of $\mathcal{N}$, although slowed down by a constant factor linear in the bandwidth of the input network.   □

6.4.3 Equivalence of Cellular Automata and Neural Networks

Theorem 6.8 is effectively a converse of Theorem 6.4 for (possibily infinite) networks of finite bandwidth. It follows from results in the previous section and the following lemma.

Lemma 6.16 *Given an activation set A with an identity element, there is a universal neural network (over an expanded set) for finite bandwidth neural networks over A, whose underlying digraph is a subdigraph of a euclidean lattice.*

Proof (of Theorem 6.8). The proof is now straightforward from the construction in the previous section, For suppose $\mathcal{P}$ is a dynetic problem solved by the global dynamics T of a neural network $\mathcal{N}$ over an activation set A. This means that for every instance x, T eventually stabilizes at $\mathcal{P}(x)$. By Lemma 6.16 there is a universal neural network $\mathcal{U}'$ for networks over A that can be embedded in a euclidean lattice. Hence, by Lemma 6.15, $\mathcal{U}'$ can be implemented by a cellular automaton. Since $\mathcal{U}'$ implements $\mathcal{N}$, T can be implemented by a euclidean celular automaton.

This does not complete the proof, however, since by definition, $\mathcal{U}$ *is required to stabilize* on its input if it is to solve the associated dynetic problem. Unfortunately, the euclidean embedding of $\mathcal{U}'$ as given does not quite stabilize on input $\langle \mathcal{N}, x \rangle$ when $\mathcal{N}$ does. It is necessary to enable $\mathcal{U}$ to detect whether $\mathcal{N}$ has stabilized, and, when this happens, have it remove the residual activity in auxiliary cells. It is indeed possible to add what essentially amount to a stack of bandwidth height at each io-cell in the design of $\mathcal{U}'$ to obtain a third universal neural network $\mathcal{U}$ that compares the content of the stack and of all stacks together using some global ANDs (locally implemented) so as to detect stability of the input network on every configuration of finite support, and then halt its simulation as well. Its layout is shown in Fig. 6.1.   □

6.4.4 Equivalence of Neural Networks and Random Networks

In order to finish the proof of Theorem 6.9 it remains only to prove that $\mathbf{RN_0} \subseteq \mathbf{NN_0}$.

Proof (of Theorem 6.9). A random boolean network $\mathcal{A}$ has, by definition, only finitely many distinct types of finite-state machines at its cells. By the argument proving Lemma 6.15 one can assume without restriction that all cells of $\mathcal{A}$ are driven by copies of the same finite-state machine M. Thus the argument in the

proof of Theorem 6.4 can be applied to obtain a neural network on the same
digraph that emulates the action of $\mathcal{A}$.   □

6.4.5 The Stability Problem is Neurally Unsolvable

Every pair consisting of a finite bandwidth network $\mathcal{N}$ and its input x has been
encoded as a configuration $\langle \mathcal{N}, x \rangle$ of the universal network $\mathcal{U}$ in section 6.4.3.
This encoding has the following properties:

1. the universal network $\mathcal{U}$ can simulate $\mathcal{N}$ on x when given $\langle \mathcal{N}, x \rangle$ as initial
   configuration.

2. it preserves finiteness of $\mathcal{N}$, i.e., it has finite support if $\mathcal{N}$ has only finitely
   many cells.

3. $\mathcal{N}$ stabilizes on input x if and only if $\mathcal{U}$ stabilizes on input $\langle \mathcal{N}, x \rangle$.

Now consider the dynetic problem of membership in the set of configurations

$$L_s(A) \;:=\; \{\langle \mathcal{N}, x \rangle : \;\; \mathcal{N} \text{ has finite bandwidth and}$$
$$\text{does } not \text{ stabilize on } x\} \,.$$

The complement of this set is the STABILITY PROBLEM for neural networks of
finite bandwidth. Recall that a neural network weakly solves this problem if it
stabilizes on $\langle \mathcal{N}, x \rangle$ if and only if $\langle \mathcal{N}, x \rangle \in L_s(A)$.

Proof. (of Theorem 6.10). If the stability problem were weakly solvable by a
network of finite bandwidth $\mathcal{N}_0$ over A, it would also be weakly solvable (after
suitable recoding) by the universal network $\mathcal{U}$ by Theorem 6.8 (see the end of
its proof). But, on input $\langle \mathcal{N}_0, x_0 \rangle$, $\mathcal{U}$ either

1. stabilizes, i.e., $\langle \mathcal{N}_0, x_0 \rangle \in L_s(A)$, hence $\mathcal{N}_0$ does not stabilize on x_0, and
   neither will $\mathcal{U}$ stabilize, a contradiction.

2. or, if $\mathcal{U}$ does not stabilize on x_0, then $\langle \mathcal{N}_0, x_0 \rangle \notin L_s(A)$. Hence $\mathcal{N}_0$ does
   stabilize on x_0, and so also $\mathcal{U}$ should stabilize on $\langle \mathcal{N}_0, x_0 \rangle$, a contradiction.

This is once again the well-known diagonalization argument: such a network $\mathcal{N}_0$
does not exist.   □

Finally, Theorem 6.12 follows since the network $\mathcal{U}$ stabilizes if and only if
its input network stabilizes on its initial configuration.

6.5 Problems

Problems marked * may need to be looked up in the literature.

LOCAL DYNAMICS

1. Show that a 2D (or higher dimensional) cellular automaton does not have finite bandwidth.

2. Show that there exists uncountably many neural nets. *

GLOBAL DYNAMICS

3. Show that there exists uncountable many distinct global dynamics of neural nets. (In particular, uncountably many neural nets.) *

4. Prove in detail that membership in every string language is a solvable dynetic problem.

5. Show that every dynetic problem with a finite number of instances that can be formulated as a set of configurations of finite support is neurally solvable. Show this statement is not necessarily true for arbitrary dynetic problems with finitely many instances.

6. Prove by a cardinality argument that most dynetic problems are unsolvable.

7. Is it true that a dynetic decision problem is solvable if it and its complement are weakly solvable?

6.6 Notes

Neural nets have experienced an explosive rebirth in the late 1980s and early 1990s. They have been successfully applied to create computational models of a cognitive universe and their phenomena (associative memory, learning and adaptation, pattern recognition, speech [S-R], artificial life [L], etc. – see Hartley–Szu [H]). Their similarity to vertebrate nervous systems appeals to cognitive intuition. Neural nets appear, in many respects, just as suitable for cognitive modeling as cellular automata are for physical modeling as shown in Toffoli–Margolus [T-M] and Wolfram [W]. In addition, combinatorial problems that are hard in the classical sense can be given fast, approximate solutions using neural nets, as pointed out by Hopfield [Ho1, Ho2].

The possibility of simulating a Turing machine with neural nets was pointed out early on by McCulloch and Pitts [M-P]. Technically speaking, cellular automata simulations are also neural net implementations. A direct, explicit simulation appears in Franklin–Garzon [F-Ga1]. Some basic analysis of boolean neural nets from the point of view of classical computation appear in Parberry's primer on neural nets [P]. Most of the results in this chapter are based on results by Garzon–Franklin [G-Fr2]. The proof of the auxiliary result Lemma 6.14 appears in Babai–Šos [B-S, Theorem 2.1] and Godsil–Imrich [G-I, Theorem 2.1]. A full-fledged description of a universal neural network for networks of finite bandwidth appears in Garzon–Franklin–Baggett–Boyd–Dickerson [GFBBR].

Finite random networks have been recently applied in modeling of biological phenomena (particularly selection, self-organization, metabolism, functional integration, cell differentiation, and the origin of life and order) in the work of S.A. Kaufmann [K]. Some of these results will be described in the next chapter. A wealth of material about finite networks can be found in Goles-Martinez [G-M]. Additional references can be found in Fogelman-Soulié [Fo].

Turing's concept of computability has been sometimes confused with the notion of *constructability* introduced by Felix Browder. A discussion of the distinction has been made in Franklin–Garzon [F-Ga4]. Constructivists do not accept an existence proof of an object (such as π or the existence of a universal Turing machine) until an actual *construction* of the object has been provided. Now, most proofs in computability are actually done by construction of the appropriate algorithm. The naive tendency is to identify computability and constructibility.

There is a sharp contrast between the two positions, however, that one can perceive upon a moment's reflection. A proof of unsolvability *cannot* be constructive since it calls for no object to be constructed. Such a proof is only possible through an indirect argument that the existence of the object in question would lead to contradiction. But even in existence proofs, the classical theory may perfectly well assert and prove the existence of an algorithm to solve a problem without actually providing a construction of it. For instance, a long-standing conjecture such as the Riemann Hypothesis is trivially algorithmically solvable. For the problem consists of a single question whose answer is either 'yes' or 'no' Certainly there exists a machine that provides the correct answer to this question. Build two machines M_{yes}, M_{no}, one simply saying 'YES', the other 'NO'! One may never find out which machine provides the correct answer, but we certainly have in our hands one that does effectively solve the problem. Another, less trivial, instance is the well-known Seymour–Robertson result on graph minors which implies the existence of several polynomial time algorithms, none of which have yet been (or perhaps will ever be) found. Therefore computable entities are not necessarily constructible in the mathematical sense of the word.

On the other hand, constructivists assume that entities such as the real number $\sqrt{2}$ are accessible to the human mind. This despite the fact that, given physical measurement constraints, a computer, or anyone for that matter, would be hard-pressed to present us with a whole *finished construct* of anything like $\sqrt{2}$ or a reasonable resemblant thereof *in finite time*, not just a *name* for it. Constructible operations may not be possible, or even consistent with the nature of Turing machines constructions. One may then conclude that there is very reasonable doubt that constructible entities are, in general, Turing computable.

Finally, a more systematic discussion of issues related to hardware implementation of cellular automata and neural nets can be found in Prince [Pri], Franklin–Garzon [F-Ga3, G-B] and Garzon–Botelho [G-B]. The less theoretical question of *physical* implementation of the models considered here has been very much ignored by the orthodox treatment of theoretical computation. That this is, in fact, a very important issue is further argued, from a different perspective, by Fields in [F]. Motivated by ultimate fundamental limitation to the size and speed of circuits, more concrete physical models, where communication is based on actual physical contact at nanoscales instead of wires, has been contemplated by several authors, see, e.g., Prince [Pri] and Biafore [Pri, B].

References

[B]   M. Biafore: Universal computation in few-body automata. Complex Systems **7**:3 (1993) 221–239

[B-S]   L. Babai, V.T. Šós: Sidon sets in groups and induced subgraphs on Cayley graphs. Europ. J. Combin. **6** (1985) 101–114

[Bi]   E. Bishop: Constructive analysis. Springer-Verlag, Berlin, 1985

[B-S-S]   L. Blum, M. Shub, S. Smale: On a theory of computation over the real numbers; **NP**-completeness, recursive functions and universal machines. Bull. of the Amer. Math. Soc. **21** (1989) 1–46

[C-M]   H.S.M Coxeter, W.O. Moser: Generators and relations for discrete groups. Springer-Verlag, New York, 1972

[F]   C. Fields: Consequences of nonclassical measurement for the algorithmic description of continuous dynamical systems. J. Exper. and Theoret. Artif. Intelligence **1** (1989) 171–178

[F-Ga1]   S. P. Franklin, M. Garzon: Neural computability. In Progress in Neural Networks, Omid Omidvar (ed.). Ablex, Norwood NJ 1990, pp 127–145

[F-Ga2]   S. P. Franklin, M. Garzon: Global dynamics in neural networks. Complex Systems **3**:1 (1988) 29–36

[F-Ga3]   S.P. Franklin, M. Garzon: On stability and solvability (Or, when does a neural network solve a problem?). Minds and Machines **2** (1992) 71–83

[F-Ga4]   S.P. Franklin, M. Garzon: Computation by discrete neural nets. In: Mathematical perspectives on neural nets, P. Smolensky, M. Mozer, D. Rumelhart (eds.). Lawrence Erlbaum Publishers, in press

[Fo]   F. Fogelman-Soulie, Y. Robert, M. Tchuente: Automata networks in computer science. Princeton University Press, Princeton NJ, 1987

[G-J]   M.R Garey, D. Johnson: Computers and intractability: a guide to the theory of NP-completeness. W.H. Freeman, San Francisco, 1978

[G-B]   M. Garzon, F. Botelho: Stability and Observability. In: Proc. 6th Neural Inf. Proc. Conf. Morgan Kaufmann, San Mateo CA, 1994, pp 455–462 and 1171–1172

[G-Fr1]   M. Garzon, S. P. Franklin: Global dynamics in neural networks II. Complex Systems **4**:5 (1990) 509–518

[G-Fr2]   M. Garzon and S. Franklin: Neural Computability II. Ext. abs. in: Proc. 3rd Int. Joint Conf. on Neural Networks, Washington D.C. **I** (1989) 631–637. Final version in J. of Artificial Neural Systems **1** (1994).

[GFBBR]   M. Garzon, S.P.Franklin, W. Baggett, W. Boyd, D. Richardson: Design and testing of a general purpose neuro-computer. J. Parallel Distrib. Comput. **14** (1992) 203–220

[G-I]   C.D. Godsil, W. Imrich: Embedding graphs in Cayley graphs. Graphs and Combinatorics **3** (1987) 39–43

[G-M]   E. Goles, S. Martinez: Neural and automata networks (dynamical behavior and applications). Kluwer, Dordrecht, 1990

[H-S]   R. Hartley and H. Szu: A comparison of the computational power of neural network models. Proc. IEEE First Int. Conf. on Neural Networks **III** (1987) 17–22

[H]   S. Haykin: Neural networks: a comprehensive approach. Macmillan, New York, 1994

[H-U]   J.E. Hopcroft, J.D. Ullman: Introduction to automata theory, languages and computation. Addison-Wesley, Reading MA, 1979

[Ho1]   J.J. Hopfield: Neural networks and physical systems with emergent collective computational abilities. Proc. Nat. Acad. Science **79** (1982) 2554–8

[Ho2]   J.J Hopfield and D.W. Tank: Computing with neural circuits: a model. Science **233**: 4764 (1986) 625–633

[K]   S.A. Kauffman: The origins of order (self-organization and selection in evolution). Oxford University Press, Oxford, 1993

[L]   S. Levy: Artifical life. Jonathan Cape, London, 1992

[M-P]   W.S. McCulloch, W. Pitts: A logical calculus of the ideas immanent in nervous activity. Bull. Math. Biophys. **5** (1943) 115-133

[P]   I. Parberry: A primer on the complexity of neural networks. Pitman and Hall, London, 1988

[Pri]   J.L. Prince: Very Large Scale Integration. Springer-Verlag, New York, 1980

[R-H-M]   D. E. Rumelhart, G. Hinton, J. L. McClelland: A general framework for parallel distributed processing. In: Parallel Distributed Processing, vol. 1. D. E. Rumelhart, J.L. McClelland, et al. (eds.). MIT Press, Cambridge MA, 1986

[S-R]   T.J. Sejnowski, C.R. Rosenberg: Parallel networks that learn to pronounce english text. Complex Systems **1** (1987) 145–68

[T-M]   T. Toffoli, N. Margolus: Cellular automata machines. MIT Press, Cambridge MA, 1987

[V]   G. Vichniac: Simulating physics with cellular automata. Physica **10D** (1984) 96. Reprinted in: Cellular Automata, Proc. Los Alamos Conference. North-Holland, Amsterdam, 1983

[W]   S. Wolfram: Theory and applications of cellular automata. World Scientific, Singapore, 1986

7. General Properties

Perhaps the most fundamental problem in the study of cellular automata concerns the classification of local rules according to a natural notion of equivalence related to their long-time behavior. One of the difficulties in answering this question is precisely the fact that there is not a single most dominant aspect in the long-term evolution of a global map. Previous chapters have dealt with particular types of local rules and their properties. Even totalistic rules of the type considered in Chapter 4 show that a bottom-up approach to the classification of cellular automaton dynamics in terms of our ability to predict their long-term behavior is not fine enough to constitute a satisfactory categorization.

A satisfactory classification requires the formalization of global rules as abstract objects that can then be identified under a certain notion of equivalence. The first requirement has been accomplished in Chapter 2 with the notions of cellular space, configuration space, dynetic problem, and global dynamics.

The second requirement, namely the appropriate notion of equivalence, appears more elusive. In the strictest sense, equivalence can be interpreted as identity of global dynamics. In this case, the problem becomes that of distinguishing those global transformations of configuration space that arise from local maps defined by cellular automata and its generalizations. In a more relaxed sense, equivalence can be interpreted as equivalence of the dynamical systems generated by global rules under successive iteration. In fact, the question of classifying dynetic tasks solvable by *dynamical* systems upon iteration of local rules of cellular automata and/or neural networks is a difficult unsolved problem of current interest.

There are other alternatives to a classification of global dynamics based on a computational viewpoint. The strict notion of equivalence mentioned above gives rise to a class of self-maps of configuration space computable in a *single* step (equivalently, *constant time*) by a cellular network. The second notion of equivalence is also interesting form a computational point of view since the computational power of cellular automata on infinite configurations goes beyond that of the classical models such as Turing machines.

The purpose of this and the following two chapters is to present some results concerning the classification problem under these notions of equivalence. In this chapter we explore the nature of self-maps of configuration space arising from local rules. This type of classification proceeds top-down, as opposed to the bottom-up approach in Chapters 5 and 6. It is thus reasonable to expect that somewhat more powerful tools than the ones we have used so far will become necessary. Although the right tools are old ones in the mathematician's tool-kit (the ideas of continuity and convergence that constitute the area of topology), no familiarity even with the basic necessary ideas is assumed on the part of the reader. As a byproduct of the main characterization, there follow a number of interesting corollaries on the relationships between injectivity and surjectivity for global rules that show that configuration space is encouragingly akin to a finite set when it comes to local maps.

7.1 Metric Preliminaries

In an abstract sense, a global function is simply a self-map $T : \mathbf{C} \to \mathbf{C}$. Obviously, not all functions T of this type are induced by a local rule δ (see Problems 2.21 and 3.12). In order to find a number of properties that will characterize such self-maps induced by local rules we begin by observing that they must satisfy the following two conditions:

1. $T(O) = O$, since no spontaneous activity can take place in a cellular automaton.

2. T commutes with all shifts S_k, i.e.,

$$\forall x \in \mathbf{C}, T(S_k(x)) = S_k(T(x)).$$

Thus, in order to distinguish arbitrary self-maps of $\mathbf{C}$ from those induced by local rules, we will call the latter ones *local transformations* or *local dynamics*. Although strictly speaking this reference may be confusing δ and T, the notation and context will keep us straight.

There is a third crucial condition that must be satisfied by a local rule. It is simply that no *instantaneous* action-at-a-distance can take place in the evolution of any configuration. There is no spontaneous activity and the only way that a quiescent cell can change its state is if a neighboring cell becomes nonquiescent. This is not to say that in the case of an infinite input configuration only finitely many cells may change states at once under the action of the local rules (try the all-ones configuration under SUM_MOD5 in the von Neumann neighborhood in 1D euclidean space.) In order to state this condition we need to give a precise meaning to the basic idea embodied in the words *near, close to,* and *distance*. There is an area of mathematics that deals with these ideas in a precise fashion. It is called *topology* and we will take a brief excursion into it.

7.1.1 Metrics and Topologies

The problem with the idea of nearness is that it is a very *relative* idea and it can be interpreted in many different ways. The simplest idea of nearness is probably physical closeness, in the sense that the distance between the two objects that are *near* each other is small (close to 0). In order to explain this simple idea of *nearness* we have to use another object, the ordinary set of real numbers and their order. We also speak of *closeness* in the context of human relations and feelings, but even there it is clear that one needs a reference frame that defines *near*. Since there is no absolute notion of nearness, we must search for a relative definition. Geometrically speaking, one can determine that two points are near on the euclidean plane because we can associate with them a real number, a *distance* between them, which is a small number. In general we do not have to be constrained to points in the plane. As long as there is a definite way to associate a number with every pair of objects, one can speak of them being near. However, some minimal requirements are to be observed because otherwise everything may become near to everything else, which is probably undesirable.

Definition 7.1 *A* distance *or* metric *on an arbitrary set Y is a mapping*

$$|*, *| : Y \times Y \to \mathbf{R}_{\geq 0}$$

that assigns a nonnegative real number $|x, y|$ to every pair of points $x, y \in Y$ so that

1. *$|x, y| = 0$ if and only if $x = y$;*

2. *$|x, y| = |y, x|$;*

3. *The triangle inequality: $|x, y| \leq |x, z| + |z, y|$.*

for all $x, y, z \in Y$. In this case, the pair $\langle Y, |*, *| \rangle$ is called a metric space. *Two metric spaces Y, Y' are called* isometric *(or* isometrically equivalent*) if it is possible to establish a bijective correspondence $\rho : Y \to Y'$ such that $|x, y| = |\rho(x), \rho(y)|$ for all $x, y \in Y$.*

A coarse way to define a metric is thus to only assign two values to $|*, *|$, namely 0 (to equal points) or 1 (to distinct points). A second example is the absolute value $|\alpha - \beta|$ function of a difference between two real numbers α, β. A more interesting example for us is a distance on a 1D euclidean cellular space. Define the distance between two configurations $x, y \in \mathbf{C}$ as follows

$$|x, y| := \sum_{i \in \Gamma} \frac{|x_i - y_i|}{2^{|i|}} . \tag{7.1}$$

It is an exercise in elementary calculus to verify that this series converges for every x, y and that the numbers thus assigned to the pairs x, y indeed satisfy properties 1–3 above. Likewise, one can define a metric on any other configuration space. Simply fix an *enumeration* $i_1, i_2, \ldots$ of the underlying cellular space Γ so that points closer to the origin in Γ appear earlier than points farther away, and define the distance between two configurations by exactly the same formula (7.1). The same argument given before establishes that $\mathbf{C}$ is again a metric space. This space is sometimes referred to as the *product metric*, or even

the *astronomer's metric* on **C**. Note that according to this metric, two configurations are near each other, i.e., the distance between them is small, if cells
near the origin of the cellular space have identical states and they differ only far
away from the origin. In other words, to an observer located at the origin (e.g.,
the astronomer), close configurations will look identical, since the differences
are so far away that she "will not be able to see" them. (Thus, in some sense,
this metric is psychologically closer to our daily experience with distance!)

Observe that, unlike the arbitrarily large distance that one can find between
two real numbers, cellular space configurations cannot be farther than 3 apart.
This is not surprising when one realizes that a similar bound also occurs with
the 2-dimensional surface of a sphere of finite radius.

7.1.2 Convergence and Continuity

Once a metric space is established as a frame of reference, it is easy to speak of
nearness, but we still need to determine *how near* by giving a threshold distance
(to tell the difference near-far) and an object to be near to. The ball of center
a and radius r in a metric space Y is the set

$$B[a, r] := \{y \in Y : |y, a| \le r\},$$

or simply $B_r[a]$, of all points within a distance r from the point a. Once we get
an idea of distance we get an idea of *convergence* of a sequence of points in the
metric space to a point a in it, namely the idea that the distance of the points
in the sequence to a decreases to zero. And once we have acquired the idea of
convergence we would like to consider transformations of the host metric space
that somehow "preserve" convergence in the following sense.

Definition 7.2 *A sequence (x^n) of points in a metric space Y converges to a
point $a \in Y$ (denoted $x^n \to a$) if the sequence of distances $|x^n, a|$ tends to zero in
the usual sense. A function $f : Y \to Y'$ between two metric spaces is* continuous
*if it does not disturb convergence, that is, $f(x^n) \to f(a)$ in Y' whenever $x^n \to a$
in Y. In symbols, f is* continuous *if*

$$\lim f(x^n) = f(\lim x^n),$$

where $\lim$ *denotes the limit of a sequence (which is always unique in a metric
space). A* cluster *or* accumulation point *of a sequence $(x^n)_{n \ge 0}$ is a point a such
that some subsequence $(x^{n_t})_t$ converges to a.*

In the case of configuration space, that a sequence of configurations (x^n)
converges to a limit configuration a means that, as we run through the configurations $x^1, x^2, x^3, \ldots$, they begin to match *cellwise* with a in larger and larger
vicinities of the origin of the cellular space. (That's why this topology may be
called the 'zipper topology': if we imagine the cells in the cellular space as points

of a one way infinite zipper spiraling from and away from the origin, convergence may be thought of as a "zipping process".) And a cluster configuration of a sequence (x^n) is a configuration a that sporadically but surely matches the configurations x^n in arbitrarily large neighborhoods of the origin of the cellular space *infinitely often* (although the matching may be destroyed before it reappears again). It should be clear that if a sequence converges in a metric space then it has a *unique* cluster point – namely its limit – but not conversely (look at $1, -1, 2, -1, 3, -1, \ldots$ over the real numbers).

Although strictly speaking each choice of the possible enumerations of Γ in equation (7.1) gives rise to a different metric, all of them will define the same notion of convergence and continuity, so we do not bother to specify it very carefully. The notions of convergence and continuity thus determined are called the *product topology*.

Remark. The term *topology* also has a precise technical meaning that need not concern us here. Intuitively speaking, a *topology* is essentially an abstraction of the idea of a metric without a reference to a distance function that enables one to speak about nearness without having to use a yardstick. It requires substitution of a metric by analogs of balls $B[a, r]$ with a minimum set of properties to make them resemble metric spaces.

There are plenty of examples of continuous self-mappings of configuration space. In fact, we can now state a third basic property of local transformations:

3. a global dynamics T induced by a local rule is continuous in the product topology.

This property is obvious once we realize that by the way we chose the enumerations of Γ to define the metric on $\mathbf{C}$, the contribution to the distance between two configurations decreases rapidly with distance of the cells from the origin. Thus if a sequence $x^n \to a$, large values of n give configurations that match a within larger and larger cellular neighborhoods about the origin of the cellular space. Since T is local, the sequence $T(x^n)$ must behave analogously.

At this point one may wonder exactly what type of conditions a continuous self-map of configuration space must satisfy in order to be specifiable by a local rule of a cellular automaton. This is the type of question dealt with in the following two sections.

7.2 Basic Results

That the three necessary conditions stated in the previous section are, in fact, sufficient to guarantee locality is a striking fact.

Theorem 7.3 *A self-mapping $T : \mathbf{C} \to \mathbf{C}$ is a local dynamics of a cellular automaton if and only if it satisfies*

1. $T(O) = O$;

2. T commutes with shifts;

3. T is continuous.

   This section presents the proof and a number of apparently unrelated consequences. Although it is possible to give other more topological and terser account (see Problem 8), we have chosen to use the original ideas based on results of Moore and Myhill because they are of interest in themselves.

7.2.1 The Moore–Myhill Theorem

One can think of a cellular space as an 'imaginary' world where configurations are total states of the whole universe. Each line in the table of a local rule of cellular automaton can then be thought as a 'physical' law that determines the next local state of the universe under given local conditions. The local rule itself is thus a complete ensemble of physical laws governing the deterministic evolution of this world. On thinking about such a universe, one naturally wonders whether there are certain states of the world that cannot be arrived at by evolution no matter what initial configuration one may start with. In the same spirit, one can ask the apparently unrelated question of whether or not the evolution of this world is reversible *by the same type of localistic law.* An obvious necessary condition is that no two different states of the universe be collapsed together by the global transformation. The results in this section establish that these two properties of a cellular automaton are essentially equivalent for a very general type of cellular space that includes all euclidean spaces and, more generally, Cayley graphs of nonexponential growth.

   These questions can be made rigorous with the following notational precisions. Recall that an *array* of a cellular space is the restriction of a configuration to a *finite* subgraph F of the cellular space (which is called the *support* of the array). Every array a is the restriction $T|_F$ of a configuration quiescent outside F and with values identical to the array in F (and will be referred to, in this chapter, as the *configuration* a as well).

Definition 7.4 *Two arrays (or a configuration and an array) agree if they have the same state at every cell in the intersection of their supports. A pattern over* **C** *consists of all arrays that differ by a shift (i.e., are equivalent under translation).*

   An array B contains a copy of an array a if, under some shift, the image of a agrees with B. An array B contains n copies of an array a if the supports of the corresponding shifted arrays are disjoint. The notation includes subgraphs of the underlying graph as quiescent arrays. For a given neighborhood N, B^+ (respectively $B^-, B^\pm$) consists of

$$B^+ := \bigcup_{i \in B} N_i$$

$$B^- := \bigcup_{\substack{i \in B \\ N_i \subseteq B}} N_i$$

$$B^\pm := B^+ - B^-$$

An array a is a garden of Eden restriction *of a configuration c (with respect to T) if the predecessor problem*

$$T(x) = a$$
$$\text{subject to } x|_F = c|_F$$

has no solution. Finally, two arrays a, b are mutually erasable *(or just mea) if whenever two configurations x, y that respectively agree with a, b in F are identical outside F, they are collapsed to the same configuration $T(x) = T(y)$.*

Observe that the notions concerning arrays are independent of T, but gardens of Eden and mea's are not.

Proposition 7.5 *If two finite subgraphs F and B of a Cayley graph Γ satisfy*

$$|Q|^{|B|-|B^\pm|} > (|Q|^{|F|} - 1)^n |Q|^{|B|-n|F|} \tag{7.2}$$

and if B^- contains at least n copies of F, then

(i) *If F is the support of two erasable patterns, then B^- is the support of a garden of Eden pattern;*

(ii) *if F is the support of a garden of Eden pattern, then B^+ is the support of two erasable patterns.*

The proof is a direct extension of the original proofs by Moore and Myhill for the euclidean case.

Proof. (i) Let R be an equivalence relation that holds for two arrays a, b with support F in case $a = b$ or a, b are mutually erasable. The number of equivalence classes of R is at most $|Q|^{|F|} - 1$ if there exists garden of Eden arrays with support F. Let R^* be the equivalence relation induced on pairs of arrays having support B as follows: they are identical outside all copies of F, and their restrictions to every two copies in B are in relation R. The number of classes in R^* is at most the right-hand side in (7.2). Plainly, two arrays with support B that are R^*-equivalent are either equal or mea. By definition of B^-, the restrictions of every two successors $T(x), T(y)$ of extension configurations x, y of two mea on B are identical on B^-. The number of such arrays on B^- is thus at most the number of classes in R^* and the number of arrays supported by B^- is the left-hand side of (7.2). The given inequality thus implies that there must exist an array supported on B^- which is not a restriction of type $T(x)|_F$, i.e., a garden of Eden array.

(ii) Let n' be the number of nongarden of Eden arrays with support B. We claim that n' separates the two sides in (7.2), that is

$$|Q|^{|B|-|B^{\pm}|} \leq n' \leq (|Q|^{|F|} - 1)^n |Q|^{|B|-n|F|}.$$

In effect, because the right-hand side counts the number of arrays supported by B not containing at least one copy of the garden of Eden (which clearly are garden of Eden arrays), it is an upper bound on n'. Call T *injective* on an array B if whenever two distinct arrays are extended identically to two configurations, their images under T remain distinct. If statement (ii) were not true and T is injective on B^-, then the number of restrictions of images of T to B, i.e., n', would be at least the number of distinct patterns on B^- (the left-hand side of (7.2)). Therefore one would have the said separation, contrary to the hypothesis. Finally, to verify the said injectivity on the hypothesis that there are no mea on B^+, assume contrarily that two configurations x, y differ on B^- but agree on $B^{\pm}$ and have the same image $T(x) = T(y)$. Define a third configuration z that agrees with y on B^+ and with x elsewhere. Thus $T(z)|_B = T(y)|_B = T(x)|_B$. Since one can easily verify that $\Gamma - B \subset (\Gamma - B^-)^-$, x and z must also agree on $\Gamma - B$, a contradiction. $\square$

Now we need to examine in which cellular spaces one can produce the hypotheses of Proposition 7.5. The next proposition shows a sufficient condition in terms of the proportion of the size of successively larger balls in their Cayley graphs to the number of sites on their boundaries.

Proposition 7.6 *Let F be a subgraph of Γ. If there exists a real number $\varepsilon > 0$ and a sequence $B_1, B_2, \ldots$ of finite subgraphs of Γ such that each B_n contains at least $\varepsilon \frac{|B_n|}{|F|}$ copies of F, where*

$$\liminf_{n \to \infty} |B_n|/|B_n - B_n^{\pm}| = 1, \tag{7.3}$$

then

(i) *If F is the support of two erasable patterns, then some B_n^+ is the support of a garden of Eden pattern;*

(ii) *if F is the support of a garden of Eden pattern, then B_n^- is the support of two mutually erasable patterns.*

Proof. It suffices to establish the conditions of proposition 7.5 for some $B = B_n$. Let m_n be the number of copies of F in B_n. Taking logarithms shows that (7.2) is equivalent to

$$\frac{|B_n|}{|B_n| - |B_n^{\pm}|}[1 + \frac{m_n|F|}{|B_n|}(\frac{\log(|Q|^{|F|} - 1)}{|F|} - 1)] < 1.$$

Since both factors $\frac{m_n|F|}{|B_n|} \leq 1$ and $\frac{log(|Q|-1)}{|F|} < 1$, the quantity in brackets is positive and less than 1. Since $\frac{m_n|F|}{|B_n|} > \varepsilon$, this quantity is bounded above by

$$\tau := 1 + (\frac{\log(|Q|^{|F|} - 1)}{|F|} - 1).$$

Hence the desired n exists whenever

$$\liminf \frac{|B_n|}{|B_n| - |B_n^{\pm}|} < \frac{1}{\tau},$$

which holds under the given condition on the $\liminf$. $\qquad \square$

This growth condition on the relative size of subgraphs B is a condition satisfied by great numbers of Cayley graphs. It amounts to an upper bound on the number of vertices in a ball that lie outside of copies of F.

Definition 7.7 *A graph Γ is* homogeneous *if for all finite subgraphs F there exists a real number $0 < \varepsilon \leq 1$ and an integer n_0 such that the number m_n of (disjoint) copies of F in every ball $B_n := B[i, n]$ $(n \geq n_0)$ of Γ is at least*

$$m_n \geq \varepsilon \frac{|B_n|}{|F|}. \tag{7.4}$$

Lemma 7.8 *Cayley graphs are homogeneous.*

Proof. Without loss, assume that F is itself a ball of Γ of radius m, say $B_1 := B[i_1, m]$. First, obtain a sequence of pairwise disjoint balls $B_n := B[i_n, m]$ of radius m such that the balls $B[i_n, 2m]$ cover Γ as follows. Let $i_2, i_3 \cdots$ be the points at distance $m + 1$ on Γ from B_1 (and hence at distance $\geq 2m + 1$ from i_1). The ball $B_2 := B[i_2, m]$ is disjoint from B_1. Renumber so that i_3 is now the next vertex at a distance $m + 1$ from $B_1 \cup B_2$. Continuing on in this fashion one obtains disjoint balls $B_1, B_2, \ldots, B_t$. Reiterate the procedure with the union of these balls in place of B_1, and so on, *ad infinitum*.

Now consider an arbitrary ball $B[i, n]$ of radius n and let m_n be the number of copies of B_1 contained in it. Let γ_n be the growth sequence of the B_n (see Problem 2.5). Clearly m_n is at least equal to the number of i_l's contained in $B_n[i, n - m]$. Every $j \in B[i, n - 3m]$ belongs to some $B[i_l, 2m]$, which is thus in $B[i, n - m]$, as a picture immediately shows. Therefore

$$
\begin{aligned}
\gamma(n - 3m) &= |B[i, n - 3m]| \\
&\leq |B_{i_l} \cap B[i, n - m]| \gamma(2m). \\
&\leq m_n \gamma(2m)
\end{aligned}
$$

We know that $\gamma(r + s) \leq \gamma(r) \gamma(t)$, so it follows that $\gamma(n - 3m) \gamma(3m) \geq \gamma(n)$ and hence that

$$m_n \geq \frac{\gamma(n)}{\gamma(2m)\gamma(3m)} = \frac{\gamma(m)}{\gamma(2m)\gamma(3m)} \frac{\gamma(n)}{\gamma(m)}.$$

Thus one can take $\varepsilon := \frac{\gamma(m)}{\gamma(2m)\gamma(3m)}$ in the homogeneity condition.   □

This result puts us within walking distance of a generalization of the original Moore–Myhill theorem to a large family of Cayley graphs.

Theorem 7.9 *Over a finitely generated cellular space of nonexponential growth, a cellular automaton has garden of Eden patterns if and only if it has mutually erasable patterns.*

Proof. With the notation as before, it suffices to show that if the inequality (7.4) in the hypothesis of proposition 7.6 is violated, i.e., if

$$\alpha := \liminf \frac{\gamma(n)}{\gamma(n-1)} > 1,$$

then γ has exponential growth. For all $\varepsilon > 0$ there exists n_0 such that

$$\gamma(n) \geq (\alpha + \varepsilon)\gamma(n-1),$$

for all $n \geq n_0$. Therefore

$$\gamma(n_0 + m) \geq (\alpha + \varepsilon)^{m+1}\gamma(n_0 - 1)$$

for all $m \geq 0$. Hence

$$\gamma(n) \geq (\alpha + \varepsilon)^n \frac{\gamma(n_0 - 1)}{(\alpha + \varepsilon)^{n_0 - 1}}$$

for all $n \geq n_0$. The second factor is the desired constant.   □

This result no longer holds for all graphs of exponential growth (see Problem 13).

7.2.2 Nondeterministic Cellular Automata

Richardson's theorem can be generalized to nondeterministic transition roles and more general (partially defined) relations, which we continue to denote T on $\mathbf{C}$.

As discussed in Chapter 1, nondeterminism is a natural way to generalize computational models and a cellular automaton is one of them. Assume hence that the fsm at each site of the cellular space is a nondeterministic fsm, that is, its local dynamics δ now assigns a finite list of *possible* moves to each pair (p, x_i) in its domain. Formally, a nondeterministic fsm M is a multivalued function, i.e., a *relation*

$$\begin{aligned}
\delta : Q \times \Sigma &\rightarrow Q \\
(p, x_i) &\mapsto \{q_1, q_2, \ldots, q_r\}
\end{aligned} \tag{7.5}$$

On an input string, M can go a number of ways *by making a choice* (also called a *guess*) from the set in the right-hand side at every move (if empty, M gets "stuck"). These choices are the possible ways in which M can identify the relevant property in the input string x. Thus, different runs of M on the same input string may produce a different orbit in the state set. But as before, M is said to *accept* a string if *any* one run of M (i.e., with the machine M "guessing" at the top of its cunning, in the best of all possible worlds) ever enters a final state. One can define a nondeterministic Turing machine in a similar way. In both cases, one can show that this guessing ability at these extremes of the spectrum does not contribute anything substantial to their computational power, because one can prove that there exist algorithms that construct equivalent *deterministic* fsm's or Turing machines that perform exactly the same function as the given one. The reason nondeterministic models are of interest in the classical case is that intermediate machines (like pushdown-stack machines or resource-bounded machines such as polynomial time) do *not* or are not known to have this property.

In the context of cellular automata, however, nondeterministic machines do make a difference, at least as far as global dynamics are concerned. If the sites of a cellular space is occupied by copies of a nondeterministic fsm, the global dynamics is no longer a self-mapping of $\mathbf{C}$ but rather a *relation* between configurations: a configuration x is related to configuration y, denoted xTy in case there is a possible run of the cellular automaton under the local rule (7.5) that transforms x into y in a single time step. In this case, y is also called a *successor* of x. Notice that x may well have no successors, so that T may also be partially defined. This relation will continue to be called the *global dynamics* of the automaton in this section. We maintain the restriction of no spontaneous activity, that is $\delta(0\cdots0) = \{0\}$, as well as the previous notation, for nondeterministic local rules.

As before, an arbitrary relation T on $\mathbf{C}$ satisfies the following properties:

(A) OTx if and only if $x = O$;

(B) T is *shift-commuting*:

$$xTy \text{ if and only if } S_k(x)TS_k(y);$$

(C) T is *(forward) continuous* in the following sense: if x^∞ is the unique cluster point of a sequence $x^1, x^2, x^3, \cdots$ then $x^\infty Tx$ iff x is a cluster point of some sequence (y^n) satisfying $\forall n\, (x^n Ty^n)$;

(D) T satisfies the *independence condition*: if for arbitrary sites $i_1, \ldots i_n$ on the cellular space in states $q_1, \ldots, q_n$ there exists a "partial" successor y^n (that may depend on x and the i_l's) such that

$$xTy^l \text{ and } \forall l(y^n_{i_l} = q_l),$$

then there exists a single configuration y such that

$$xTy \text{ and } \forall l(x_{i_l} = y_{i_l} = q_l).$$

Note that if T is deterministic, conditions (C) is equivalent to continuity and (D) is automatically satisfied. Richardson's theorem asserts that these four conditions, in fact, constitute a characterization of binary relations on $\mathbf{C}$ induced by (nondeterministic) local dynamics. We proceed to a proof of this result.

Lemma 7.10 *Every infinite sequence of configurations $(x^n)_n$ in $\mathbf{C}$ has a cluster point.*

Proof. Let (i_n) be an enumeration of all cells in the cellular space so that some sequence (a^n) of arrays satisfies

(i) a^n is nonquiescent exactly up to site i_n;

(ii) a^n agrees with (x^l) at infinitely many i_n's;

(iii) a^n agrees with a^{n+1}.

The point $x^\infty := \lim_n a^n_{i_n}$ is a cluster point of (x^n). $\square$

Theorem 7.11 *A binary relation T is defined by a local rule if and only if it satisfies conditions (A), (B), (C), and (D).*

Proof. Assume first that T is local. Conditions (A), (B), and (D) are obvious. To prove (C), let $x^n T y^n$ for all $n \geq 1$ and let y^∞ be a cluster point of (y^n). To show that $x^\infty T y^\infty$ it suffices to show that $x^\infty_i = y^\infty_i$ for each i. Since $y^n = d^\infty$ for infinitely many n, $\delta(x^n_i, *)$ must contain y^∞ for infinitely many n. Since x^∞ is the unique cluster point of (x^n), it follows that $y^\infty_i \in \delta(x^n_i, *)$, hence $x^\infty T d^\infty$. Conversely, if $x^\infty T y$, define (y^n) as follows: for each n and i, put $y^n_i = y_i$ if $y_i \in \delta(x^n, *)$. Otherwise let y^∞ be an arbitrary state in $\delta(x^n_i, *)$. (There must be at least one such state since x^n is in the domain of T). Clearly $x^n T y^n$ for all n. To show that y is a cluster point of (y^n), let a be a finite configuration that agrees with y on its support $i_1, ..., i_r$. We need to show that a agrees with infinitely many y^ns, i.e., that each y^n_i belongs to $\delta(x^n, *)$ for infinitely many n's. After a moment's reflection, this follows from the fact that x^∞ is a cluster point of (x^n).

In order to prove the converse, assume T has properties (A)–(D). Let $q_0 = 0, ..., q_r$ be all the states of the nondeterministic automaton T and let δ be the local rule defined as follows:

$$q \in \delta(x_i, x_{i+N}) \quad \text{iff} \quad \exists x, y \in \mathbf{C}, \exists i \, (x_i = p \,\&\, xTy \,\&\, y_i = q).$$

Now let a global rule Δ be defined as follows:

$$x \Delta y \quad \text{iff} \quad \forall i \exists y^i \, (xTy^i \,\&\, y^i_i = y_i).$$

Assume $x \Delta y$. Since T has the independence property (D), if a finite pattern a agrees with y, then some configuration y^a agrees with a and satisfies xTy^a.

Find a sequence (y^a), by Lemma 7.10, that has y as cluster point. Since T is continuous, it follows that xTy. This proves that if $x\Delta y$ then xTy. The converse is obvious, so that T and Δ are identical whenever x belongs to the domain of T. However, it is necessary to prove also that Δ is a local rule.

Linearly order all possible neighborhoods by size and by inclusion for neighborhoods of the same size. For a neighborhood N, let δ_N be the local rule defined as follows:

$$\delta_N(p, x_{i+N_+}) \quad \text{iff} \quad \exists x, y \in \mathbf{C}\left(xTy \,\&\, \exists i(y^i = q \,\&\, \forall j \in N_\bullet\,(x_{i+j} = x_j))\right).$$

The last condition implies that, in particular, $x_i = x_0 = p$. Also, in the particular case $N_\bullet = \{0\}$, $\delta_N = \delta$. Moreover, $x\Delta_N y$ whenever xTy because Δ_N agrees with T in its domain.

It suffices to prove that if N is large enough, δ_N allows *only* the all-quiescent configuration O in its domain as the next state of O. This follows by taking a neighborhood N larger than the neighborhoods N_1 and N_2 from the next two lemmas. □

Lemma 7.12 *For some neighborhood N_1, if N_1 strictly contains N then Δ_N is local.*

Proof. If not so, some sequences $(x^n), (y^n)$ satisfy $x^n = 0$ and $x^n \Delta_{N_n} y^n$, but for some q, $y^n_{i+j} = q$ for all $j \in N_n$. By invariance under shifts, we can assume that $i = 0$. Therefore O would be the only cluster point of (x^n) because each N_{n+1} is strictly larger than N_n, and (y^n) would have a not all-quiescent cluster point d^∞. This would violate continuity because we should then have OTy^∞, which contradicts condition (A). □

Lemma 7.13 *There is a neighborhood N_d such that if N properly includes N_d then for all x in the domain of T, xTy iff $x\Delta_{N_d} y$.*

Proof. It was already pointed out that Δ_N agrees with T. Suppose, for the sake of contradiction, that the converse is false. If so, there are sequences $(x^n), (y^n)$, with each x^n in the domain of T, such that $x^n \Delta_n y^n$ but not $x^n T y^n$. Hence, for each n there is some i_n such that $x^n \neq y^n$ whenever $x^n_{i_n} T y^n_{i_n}$. (Otherwise we use properties (C) and (D) of T to get $x^n T y^n$). By property (B), assume $i_n = 0$.

Now, y^n_0 has only finitely many possibilities, so x^n_0, y^n_0 can be both further assumed to satisfy $\forall n(y^n_0 = q)$ for some fixed state q. Since any sequence of configuration contains a subsequence with a unique cluster point by Lemma 7.10, further assume (x^n) has a unique cluster point. By properties (C) and (D), there exist sequences $(\bar{x}^n), (\bar{y}^n)$ such that for all n, $\bar{x}^n T \bar{y}^n$, $\bar{x}^n$ agrees with x^n in N_n, and $\bar{y}^n_0 = q$. The unique cluster point x^∞ of x^n is also a unique cluster point of $(\bar{x}^n)$. If y^∞ is a cluster point of $(\bar{y}^n)$, it also holds that $x^\infty T y^\infty$ and $y^\infty = q$. By property (C), some sequence $(\tilde{y}^n)$ satisfies $\bar{x}^n T \tilde{y}^n$ for all n, and, moreover has y^∞ as a cluster point, which contradicts (B) because then some $\tilde{y}^n_0 = q$. □

7.3 Injectivity, Surjectivity and Local Reversibility

Richardson's theorem, can be used, in conjunction with the Moore–Myhill result, to establish a number of interesting results about injectivity and surjectivity of global dynamics. To that end, the latter can be conveniently formulated as follows.

Theorem 7.14 *If T is a deterministic local rule and T_0 its restriction to C_0 then T_0 is one-one iff for any array a there exists a configuration x in the range of T that agrees with a.*

Corollary 7.15 *If T is one-one then T^{-1} (possibly a nondeterministic rule) is defined by a local transformation.*

Proof. The inverse T^{-1} has properties (A) and (B). Property (D) holds since T^{-1} is single-valued. To prove (C) let (y^n) be a sequence in the range of T with a unique cluster point y^∞. Let (x^n) be a sequence such that $x^n T y^n$ for all n. Let x^∞ be a cluster point of (x^n). $x^\infty T y^\infty$ since T has property (C). Therefore T^{-1} is continuous, i.e., it has property (C). The converse is easy once T is proved onto as in the proof of Corollary 7.16 below.   □

Corollary 7.16 *If T is deterministic and defined by a local transformation then the following relations hold*

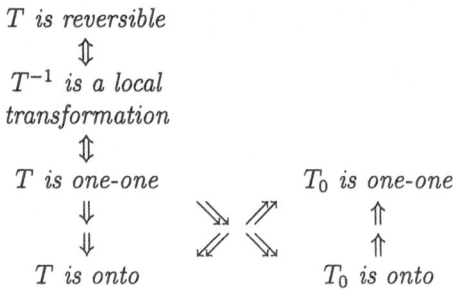

Proof. The proof proceeds bottom-up, right to left, in the diagram above.

⇑: If T_0 is onto then T_0 is one-one by the Moore–Myhill theorem.

⤢: If T_0 is one-one, let y be any configuration and let (b^n) be a sequence of arrays so that y is the only configuration that agrees with all b^n's. By the Moore–Myhil theorem, there are sequences $(x^n), (y^n)$ such that $T(x^n) = y^n$ and y^n agrees with b^n for each n. If we let x^∞ be a cluster point of (x^n), then $T(x^\infty) = y$. This proves that T is onto. Note that since x^∞ need not be finite, it does not necessarily follow that T_0 is onto.

$\nearrow$: Conversely, if T is onto, given any array a, there is a configuration in the range of T that agrees with a. Since a has finite support, there is also a configuration in the range of T_0 that agrees with a. Since a is arbitrary, T_0 is one-one by the Moore-Myhill Theorem 7.14.

$\searrow$: If T is one-one, obviously T_0 is one-one, so T is onto and bijective. By Corollary 7.15, T^{-1} is defined by a (deterministic) local transformation. So, if y is any finite configuration, there exists $x \in \mathbf{C}$ such that $T(x) = y$. Since T^{-1} is local, x must be finite, hence $T_0(x) = y$. Therefore T_0 is onto.

$\Downarrow$: follows from the above three.

$\Updownarrow$: The upward implication follows from Corollary 7.15 and the fact that T is onto. For the converse, note that T must be onto, hence one-one, so T_0^{-1} is deterministic. But then so must be T^{-1}, whence T is one-one.

$\square$

The following example, due to D. E. Muller, shows that if T is surjective, it does not necessarily follow that T is one-one.

Consider the local dynamics δ defined by the table

00	10	20	01	02	11	22	12	21	*0
0	1	2	0	0	1	1	2	2	*

Any configuration y has a predecessor x which has quiescent cells wherever y does and the rest is uniquely determined by the above table. The configuration $x = \cdots 12112 \cdots$ has as successor $T(x) = T(O) = O$. Likewise, the Moore–Myhill statement does not hold on graphs of exponential growth (see Problems 13 and 14).

7.4 Some Generalizations

The elegance of the characterization of local computation in terms of continuity in the product topology of configuration space leads naturally to similar questions for generalizations such as neural and random networks. In this section se explore these generalizations, as well as combinatorial generalizations of a different type.

7.4.1 Neural and Random Networks

Certainly one cannot expect natural conditions involving the shift for arbitrary networks since the communications digraphs can be completely asymmetric. On the other hand, neural nets appear close enough to linear maps that characterization in terms of some sort of semilinearity condition appears plausible. This is, in fact, the case if another simple ingredient is allowed. Let's call a self-map of configuration space F *strictly local* if the values of F at any pixel do not change upon changing pixels *elsewhere*, i.e., if $F(x)_i = F(x_i e^i)_i$ for all

x and i. This condition is equivalent to the existence of maps $f_i : Q \to Q$ such that $T(x)_i = f_i(x_i)$ for each i (see Problem 18). Also let π_i be the projection $\pi_i : \mathbf{C} \to Q$ onto the ith component.

Theorem 7.17 *A self-map $T : \mathbf{C} \to \mathbf{C}$ is realizable over a finite activation set A with unity 1 as an activation global dynamics of a neural network if and only if*

   1. $T(O) = O$;

   2. T is continuous;

   3. $T(e^k)$ has finite support for each pixel configuration e^k; and

   4. $T = F \circ L$, where L is a linear self-map of $\mathbf{C}$ and F is strictly local.

In this case, one can say that such a self-map T is *implemented* or *realized* by the corresponding network. Note that condition 4 holds with a linear activation map F if and only if T is linear.

The interesting feature of this result, which is already impliciy in Richardson's theorem, is that the underlying digraph is not mentioned anywhere in conditions 1–4. In fact, condition 2 allows the recovery of the underlying network structure. Hence the architecture of a network is *encoded* in its global dynamics. Condition 3 reflects the local finiteness of the network. Condition 4 shows that networks are perturbations of linear maps by local perturbations of the states.

Proof. We first show the necessity of the four conditions. The first three follow immediately from Theorem 7.18 (shown next) since every neural network is a random boolen network. For the fourth condition, let L the net activation map *net*. L is clearly linear and $T = F \circ L$, where F is the product of the activation functions. One readily checks that F is strictly local.

For sufficiency, suppose that the four conditions are satisfied. Choose a countably infinite set V to serve as the vertices of the underlying digraph. The continuity of T along with $T(O) = O$ provides the links in the digraph. Links constructed in this way yield a graph of finite in-degree. Condition 3 implies finite out-degree, and thus the digraph is locally finite. To completely specify the neural network, it remains to determine the weights on the links and the activation functions. F will supply the activation functions and L the weights. Define the weight on the link from vertex j to vertex i by $w_{ij} := L(e^j)_i$. Finally, define the activation function for each vertex i by $f_i(r) := F(re^i)_i$. Clearly $f_i(0) = 0$ by conditions 1 and 4, so that each f_i is an activation function. It remains only to show that T is the global dynamics of this neural network. The following calculation uses successively the definitions of w_{ij} and f_i, the linearity of $\pi_i \circ L$, and the strict locality of F.

$$f_i(\sum_j w_{ij}x_j) \; = \; f_i[\sum_j x_j(\pi_i \circ L(e^j))] = f_i \circ \pi_i \circ L[\sum_j x_j e^j]$$

$$= \; \pi_i \circ F[(L(\sum_j x_j e^j))_i e^i] = [F \circ L(\sum_j x_j e^j)]_i$$

$$= \; T(\sum_j x_j e^j)_i \,. \; \Box$$

A similar characterization can be given for random boolean networks. Here we let $\mathbf{C} := \prod_{i \in V} Q_i$ be an encoding of the configuration space. Each Q_i is a finite set of states contained in a common finite set Q and V is a countably infinite index set. Also a pixel j is said to T-*influence* another pixel i if there exist two configurations $x, y \in \mathbf{C}$ there such that $y_k = x_k$ for all $k \neq j$ but $T(y)_i \neq T(x)_i$.

Theorem 7.18 *A self-map* $T : \mathbf{C} \to \mathbf{C}$ *is realizable as the global dynamics of a random boolean network if and only if*

1. $T(O) = O$;

2. T *is continuous*;

3. *each* $j \in V$ T-*influences only finitely many* $i \in V$.

Proof. It is easy to see that each of the three conditions is necessary. The first is immediate since each δ_i sends quiescent input to 0. The continuity of T follows from that of each $\pi_i \circ T = \delta_i$ since T is the product of the δ_i. (δ_i can be considered to be a function on the larger domain $\mathbf{C}$ by computing its value at an arbitrary $x \in \mathbf{C}$ by extending x to be quiescent elsewhere.) For the third condition note that, unless j is a vertex indexing the domain of δ_i, it cannot T-influence i. Since j has finite out-degree, it can contribute to the definition of only finitely many such domains.

For sufficiency, suppose that $T : \mathbf{C} \to \mathbf{C}$ satisfies the three conditions under the encoding $\mathbf{C} = \prod_{i \in V} Q_i$ as above. Let V be the vertex set, and Q_i the state set of the ith finite-state machine at site i.

The continuity of T will allow the recovery of the links between the vertices in V, completing the underlying digraph. For each vertex $i \in V$ and for each state $q \in Q_i$, let $X^{i,q} := \{x \in \mathbf{C} \mid x_i = q\}$. The set $X^{i,q}$ is both open and closed in the product topology (see Problem 2). Since T is continuous $T^{-1}(X^{i,q})$ is open and hence the union of basic open sets of $\mathbf{C}$, i.e., sets that restrict only finitely many components (see Problems 5-6). Since $\mathbf{C}$ is compact and $T^{-1}(X^{i,q})$ is closed, it is the union of finitely many such basic open sets. Let $V^{q,i}$ be the set of vertices indexing the restricted components. Assume also that each restricted coordinate is necessary in the sense that making it unrestricted would drive its basic neighborhood outside of $T^{-1}(X^{i,q})$. Let $V^i = \bigcup\{V^{q,i} \mid q \in Q_i\}$. For each $i \in V$, V^i is a finite set. For each $j \in V^i$, include a link form j to i in the underlying digraph. Thus each $i \in V$ has finite in-degree.

If j T-influences i, there is an $x \in \mathbf{C}$ satisfying the defining condition above. If $T(x)_i = q$, then $x \in T^{-1}(X^{i,q})$, and thus x belongs to some one of the finitely many basic neighborhoods covering $T^{-1}(X^{i,q})$. If none of these basic neighborhoods restrict j, the condition would be contradicted. Thus there is a link from j to i. Likewise, if there is such a link, $j \in V^{q,i}$ and thus is restricted in some basic neighborhood $B \subseteq T^{-1}(X^{i,q})$. Any $x \in B$ will serve to show that j T-influences i. Thus j T-influences i just in case there is a link from j to i. The third condition of the theorem now implies that each vertex has finite out-degree, and so the underlying network is locally finite.

Take V^i to define the domain of δ_i. Given any input states to i extend them with zeros to a configuration x and define the value of δ_i to be $T(x)_i$. It is easy to see that the choice of x does not affect the value. This defines, for each i, a finite state machine in such a way as to yield T as global dynamics of the network.   □

7.4.2 Combinatorial Generalizations on Euclidean Spaces

One can also regard the previous results of this chapter as generalizations of the obvious fact that a self-mapping of a *finite* set is one-one iff it is onto iff it is bijective. The consequences above say that in going to cellular automata on *infinite* euclidean spaces one only loses one of the implications. There are also generalizations along these lines in relation to linear self-maps. Linear cellular automata have the obvious property that the number of configurations collapsed by a global dynamics to a given configuration b is the same regardless of b (since it obviously equals the size of the kernel of T as a linear transformation; see Problem 3.18). Note that T can be extended to a mapping of patterns to patterns by invariance with respect to shifts. We finish this chapter by stating, without proof, several combinatorial formulations of the familiar concepts of injectivity and surjectivity in 2D spaces, and some of their variants.

Definition 7.19 *A global dynamics T in 2D euclidean space is n-balanced if every pattern of size $n \times n$ has exactly $\frac{|Q|^{(n+1)^2}}{|Q|^{n^2}}$ predecessor patterns. T is balanced if it is balanced for every $n \geq 1$. The center of depth m of a $(2n+1) \times (2n+1)$ pattern is the pattern of the central array of size $(2m+1) \times (2m+1)$ left by removing the remaining rows and columns. Finally, T is hard if there exists an integer $m \geq 0$ such that for any $n > 2m$ and any pattern a of size $n \times n$, every pattern mapped by T to a has a unique center of depth m.*

Theorem 7.20 *If T is a local transformation, then*

1. *T_0 is one-one if and only if it is balanced.*

2. *T is one-one if and only if it is hard.*

It follows, in particular, by the results in Sect. 7.3 that, if T_0 is onto, then it is balanced.

The notions injectivity and surjectivity can be generalized as follows.

Definition 7.21 *Two configurations are R-equivalent if and only if they differ at finitely many places. Let Ra denote the equivalence class of $a \in \mathbf{C}$. Different global transitions may induce the same given global transition $T : Ra \to RT(a)$. A local rule $T : \mathbf{C} \to \mathbf{C}$ is weakly R-injective (respectively, subjective) if the induced transition $T : RC \to RT(a)$ is injective (surjective) for some $a \in \mathbf{C}$. T is strongly R-injective (-surjective) if $T : Ra \to RT(a)$ is injective (surjective) for all $a \in C$.*

Theorem 7.22 *The following relations hold:*

$$T \text{ is one-one} \quad \Rightarrow \quad T \text{ is strongly } R\text{-surjective}$$
$$\Updownarrow$$
$$T_0 \text{ is } R\text{-surjective}$$
$$\Updownarrow$$
$$T \text{ is weakly } R\text{-surjective}$$
$$\Updownarrow$$
$$T \text{ is onto} \quad \Rightarrow \quad T \text{ is strongly } R\text{-injective}$$
$$\Updownarrow$$
$$T_0 \text{ is weakly } R\text{-injective.}$$

7.5 Problems

Problems marked * require some substantial effort to establish and may need to be looked up.

TOPOLOGY
The following are well known facts in metric spaces. Verify them in the case of configuration space.

1. An open ball $B(a, r) := \{x \in Y : |x, a| < r\}$ is an open set (that is, it actually contains an open ball of positive radius centered at each of its point $x \in B(a, r)$) in every metric space Y.

2. A basic neighborhood $X^{i,q} := \{x \in \mathbf{C} \mid x_i = q\}$ is an open set. It is also closed (i.e., its complement is open.)

3. An intersection of a finite number of basic neighborhoods is an open set.

4. The union of an arbitrary number of open sets is open. Likewise, the intersection of an arbitrary number of closed sets is closed.

5. Every open set is a union of open balls.

6. A global map is continuous iff the inverse image of every open set is an open set. Likewise for closed sets.

7. Show that the ternary Cantor set is compact, .i.e., a subset every cover of which consisting of open sets always contains a finite subcover. (A *cover* of a space X is a family $\{U_l\}_{l \in \alpha}$ of subsets whose union spans all of X.) *

8. Prove the sufficiency of the conditions in Richardson's theorem topologically. [Use the compactness of configuration space on the cover given by the inverse image of the sets of configurations that map the sites to various states to obtain a local rule at the origin; then use commutation with the shift for the other cells.] *

9. Show that any two configuration spaces are homeomorphic to the ternary Cantor set (and hence to each other). * (Two metric spaces X, Y are *homeomorphic* if they are topologically identical, i.e., it is possible to establish a bijective correspondence between their elements which is continuous and has a continuous inverse.) [A proof can be found in [Ho].] *

GLOBAL DYNAMICS

10. Show that the Moore–Myhill theorem holds for finite cellular spaces.

11. Calculate the smallest possible value of the constant ε for square arrays in euclidean universes in Definition 7.7.

12. Find a local transformation with mea's which is not injective.

13. Find a counterexample to Moore's theorem on universes of exponential growth. [Consider a game $\delta(x_i x_{i+A} x_{i+B} x_{i+C})$ of VOLATILE-LIFE given by

$$(1 - x_i)[x_{i+A} + x_{i+B} + x_{i+C}] + x_{i+A} x_{i+B} + x_{i+A} x_{i+C} + x_{i+B} x_{i+C}$$

with the von Neumann neighborhood on the Cayley graph of the modular group given by two generators A, B subject to

$$\langle\, A, B \mid 2A = 3B = 0 \,\rangle.$$

Recall that the operation on site coordinates is not necessarily commutative; see Problem 2.2.]

14. Find a counterexample to Myhill's theorem on universes of exponential growth. [Consider the cellular space rule

$$\langle\, A, B, C \mid 2A = 2B = 2C = 0 \,\rangle$$

on the state set the Klein group $Q := \{0, 1, 2, 3\}$ and the linear global dynamics of Problems 3.21–22.]

15. Construct a 2D universal cellular automaton. [Use the same idea for a universal cellular automaton in the proof of Theorem 6.7.]

16. Prove that every local rule of the type

$$x_i(t+1) = g(x_i(t), x_{i+N}(t)) - x_i(t-1)$$

which stores the previous state and subtracts from an arbitrary new state gives rise to a reversible rule.

17. Let N_i denote the cells that influence a given cell i. Show that a linear cellular automaton is continuous if and only if every N_i is finite, where N_i denotes the set of cells that influence a cell i.

18. Show that a map $F : \mathbf{C} \to \mathbf{C}$ is strictly local if and only if there exist maps $f_i : Q \to Q$ such that $T(x)_i = f_i(x_i)$ for each i.

19. Show that if a local map T is open, then it is surjective.

20. Show that the problem of 1D OPENNESS is algorithmically solvable.

7.6 Notes

The Moore–Myhill theorem actually consists of two halves. The first one was proved by Moore [Moo] and the (modified) converse by Myhill [My] for euclidean cellular spaces. They were probably the first general theorems on cellular automata after von Neumann's definition. The generalization to Cayley graphs whose growth function is nonexponential is due to Machi–Mignosi [M-M], which the presentation in Section 7.4.2 follows closely (in content although not in notation, particularly for the notions of 'array' and 'pattern').

The idea of considering the global dynamics from a topological viewpoint was originated in Curtis and Hedlund [H], where the 1D version of Richardson's theorem was first established via shifts of finite type. The natural follow-up on other topological approaches is considered in Problem 8. A combinatorial proof of the decidability of openness for 1D cellular automata in Problem 20 can be found in Willson [W].

Richardson's theorem and its consequences in this chapter appear in his celebrated [R]. Extensions of Richardson's theorem to generalizations of cellular automata such as neural networks and automata networks appear in Garzon–Franklin [G-F]. Another characterization of global dynamics of neural nets in terms of net input (as opposed to activation) functions appears in Franklin–Garzon [F-G]. Results similar to Theorem 7.18 that the input/output map determine essentially the internal structure (architecture and weights) for finite neural networks with analog activations have been shown by Albertini–Sontag [A-So1, A-So2]. Related results about identifiability and observability for discrete and analog networks will be discussed later in Chapters 9 and 11.

The conditions for injectivity and surjectivity in terms of hard and balanced global dynamics of Theorem 7.20 are due to Maruoka–Kimura [M-K1]. Further conditions in Theorem 7.22 appear in Maruoka–Kimura [M-K2]. What effect the restriction to a subfamily of configurations has on the truth of Corollary 7.16 appears to be an interesting open problem.

Nondeterministic cellular automata are used in this chapter as they were in the original proof by Richardson. They were considered later on by Yaku [Ya1, Ya2]. He proves in [Ya1], that the range of a cellular automaton may be a nonrecursive set, even if the inputs are just finite configurations, while in [Ya2] he establishes the undecidability of surjectivity for nondeterministic euclidean automata in dimension $d \geq 2$ restricted to recursive (or even just finite) configurations. However, very little is known at present about their computational power, particularly in relation to deterministic cellular automata.

That reversible cellular automata are capable of universal computation was first shown by Toffoli [M-T]. More efficient universal automata were constructed by Margolus at the expense of uniformity in time and space (different rules apply at even/odd time steps with two different neighborhoods, such as the second-order rules in Problem 16). They have been used to produce digital models of several physical phenomena – see, for instance, Margolus–Toffoli [M-T, Sect. 14.2] for examples. These nonuniformities in time and space have been recently removed in Morita [Mo1] by giving a partitioned structure to the state set. The same type of automaton has been used to provide a characterization of linear-bounded and a generalization of context-sensitive languages in the euclidean plane [Mo2]. The same techniques are used to produce a 1D euclidean computation universal reversible cellular automaton in his more recent works [M-H, Mo3].

References

[A-So1]   F. Albertini and E.D. Sontag: Identifiability of discrete time neural networks. Proc. European Control Conference, Groningen, The Netherlands, 1993

[A-So2]   F. Albertini and E.D. Sontag: For neural networks, function determines form. Neural Networks 6:7(1993), 975-990

[F-G]   S. Franklin, M. Garzon: Global dynamics in neural networks. Complex Systems 3:1 (1989) 29–36

[G-F]   M. Garzon, S. Franklin: Global dynamics in neural networks II. Complex Systems 4:5 (1990) 509–518

[H]   G.A. Hedlund: Endomorphism and Automorphism of the Shift Dynamical System. Math. Syst. Theory 3 (1969) 320–375

[Ho]   J.G. Hocking, G.S. Young: Topology. Addison-Wesley, Boston MA, 1969

[M-M]   A. Machi, F. Mignosi: Garden of Eden configurations for cellular automata on Cayley graphs on groups. SIAM J. Discr. Math. 6:1 (1993) 44–56

[M-T]   N. Margolus, T. Toffoli: Cellular Automata Machines (A new environment for modeling). MIT Press, Cambridge MA, 1987

[M-K1] A. Maruoka, M. Kimura: Condition for Injectivity of global maps. Inf. and Control **32** (1976) 158–162

[M-K2] A. Maruoka, M. Kimura: Injectivity and surjectivity of parallel maps for cellular Automata. J. Comput. Syst. Science **18** (1979) 47–64

[M-K3] A. Maruoka, M. Kimura: Strong injectivity is equivalent to C-injectivity. Theoret. Comput. Science **18** (1982) 269–277

[Mo1] K. Morita: Computation-universal models of two-dimensional 16-state reversible cellular automata. IEICE Trans. Inf. and Syst. Vol. E75-D:1 (1992) 141–147

[Mo2] K. Morita: Parallel generation and parsing of array languages using reversible cellular automata. Preprint

[M-H] K. Morita, M. Harao: Computation universality of one-dimensional reversible (injective) cellular automata. Trans. IEICE E **72** (1989) 758–762

[Mo3] K. Morita: Computation universality of one-dimensional two-way reversible cellular automata. Inf. Proc. Lett. **42**:6 (1992) 325–329

[Moo] E.F. Moore: Machine models of self-reproduction. In: Proc. Symp. in Applied Mathematics. American Mathematical Society, Providence RI, 1962

[My] J. Myhill: The converse of Moore's garden-of-Eden theorem. Proc. Amer. Math. Soc. **14** (1963) 658–686. Revised version in: Essays on cellular automata. A.W. Burks (ed.): University of Illinois Press, Chicago, 1970

[R] D. Richardson: Tessellation with local transformations. J. Comput. Syst. Sci. **6** (1972) 373–388

[T] T. Toffoli: Computation and construction universality of reversible cellular automata. J. Comput. Syst. Sci. **15** (1977) 213–231

[W] S. Wolfram: Theory and applications of cellular automata. World Scientific, Singapore, 1986

[Ya1] T. Yaku, The constructibility of a configuration in a cellular automaton. J. Comput. Syst. Science **7** (1973) 481–496

[Ya2] T. Yaku: Surjectivity of nondeterministic parallel maps induced by nondeterministic cellular automata. J. Comput. Syst. Sci. **12** (1976) 1–5

[W] S.J. Willson: Decision procedures for openness and local injectivity. Complex Systems **5** (1991) 497–508

8. Classification

By indirections finds directions out.
Hamlet

Cellular automata and discrete neural networks constitute very simple and general models that seem to capture the fundamental features of a variety of highly complex systems. Their study offers the possibility of obtaining some understanding of the most important and characteristic properties of complex and self-organizing systems, the evolution of which currently appears chaotic, disorganized and beyond the scope of known laws of nature.

Oddly enough, the level of explanation provided by such simple local rules appears at least one level deeper than the orthodox explanations provided by well known physical and mathematical theories. Thus, for the first time, a systematic classification of local rules becomes, in fact, a common classification of complex discrete dynamical systems modeling a wide spectrum of physical, biological and computational phenomena. This is in contrast with received views that laws and equations governing physical and biological phenomena, for example, are of an entirely different nature.

Classification means arranging systems into groups or classes, each consisting of systems bearing some common and characteristic properties. One usually classifies things into three great kingdoms: animal, plant, and mineral. This is essentially a classification in terms of degree of "liveliness". By virtue of the fact that phenomena from different kingdoms can be modeled by identical cellular automata or neural networks, the classification of these models falls transversal to established criteria. Hence, it is unquestionable that a satisfactory classification of cellular automata and neural networks requires a precise definition of *equivalence*, i.e., isomorphism, or membership in the same class.

Current classifications of neural networks and cellular automata are made in terms of their local dynamics or the architecture of the communications digraph, activation values or activation functions (e.g., linear or totalistic rules, linear-thresholds networks, feed-forward neural networks, etc.). It is clear, however, that the real value of these models lies in their ability and potential to serve as models of evolution and complex behavior over time, either as dynamical or computational systems. Thus, the most important classification is one made on the basis of their *emergent* long-term behavior. At face value, this is a most difficult problem, the complete solution of which requires, in the words of S. Wolfram, the pioneer of this type of classification, "a multitude of questions to be asked and answered."

It would be ideal to have a classification of cellular automata and neural networks in terms of well-known mathematical invariants of their transition tables, say of their weights and/or activation vectors, the way classification is ordinarily done in mathematics. Unfortunately, it is easy to see that networks quite different by these standards may have identical or very similar long-term behavior. Vice versa, very similar models by these standards may have very different behavior. On the other end of the spectrum, a classification can be made from a purely anthropomorphic point of view, i.e., in terms of a so-called *functional isomorphism* in which the number of cells or the exact transitions are not as important as the satisfaction of an overall function that the corresponding dynamical systems are supposed to accomplish. For instance, a primitive organism's brain may be regarded as a neural network of some sort, and two organisms of the same species are considered functionally identical because they exhibit very similar *behavior*. Although important from a biological viewpoint, this criterion is just rising on the research horizon, and so it will not be pursued here.

There are essentially two ways to study a cellular model: as a discrete dynamical system in the classical mathematical sense, or as a parallel computer. These two approaches are not contradictory, but in fact jointly provide an enhanced understanding of the model in question. For instance, the brain-as-a-computer paradigm in contemporary models of the mind ushered in by recent developments in computation and complexity has proved extremely useful in the analysis of old psychological and philosophical problems such as the mind–body problem. Dynamical systems, on the other hand, offer a quantitative handle and a set of tools that have been successful for the study of the long-term evolution of a system under given initial conditions and evolution laws, albeit in the area of continuous dynamical systems.

The purpose of this chapter is to give an exposition of the progress and known results on classification of cellular automata and neural networks as dynamical systems. The nature of the problem is somewhat different for finite networks, where one must only be concerned with questions of complexity, than for infinite networks, where infinities and ensuing issues about convergence and complexity are also involved. Thus Section 8.1 will deal with finite networks and the remaining sections will deal with infinite networks. Section 8.2 describes Wolfram's classification, the first attempt at a comprehensive classification based on purely experimental evidence in the absence af any analytical tools to attack the problem. Section 8.3 presents attempts at anayltic definitions of Wolfram classes and the ensuing problems. Section 8.4 deals with a preliminary approach to the classification problem via mean field theory and statistical mechanics. Finally, Section 8.5 deals with a more general, probabilistic approach, the so-called *local structure theory*.

8.1 Finite Networks

In this section we consider the classification problem of *finite* neural networks, and in particular, cellular automata. The following sections will consider infinite neural networks. In order to make the notion of equivalence precise, it is necessary to introduce some terminology. A finite threshold neural network can be represented by a pair $\langle W, \theta \rangle$ consisting of an $n \times n$ real weight matrix $W := [w_{ij}]$ of "synaptic strengths" and a real threshold vector $\theta \in \mathbf{R}^n$ defining its activation functions. A threshold network is *symmetric* if and only if the weight matrix W is symmetric. A configuration of an n-cell network is just a vector $x \in \mathbf{B}^n$, where $\mathbf{B} := \{0, 1\}$. (Without loss of generality, boolean activation values will be assumed throughout.) The results in this section, being of a lower bound type, remain true for more general spaces and activation functions.

The phase space of an n-cell neural network can be naturally represented by a *dynamics digraph* with vertex set $\mathbf{B}^n$ and arcs connecting each x to $T(x)$. Each cell updates its activation values for the next time step according to

$$x_i(t+1) \quad := \quad \mathbf{1}_i(Wx(t) - b) \qquad \qquad (8.1)$$

$$= \quad \mathbf{1}(\sum_{j=1}^{n} w_{ij}x_j(t) - b_i)$$

where the ith component of the threshold function $\mathbf{1}$ is given for a vector $\mathbf{x} \in \mathbf{R}^n$ by

$$\mathbf{1}_i(\mathbf{x}) := \begin{cases} 0 & \text{if } \mathbf{x}_i < 0 \\ 1 & \text{else} \end{cases} .$$

8.1.1 The Difficulties

Different cellular automata (with different local rules) and different neural networks (with different weights and thresholds) may give rise to identical dynamics digraphs. For *finite* networks, a functional classification of long-term behavior can be achieved in terms of isomorphism of their phase spaces. These problems were specified in Chapter 5 as *strong isomorphism testing* (SIT) and *weak isomorphism testing* (WIT). It would be ideal to have a classification of isomorphism of finite networks in terms of known invariants (such as eigenvalues) of the weight matrices and activation vectors that can be verified efficiently. Unfortunately, it is easy to see that isomorphic networks may differ by more than relabeling of their cells. In particular, the weight matrices of isomorphic networks need not have the same eigenvalues. Further evidence given below suggests that a simple condition characterizing isomorphism of finite neural networks is highly impractical, if at all possible.

The best object next to a simple formula is an algorithm to test finite neural network isomorphism. Such an algorithm exists, for it is in principle possible to make an exhaustive systematic comparison of the (exponentially many) transitions in the dynamics digraphs of two given networks. Thus the fundamental question concerning a classification by isomorphism is whether it is computationally feasible. From Chapter 1, we know that when confronted with a complexity question of this kind, one has only two alternatives. Either find an *efficient* algorithm to solve the problem or prove that such an algorithm does not, or is very unlikely to, exist. As it turns out, the problem is somewhat different from the general graph isomorphism problem (GI) in at least two aspects. First, in the case of SIT one does not have to contend with a factorial number of permutations of the vertex labels, but rather with an *exponential* number of vertices in the size of the input (which is essentially the square of the number of cells in the network). On the other hand, WIT is the usual type of combinatorial isomorphism problem compounded with the basic difficulty of SIT. Moreover, in both cases, equivalent dynamics can be succinctly represented by augmented matrices which may differ by more than a permutation of rows and columns or, more generally, may not even have the same eigenvalues. For these reasons one would expect these problems to be of higher complexity than graph isomorphism.

On the other hand, finite threshold networks possess structural features that might allow to fast isomorphism testing. For instance, dynamics digraphs are *functional* digraphs of out-degree 1, and more generally, each weakly connected component is a unicyclic digraph with a unique directed shortest path between every pair of vertices. Also, symmetric network dynamics (with a symmetric weight matrix) admit an energy (or Lyapunov) function that decreases with time, which implies that they always converge to fixed points or 2-cycles (although the networks may take up to $2^{n/3}$ iterations to stabilize). Further, it is well known that every neural network on n cells is isomorphic to one with integer weights and thresholds which are expressible in $O(n \log n)$ bits. For the purpose of network isomorphism testing, instances are given as an integer weight matrix and an integer threshold vector of size polynomial in the number of cells (which can then be taken as the input size). Thus, despite appearances, the dynamics isomorphism problem is a purely combinatorial one.

8.1.2 Complexity of Classifying Finite Networks

The results in this section show that isomorphism testing of neural networks belong to families of hard combinatorial problems. Thus, a fast, efficient solution to any one of them would provide a similar solution to hundreds of problems of great importance in optimization and other areas of computation. As a byproduct of the study of the complexity of neural network isomorphism, one obtains a result on the following related problem of practical interest dealing with finite cellular automata and neural networks.

PREDECESSOR COUNTING

INSTANCE: a neural network $\langle W, \theta \rangle$ on n cells and a configuration
$x \in \mathbf{B}^n$

QUESTION: how many predecessors does x have in the dynamics of
$\langle W, \theta \rangle$?

Theorem 8.1 SIT *is coNP-complete, even if restricted to symmetric networks.*

This result is in contrast with graph isomorphism, which although known to be in **NP**, is not known to be either polynomial time or **NP**-complete.

Proof. Equivalently, *co*SIT is in **NP** as indicated above since finding the successor of a configuration from the weight matrix and threshold can be clearly done in *linear*-time in the number of cells.

The proof of **NP**-hardness for *co*SIT is by reduction from PARTITION which is known to be **NP**-complete.

PARTITION
  INSTANCE: positive integers $a_1, \ldots, a_n$
  QUESTION: is there a subset of indices S such that
$$\sum_{j \in S} a_j = \sum_{j \notin S} a_j \, ?$$

Given an instance $a_1, \ldots, a_n$ of PARTITION, construct an instance $\langle A_1, \theta^1; A_2, \theta^2 \rangle$ of *co*SIT as follows. Let A_1 be the matrix with constant rows $a_{1j} := a_j$ for all j, and $a_{ij} = 0$ otherwise. Let the first threshold in θ^1 be $\theta_1^1 := \frac{1}{2}\tau$, where $\tau := \sum_{j=1}^n a_j$. Let $A_2 := A_1$ and the first threshold in θ^2 be $\theta_1^1 := \frac{1}{2}\tau + \varepsilon$, where $\varepsilon := \frac{1}{4}$. Let the remaining thresholds in θ^1 and θ^2 be 1. This reduction can clearly be done in polynomial time.

If the given instance of PARTITION has a solution whose characteristic vector (a boolean vector whose j^{th} bit is 1 iff the jth weight is in the solution set S) is x^0, then the transitions of the networks $\langle A_k, \theta^k \rangle (\, k = 1, 2)$ differ at x^0 and are therefore not strongly isomorphic. Conversely, if $T_1(y) \neq T_2(y)$, then y defines a weight-sum $\omega := \sum_{y_j=1} a_j$ satisfying

$$\frac{1}{2}\tau \leq \omega < \frac{1}{2}\tau + \varepsilon.$$

But since τ and ω are integers and $\varepsilon < \frac{1}{2}$, this implies $\frac{1}{2}\tau = \omega$. Hence the given instance of PARTITION has a solution.

The proof for symmetric networks proceeds similarly with the networks defined by

$$A_1 := \begin{bmatrix} 0 & a_1 & a_2 & \cdots & a_n \\ a_1 & 0 & 0 & \cdots & 0 \\ & & \cdots & & \\ a_n & 0 & 0 & \cdots & 0 \end{bmatrix} \, ; \, \theta^1 := \begin{bmatrix} \frac{1}{2}\tau \\ a_1 + 1 \\ \cdots \\ a_n + 1 \end{bmatrix}$$

and

$$A_2 := \begin{bmatrix} 0 & a_1 & a_2 & \cdots & a_n \\ a_1 & 0 & 0 & \cdots & 0 \\ & & \cdots & & \\ a_n & 0 & 0 & \cdots & 0 \end{bmatrix} \, ; \, \theta^2 := \begin{bmatrix} \frac{1}{2}\tau + \varepsilon \\ a_1 + 1 \\ \cdots \\ a_n + 1 \end{bmatrix}. \square$$

Recall that if **D** is a complexity class (such as **NP**), *co***D** stands for the class of complements of problems in **D**. Therefore the question of whether SIT itself is in $\mathbf{P}$ (respectively, $\mathbf{NP}$) is equivalent to the well-known open problem $\mathbf{P} \stackrel{?}{=} \mathbf{NP}$ (respectively, $\mathbf{NP} \stackrel{?}{=} co\mathbf{NP}$). It is widely believed that $\mathbf{P} \neq \mathbf{NP}$ is the most likely event.

In contrast, the situation for WIT may be entirely different from that in Theorem 8.1 given the current belief that even $co\mathbf{NP} \neq \mathbf{NP}$.

Theorem 8.2 WIT *is* $\mathbf{NP}$-*hard, even if restricted to symmetric networks.*

Proof. The following problem is known to be $\mathbf{NP}$-complete [G-J, SP13]:

SUBSET SUM

INSTANCE: a finite set A, sizes $s(a_i) \in \mathbf{Z}^+$ for each $a_i \in A$ $(i = 1, \ldots, n)$, and an integer $B > 0$

QUESTION: is there a subset $S \subseteq A$ such that
$$\sum_{a_i \in S} s(a_i) = B?$$

Given an instance of SUBSET SUM, construct an instance of WIT in p-time as follows. Let $\tau := \sum_{i=1}^{n} s(a_i)$,

$$A_1 := A_2 := \begin{bmatrix} 0 & s(a_1) & s(a_2) & \cdots & s(a_n) \\ 0 & 0 & 0 & \cdots & 0 \\ & & \cdots & & \\ 0 & 0 & 0 & \cdots & 0 \end{bmatrix}$$

and let

$$\theta^1 := \begin{bmatrix} B \\ 1 \\ \cdots \\ 1 \end{bmatrix}, \theta^2 := \begin{bmatrix} \tau - B \\ 0 \\ \cdots \\ 0 \end{bmatrix}. \tag{8.2}$$

All cells in the first network, except the first one, go to state 0 from any configuration. Likewise, all cells in the second network, except the first one, go to state 1. In the first network, any configuration goes to $10 \ldots 0$, or directly to $00 \ldots 0$, which is a fixed point. In the second network, any configuration goes to $01 \ldots 1$, or directly to $11 \ldots 1$, which is also a fixed point. Thus the weak-isomorphism type of either network is determined by how many configurations go directly to the fixed point $10 \ldots 0$ and to the other configuration $01 \ldots 1$. In order to count these configurations, let i (respectively j) be the number of distinct *subsets* whose sums are equal to (respectively, greater than) B. Hence there are $2^n - i - j$ subsets whose sums are less than B. Therefore, the number of distinct subsets whose sums are equal to (respectively, less than) $\sum_{i=1}^{n} s(a_i) - B$ is i (respectively, j). Thus $2(i + j)$ configurations go to $10 \ldots 0$ by the first dynamics while $2j$ of them go to $01 \ldots 1$ by the second dynamics. So the two networks are weakly isomorphic if and only if $2(i + j) = 2j$, that is, if and only if $i = 0$. Thus the given instance of SUBSET SUM has no solution if and only if the two dynamics are weakly isomorphic.

The proof for the symmetric case proceeds analogously with weight matrices taken to be as in (8.2) with first columns equal to their first rows and threshold vectors

$$
b^1 := \begin{bmatrix} B \\ a_1 + 1 \\ \cdots \\ a_n + 1 \end{bmatrix} ; \quad b^2 := \begin{bmatrix} \tau - B \\ 0 \\ \cdots \\ 0 \end{bmatrix} ,
$$

where τ is as before, and it will be omitted.   □

A slight generalization of combinatorially hard problems deals with counting the number of positive answers to given instances (e.g., *how many* tours visit all cities in the traveling salesman problem?). Some problems present the same phenomenon of completeness in the corresponding class #P as observed in NP and *co*NP. (See [G-J, Sect. 7.3] for definitions and facts about #P-completeness.) The following result thus provides a measure of difficulty of the predecessor counting problem for neural networks dynamics. In view of Theorem 6.6, the results of this section also hold for finite cellular automata.

Theorem 8.3 PREDECESSOR COUNTING *is #P-complete, even if restricted to symmetric networks.*

Proof. The problem is clearly in #P by definition. The proof of #P-hardness is by reduction from the enumeration problem #PARTITION associated with PARTITION, a problem which is known to be #P-complete (see [G-J, Sect. 7.3]). Given an instance of #PARTITION, construct two instances of PREDECESSOR COUNTING as follows. The two networks are given by the networks $\langle A_k, \theta^k \rangle$ ($k = 1, 2$) obtained by the reduction in the proof of Theorem 8.1. If i (respectively, j) is the number of predecessors of $c := 10\ldots0$ in the dynamics of $\langle A_1, \theta^1 \rangle$ (respectively, $\langle A_2, \theta^2 \rangle$) then $i - j$ is the answer to the original instance of #PARTITION. The proof for symmetric networks proceeds identically with the reduction for the symmetric case in Theorem 8.1.   □

We now turn to the classification of infinite neural networks.

8.2 Wolfram Classification

Cellular automata provide a brand-new explanation of complexity in terms of simple rules. The basic difference is that, unlike classical models with a few variables that vary *continuously*, we now have a very large, in principle infinite, number of *discrete* variables (the states of the sites) that assume a very small number of values (compared to a real variable). It is then easy to understand why the analytic machinery of established fields (such as calculus and differential equations) offer little in the form of tools to read the long-term behavior of a network off the local rule of interaction. A brand new type of tools and analysis is required.

Not unlike Tycho Brahe facing the analysis of planetary motion, S. Wolfram took the only initial course of action possible to overcome this difficulty. He began a systematic observation and careful recording of the behavior of many rules on random initial configurations. His empirical tool was also the most recent invention of the time, not a telescope, but a sequential computer capable of performing the millions of operations necessary to simulate a (portion of an in)finite cellular automaton. His phenomenological approach had to overcome two basic questions. First, given a specific rule, it is possible to describe in simple terms the global evolution of a "typical" (random) initial configuration?; and second, how can one identify *similar* behavior into a small number of classes that cover the entire spectrum of behavior that cellular automata are capable of?

After thousands of hours of computer simulations of selected and random rules on disordered initial configurations, he established a purely phenomenological and very qualitative classification. Like the Chomsky hierarchy for string languages, it classifies cellular automaton rules into four basic categories, which can be described as follows.

WC1: evolution on a random initial configuration leads to a homogeneous state. They develop analogously to continuous dynamical systems with a single unique limit *point*. Precisely, every finite configuration evolves to a stable configuration in finitely many steps.

WC2: evolution on a random initial configuration leads to simple separated periodic structures. They develop analogously to continuous dynamical systems with limit *cycles*. Precisely, every finite configuration evolves to a periodic configuration in finitely many steps.

WC3: evolution on a random initial configuration leads to chaotic aperiodic patterns. This behavior is analogous to the chaotic behavior found with strange attractors in dynamical systems. More precisely, there is an algorithm to decide whether one configuration belongs to the orbit of another.

WC4: evolution on a random initial configuration leads to complex patterns of localized structures. No direct analog of this kind of behavior has been identified in continuous dynamical systems.

These classes can be regarded as an increasing hierarchy of rules of increasing complexity, **WC4** including all rules. Table 8.1 from [Wo4] shows the approximate fraction of rules falling in each of Wolfram's classes for various state sets and radii.

As mentioned above, Wolfram's classification is purely heuristic and, in fact, raised a multitude of questions that stimulated the investigation of more precise techniques for the classification problem.

Table 8.1. A sample of the distribution of rules in Wolfram classes

| # States | 2 | 2 | 3 | 3 |
Radius	$(r = 1)$	$(r = 2)$	$(r = 3)$	$(r = 1)$
WC1	.50	.25	.09	.12
WC2	.25	.16	.11	.19
WC3	.25	.53	.73	.60
WC4	0	.06	.06	.07

8.3 Classification via Limit Sets

The most natural follow-up is the search for an *analytic* definition of Wolfram's classes. In this section we describe main efforts of this type. In order to make the task manageable, we introduce a few more ideas about metrics, topological spaces and dynamical systems.

As dynamical systems, a cellular network is determined by its phase space consisting of the orbits of all the gardens of Eden. Under the action of the automaton, these orbits are organized about limit cycles (e.g., fixed points). For example, a noninjective automaton may begin to collapse orbits together (by mapping two different configurations to the same successor). In view of the topological ideas introduced in Chapter 7, the ultimate behavior of the automaton can be analyzed by looking at the *asymptotic behavior* of orbits in relation to the limit cycles and one another other.

In these terms, the first object to consider consists of those configurations that survive after arbitrarily long times, i.e., that reappear long after the rule has filtered out any peculiarities in the initial conditions.

Definition 8.4 *The* limit set $\Lambda(T)$ *of a global dynamics T is the intersection of all successive forward images of the set of all configurations, i.e.,*

$$\Lambda(T) := \bigcap_{t \geq 1} T^t(\mathbf{C}).$$

The map T is said to be nilpotent *if $\Lambda(T)$ consists of a single configuration.*

8.3.1 Culik-Yu's Classes and Ishii's Classes

One of the difficulties with Wolfram's classes is the fuzziness about their definition. They are characterized in qualitative terms by their average asymptotic dynamical behavior and so admit exceptional configurations. It would be desirable to give a precise definition and corresponding algorithms for class membership. We now present evidence that what may have been achieved in the previous section for the finite case, may be impossible under any reasonable definition for the infinite case.

Definition 8.5 *A state s is δ-stable if $\delta(ss\cdots s) = s$, i.e., $T(\mathbf{s}) = \mathbf{s}$, where $\mathbf{s}$ is the constant configuration with value s. A (respectively, finite) configuration c is s-homogeneous (respectively, s-finite) if s is the only (nonquiescent) state in c, i.e., $c_i = 0$ or s for all i ($c_i = s$ only for finitely many is). The Culik-Yu class* **CY1** *consists of all cellular automata rules that evolve every s-finite configuration to some s-homogeneous configuration. The Culik-Yu class* **CY2** *consists of cellular automata for which every s-finite configuration has a finite, i.e., eventually periodic orbit, for some stable state s.*

Note that an s-homogeneous configuration must have a finite orbit since the global dynamics preserves homogeneity. Although the state s may change under the global dynamics, iteration eventually repeats some state by the pigeon-hole principle (see Problem 2.23). The next three results are reducible from the HALTING PROBLEM.

Proposition 8.6 *The following properties are equivalent:*

1. *All configurations evolve to a homogeneous configuration.*

2. *All configurations evolve to a homogeneous configuration in constant time.*

3. *All s-finite configurations evolve to an s-homogeneous configuration for some s.*

4. *The automaton is nilpotent and the limit set only contains the s-homogeneous configuration, for some state s.*

Theorem 8.7 *Membership of cellular automata in each of classes* **CY1** *or* **CY2** *is unsolvable, even if automata are guaranteed to be in one of the two classes.*

Definition 8.8 *A cellular automaton belongs to the Culik-Yu class* **CY3** *if there exists an algorithm that decides of every pair of finite s-configurations whether one belongs to the orbit of the other. Class* **CY4** *consists of all cellular automata.*

Note that membership in class **CY4** is obviously algorithmically solvable.

Theorem 8.9 *The class* **CY3** *does not contain computation universal cellular automata. Membership in class* **CY3** *is undecidable.*

The **CY** classes are defined in terms of the dynamical behavior of finite configurations. A natural question is whether they hold if one were to consider the typical behavior of almost all configurations (as seems to have been originally intended). A similar classification based on the dynamical structure of the orbits of (not just finite but) almost every configuration has been introduced by S. Ishii. 'Typical' behavior is defined by the product measure, as extended from the uniform atomic measure on sites that gives every state the same probability $\mu(x_i) := \frac{1}{|Q|}$. Classes I (respectively, class II) consists of cellular automata in which almost all configurations – all except those in a subset of measure 0 – evolve to a homogeneous (cyclic, respectively) configuration in bounded time. They are subsets of classes **CY1** and **CY2**, and equality might hold. Automata in Ishii's class I evolve almost every configuration to an attractor which is, in a certain technical sense, cyclic. Class IV automata evolve almost every configuration to no simple attractor. Presumptively, they correspond to classes **WC4**, **WC3**, respectively, but their true status is unknown.

8.3.2 About Entropy

Entropy is a classical notion in physics, where it was introduced as a measure of *disorder*, particularly in thermodynamics. It has been mentioned that local rules can be regarded as physical laws governing the evolution of imaginary physical universes. It is hard not to think of entropy when observing the evolution of random configurations under many rules and it is thus natural to attempt to define a like concept in order to classify cellular automata.

The orbits of arbitrary configurations under a reversible automaton are kept distinct and disjoint. In general, however, they are collapsed together under the action of an automaton as a deterministic system. The notion of *entropy* quantifies the dynamical complexity of the iteration of a dynamical system f. There are at least two types of entropy for dynamical systems: *metric* and *topological*. We present the latter since it is a more general notion that does not require background in measure theory.

Definition 8.10 *Let* (X, d) *be a compact metric space and* $T : X \to X$ *be a continuous self-map of* X. *Let* n *be a nonnegative integer and* $\varepsilon > 0$. *A subset* $E \subseteq X$ *is called* (n, ε)-*spanning if for every* $x \in X$ *there exists* $a \in E$ *such that* $|T^t(x), T^t(a)| < \varepsilon$, *for every* $t = 0, 1, \ldots, n - 1$.

The smallest cardinality of an (n, ε)-spanning set is here denoted $n_{T,\varepsilon}$, or just n_ε if T is understood. Since X is compact, n_ε is finite. It usually grows exponentially fast with n. The growth rate of n_ε is measured by $\limsup \frac{1}{n} \log n_\varepsilon$.

Definition 8.11 *The* topological entropy *of* T *is given by*

$$ent(f) := \lim_{\varepsilon \to 0^+} \limsup \frac{1}{n} \log n_\varepsilon .$$

Entropy provides, at least in principle, a classification of cellular automata into a hierarchy according to dynamical evolution complexity. This classification is parametric in the sense that entropy is a real number that varies continuously from 0 possibly to infinity. The usefulness of the classification is, in general, severely limited due to the inaccessibilty of the values of the entropy, even for 1D cellular automata.

Theorem 8.12 *The following algorithmic problems are undecidable for 1D euclidean cellular automata instances T for any given $\varepsilon > 0$:*

1. *whether T is nilpotent;*

2. *whether $ent(T) = 0$;*

3. *whether $ent(T) > \varepsilon$;*

4. *to find an approximation $\hat{h}$ such that $|ent(T) - \hat{h}| < \varepsilon$.*

The proof requires the use of a special type of tiling system of the plane. Such a system t is called *nw-deterministic* if for every pair of north-west tiles $a, b \in$ t, there is at most one *matching* tile $c \in$ t that fits below b and to the right of a (such a place will be denoted $a_*^{\,b}$ and will be referred to as a *south-east se-tile* in the following proof). Thus, if t affords a valid tiling of the plane, any of its diagonals completely determines the tiling of the south-east half-plane below it. It can be proven (see [Kal]) that the problem of NW-TILING obtained by restricting 2D TILING to *nw*-deterministic systems remains unsolvable.

Proof. (1). The reduction is from NW-TILING, much in the spirit of the results in Sect. 5.3. To every *nw*-deterministic tiling system t associate a cellular automaton T whose state set is the set of tiles t enlarged with a quiescent state 0 and whose local rule is defined by

$$\delta(x_{-1}x_0x_1) := \begin{cases} c & \text{if } c \text{ fits the } se\text{-corner } a_*^{\,b} \\ 0 & \text{else.} \end{cases}$$

Since t is *nw*-deterministic, δ is well-defined and deterministic. If there is a tiling $\tau : \mathbf{Z} \times \mathbf{Z} \to$ t of the euclidean plane with t, the 1D configuration x given by $x_i = \tau(i, i)$ is a nonquiescent configuration in $\Lambda(T)$, hence T is not nilpotent. Conversely, if T is not nilpotent, it must have a limit point x with a nowhere quiescent orbit since the quiescent state 0 *spreads* under the evolution of T (i.e. $\delta(x_{-1}x_0x_1) = 0$ whenever any one of its arguments is 0). Any 1D diagonal of tiles can be extended to a tiling of the lower half-plane. Since x belongs to the limit set, by compactness, there must exist a tiling of the entire set. Therefore t affords a valid tiling of the plane iff T is not nilpotent. Hence nilpotency is undecidable.

(2) − (4). The reduction is from the HALTING PROBLEM on blank tapes. One shows an algorithmic way $\xi : \mathbf{TM} \to \mathbf{CA}$ to map each Turing machine M to a cellular automaton $\xi(M)$ with the following properties:

- $ent(\xi(M)) = 0$ or $ent(\xi(M)) \geq \log n$.

- M halts on blank tape iff $ent(\xi(M)) = 0$.

In fact, given a Turing machine M, let t be the nw-deterministic tiling used in a reduction that shows unsolvability of NW-TILING, and let δ_M be the associated rule in the construction of part (1). Define a local rule δ'_M over the state $Q' := \{0\} \cup (t \times \{0, 1, \cdots, n\})$ as follows

$$\delta'\left((x_{-1}, n_{-1})(x_0, n_0)(x_1, n_1)\right) := \begin{cases} 0 & \text{if } \delta_M(x_{-1}x_0x_1) = 0 \\ (\delta_M(x_{-1}x_0x_1), n_1) & \text{else.} \end{cases}$$

It is easy to verify that the global dynamics T'_M induced by δ'_M is nilpotent iff the one induced by δ_M is. Further, if M halts on blank tape, T' is nilpotent and $ent(T'_M) = 0$. Conversely, if T'_M is not nilpotent, it has a nowhere quiescent configuration, so that the number of its space-time windows must satisfy

$$R(w, t) \geq n^{w+t-1},$$

which implies that $ent(t'_M) \geq \log n$ (see Problem 4). In that case, M cannot halt on a blank tape. Finally, part (4) follows by the same constructing if n is chosen so that $\varepsilon < \frac{1}{2}\log n$.   □

Finally, we mention, without proof, a most general result on the overall possibility of predicting global behavior of cellular automata using sequential computers. The result is sometimes called *Rice Theorem* for cellular automata since an analogous result holds for nontrivial properties of recursive sets in Turing computation. A global property of cellular automata is *nontrivial* if some cellular automata have it and some do not.

Theorem 8.13 *Every nontrivial property of the limit sets of cellular automata is algorithmically undecidable.*

8.4 Mean Field Theory

The first *analysis* of cellular automata rule space proceeded on the analogy with physical systems with a large number of particles. Conglomerates of this type such as ideal gases and fluid dynamics have long been objects of study, for example in statistical mechanics. We know that predicting the precise long-term behavior of cellular evolution is computationally impossible by any method that would shortcut the computation performed by the automaton. Thus, any analysis of cellular models must either be about *average* behavior or make special assumptions about the type of rules for which the analysis is valid. But, from the point of view of statistical mechanics, most cellular automata are not of the reversible (nondissipative) type in physics, where the systems tend with time to a state of maximal entropy, i.e., maximal disorder. These models are more like irreversible systems in interaction with an environment, such as snowflakes and biological systems, which may evolve and self-organize from "disordered" towards more "ordered" states.

The *mean field theory* is a model of cellular evolution based on the assumption that iterative application of a rule does not introduce correlations between states of cells at different positions in the cellular space. In other words, if an initial configuration consists of more or less random state assignments to the cells in the networks, it will remain uncorrelated for the entire evolution. Although the assumption is not generally valid, it allows the derivation of a simple formula for the limit density of each possible state of a cell. In some cases this is a realistic assumption and Monte-Carlo simulations confirm that the mean-field theory may be at times a good predictor of the long-term evolution of the automaton.

One can view mean-field theory as a parametric representation of the rules of the automaton. Each value of the parameter corresponds to a subset of rules. Small changes in the parameter lead to small changes in the properties of the automaton. In particular, for the majority of rules with a given value of the parameter, rules with close parameter values will, in general have similar properties. However, sharp changes in the mean-field behavior cause, in some cases, sharp changes in Wolfram class behavior.

Precisely, the probability of a block b is defined at time $t+1$ by

$$p_b^{t+1} \quad := \quad \sum_{T(B)=b} p_B^t . \tag{8.3}$$

Thus, for example, in a 1D elementary rule with radius r, the probability that a single cell will take on the state 1 (block $b := 1$) is the sum of the probabilities of blocks B of length $2r + 1$ which lead to 1 under the rule. (Note that these probabilities can be computed from the transition table.) In general, denoting by $\#_s(B)$ ($s = 0$ or 1) the number of cells in state s and assuming the rule preserves uncorrelation as described above, the probability of a block B is just $P_B = p_1^{\#_1(B)}(1 - p_1)^{\#_0(B)}$, and hence

$$p_1^{t+1} \quad := \quad \sum_{T(B)=1} (p_1^t)^{\#_1(B)}(1 - p_1^t)^{\#_0(B)} . \tag{8.4}$$

This equation now defines a real-valued dynamical system which approximates the action of the rule on uncorrelated probability measures. In fact, the mean-field theory represents the action of the cellular automaton combined with a noise process which removes all correlations between states after rule applications.

Only blocks leading to a 1 contribute to the sum in (8.4). Blocks B and B' with $\#_1(B) = \#_1(B')$ occurring in the same sum make the same contribution. If n_l is the number of blocks which contribute to the sum and contain l cells in state 1, equation (8.3) becomes

$$p^{t+1} \quad = \quad f(n_0, \cdots, n_{2r+1}; p^t) \tag{8.5}$$

$$= \quad \sum_{l=0}^{2r+1} a_k (p^t)^l (1 - p^t)^{2r+1-l}$$

All cellular automata with the same coefficient values n_l have the same behavior in the mean-field approximation and define a class of rules that take on that parameter. Now, a fixed-point p^* solution to the equation $f(\mathbf{n}, p^*) = p^*$ gives an estimate of the large-time probability of a 1 under any cellular automaton in the class determined by $\mathbf{n} := (n_0, \cdots, n_k)$. Assuming that the probability of a block is just the product of the probabilities that each of its cells is 1, then p^* is an estimate of an *invariant measure* of the cellular automata in the class of $a := (a_0, \cdots, a_{2r+1})$. The derivative with respect to p in equation (8.5) evaluated at p^* estimates the stability of a fixed-point p^*.

These considerations lead to another attempt at a rigorous definition of Wolfram classes. Rules in class **CW1** have a trivial (only one cell state has positive probability) invariant measure, which is rapidly approached from any initial measure. Rules in class **CW2** have simple invariant measures, again rapidly approached from initial configurations (the exact nature of the measures depends on the initial one). The class **CW3** rules have complex invariant measures independent of the initial measure. **CW4** rules may not have invariant measures, and if they do, they are complex and independent of the initial configuration. In this case, convergence to the invariant measure may be arbitrarily slow. In terms of stable fixed-points, this means only trivial stable fixed points in classes 1 and 2, unique stable fixed points in class **CW3** and more than one, marginally stable fixed points in class **CW4**.

This correspondence has been experimentally observed in detail (e.g., by Gutowitz [Gut2]) for rules of various radii r in the 1D case. For $r = 1$, all but 2 of the 16 rules yield agreement with Wolfram's classification. For $r = 3$, a study of

$$f_\mu(\mathbf{n}; p) := f(0, 0, 0, \mu, \mu, 21, 0, 0; p)$$

shows that for μ below the first bifurcation point ($\mu_1 = 21.499$), O is the only stable point. For $\mu_1 < \mu < \mu_2 := 35.621$ f has two nontrivial fixed-points, the larger of which is stable. At the second bifurcation point μ_2 the derivative of f_μ passes through -1 and f_μ acquires a stable 2-cycle. This cycle persists until the next-bifurcation point $\mu_3 = 35.621$. Other slices of the parameter μ (i.e., other values of n and p) produce qualitatively similar parameters.

In summary, a parametric classification of cellular automata has been obtained. The complexity of the rules increase with the parameter μ. As μ increases, the *class* of the automaton varies from 1, to 2, to 3. Wolfram class behavior peaks near μ_2, and generally occurs for large values of μ. Other advantages of the mean-field classification include:

1. the classification is parametric, with the properties of the rule changing smoothly with the parameter.

2. membership in these classes is decidable, while membership in Wolfram's classes is at best undefined.

3. all the members of a class can be constructed. There is no method to construct all rules in one of Wolfram's class.

4. Solution of the mean-field equation produces quantitative estimates for both invariant probability measures and the stablity of these measures.

5. the classification can be systematically refined to take take longer-range correlations into account, unlike Wolfram's classification.

These refinements are known as local structure theories.

8.5 Local Structure Theory

Local Structure Theory (LST) is a generalization of mean-field theory to a sequence of analytical models of the action of a cellular automata rule on probability measures. Increasing orders of the theory take into account increasing correlation between cell states. LST has afforded empirical demonstration that there is a coupling between the values of certain coefficients and the statistical properties of the rules. The coupling becomes tighter as the order of the theory increases and refines classifications of smaller orders.

The concept of a measure is required to describe the basic ideas of local structure theory. Analogously to a topology, a *measure* is a generalization of the concepts of *length*, area, *volume*, etc. to an abstract space X. In order to have a measure it is necessary to specify certain subsets of X as *measurable sets* (technically, a σ-algebra, or *field* F) that the measure μ applies to, and the values of the measure μ for each measurable set. Like with the idea of topology before, only basic properties about length, area, and volume are required to hold in general. These properties are as follows.

Definition 8.14 *A σ-field of a set X is a set of subsets of X (called* measurable *sets) including X and closed under complementation, intersection and arbitrary countable union. A* measure *on X is a nonnegative real-valued function*

$$\mu : F \to \mathbf{R}_{\geq 0}$$

satisfying the following properties for arbitrary measurable sets E, E_l in F:

1. $0 \leq \mu(E) \leq +\infty$

2. $\mu(\bigcup_l E_l) = \sum \mu(E_l)$

The elements of F are called blocks *or* measurable sets. *If $\mu(X) = 1$ (respectively, finite), the measure is called* probabilistic (finite).

The measures of interest in LST are probabilistic and the universe X is here, of course, configuration space $\mathbf{C}$. LST regards a local rule T as a transformation of probability measures according to the equation

$$T(\mu)(E) = \mu(T^{-1}(E)) \tag{8.6}$$

where E is a μ-measurable set of configurations. Such sets can be represented as unions of *blocks* (or cylinders), i.e., contiguous sequences of cell states at specified locations. Since T is shift-invariant, one only needs to consider shift-invariant measures, for which the measure of a block b depends on sites within the block, not on its location. Therefore equation (8.6) becomes

$$P^{t+1}(b) = \sum_{|\underline{B}|=2r+1} \mathbf{1}_{T,b}(B)P^t(B), \tag{8.7}$$

where $P^t(B)$ is the probability of B at time t, and $\mathbf{1}_{T,B} = 0$ or 1 according as $T(B) = b$ or not. This recursive equation allows computation of smaller block probabilites from probabilities of larger blocks (which must be known). The system must then be iterated to obtain the probabilites of blocks of any fixed length. Mean-field theory resolves this problem by assuming that the initial configuration is made up of uncorrelated events at each cell.

On the other hand, LST inverts the procedure and attempts to use infomation about blocks of smaller size to compute probabilities of larger blocks. This approach requires an approximation of the function $\mathbf{1}_{T,*}$. The exact type of approximation determines the type of LST to be used.

8.5.1 Zeroth-order and First-order Local Structure Theories

The lower order of LST ignores all structure of a cellular automaton except one property: the number of neighborhoods which yield b upon application of the rule. In this theory equation (8.7) is radically simplified to the point where it no longer is a dynamical system: *block probabilities are assumed to only depend uniformly on their length (or size)* $|\underline{B}|$. P^t is replaced by $\frac{1}{2^{|\underline{B}|}}$ for elementary rules. $\mathbf{1}_{T,b}(B)$ now simply counts the neigborhood blocks that run to 1 under the rule. For example, for a pixel $b = 1$, only the number of blocks with support N that lead to 1 under the rule are counted. Thus, for a single pixel, equation (8.7) becomes

$$P_1 = \frac{1}{2^d} \sum_{T(B)=1} \mathbf{1}_{T,b}(B) = \frac{\lambda}{2^d}, \tag{8.8}$$

where $d := |\underline{B}|$. Therefore, the prediction of 0th-order LST for the invariant density of a rule is the density of the transition table itself. As crude as they may seem, low-order LSTs have been used with some success by Langton [L] and Li–Packard–Langton [L-P, L-P-L] to attempt to correlate long-term behavior in 1D and 2D local rules.

As it turns out, first-order LST is exactly mean-field theory. The values of $\mathbf{1}_{T,b}(B)$ here correspond to the values n_k and the more refined estimate of the probability of a block is made as in mean-field theory. Rules in the same 1*st* (mean-field) order class **n** belong to the same 0th-order class with parameter $\lambda := \sum_k n_k$, so 1st-order LST is a refinement of 0th-order theory.

8.5.2 Higher-order Local Structure Theories

In 2nd-order theory, correlations between blocks ignored in 1st-order theory are taken into account by assuming that the probability of large blocks can be computed in terms of the probability of smaller blocks that they contain.

$$P_1 = \frac{1}{2^d} \sum_{2^d} \mathbf{1}_{T,b}(B) = \frac{\lambda}{2^d}. \tag{8.9}$$

We illustrate with the 1D case. If $a := a_1 \cdots a_n$ is an n-array of binary cells, and let $P(a)$ be the probability of an n-block. If the probabilities of all 2-blocks are known, the probability of an n-block a may be estimated as

$$P(a_1 \cdots a_n) := \frac{\prod_{k=1}^{n-1} P(a_k a_{k+1})}{\prod_{k=2}^{n-1} P(a_k)}. \tag{8.10}$$

where the 2-blocks are found by summing 2-block probabilities. It is not obvious, but can be proved, that this equation assigns probabilities consistently to larger blocks and gives rise to an extension of maximum entropy of the given 2-block probabilities. The rest of the process is similar to 1st-order theory, of which 2nd-order theory becomes a refinement.

The advantages of LST are similar to those of mean-field theory. For instance, there are algorithms to construct all rules in one of the higher-order classes. Such algorithms are unthinkable for Wolfram classes themselves, unless an appropriate definition of the class is given. The definition has thus particular significance, as shown by Turing unsolvability of membership shown in Theorems 8.7 and 8.9.

8.6 Other Classifications

Other classifications of cellular automata have been attempted. A brief description of some of them is given in this section.

First, there is the algebraic approach to the classification problem based on the idea that cellular automata rules are closed under composition. The basic emphasis is more pattern-recognition oriented and aims at obtaining arbitrary configurations from a fixed set of rules and as few configurations as possible. The starting point of the approach is the following result for 1D configurations.

Theorem 8.15 *The rules* T_1, T_2 *defined by the local rules* δ_1, δ_2 *as follows:*

$$\delta_1(x_0, x_1, x_2) := \begin{cases} 0 & \text{if } x_1 = x_2 \text{ and } x_0 \neq x_1; \\ x_1 & \text{if } x_0 = x_1 = x_2 \\ x_2 & \text{if } x_1 \neq x_2 \text{ and } x_2 \neq 0; \\ 0 & \text{if } x_1 \neq x_2, x_2 \neq 0 \text{ and } x_0 = x_1; \\ x_1 & \text{if } x_1 \neq x_3, x_2 = 0 \text{ and } x_0 \neq x_1; \end{cases}$$

$$\delta_2(x_0, x_1, x_2) := \delta_1(x_2, x_1, x_0).$$

Table 8.2. Constructive chaos classification of cellular automata

Initial	Evolution type	
Conditions	Computable	Uncomputable
Uncomputable	Type I	Type III
Computable	Type IV	Type II

are bijective and for any configuration $a \in \mathbf{C}_0$ (for instance, the unit pixel) there exists some composition

$$T = T_{i_1} \circ T_{i_2} \circ \cdots \circ T_{i_k}$$

of T_1s and T_2s such that $x = T(a)$.

A similar result holds for higher dimensions with a suitable new set of building blocks. Thus one has some sort of decomposition theory for cellular automata. This theory, however, has not been actively pursued since these results appeared.

Second, there is the so-called *constructive chaos* approach to classification along related lines. Here the basic notion is *chaos* and the idea is to classify cellular automata by *how* chaotic they turn initial nonchaotic configurations, or vice versa. (The notion of chaotic is defined relative to the well known Kolmogorov–Chaitin notion; see, e.g., Li–Vitany [L-V] for a detailed treatment). Table 8.2 show the resulting four classes. For example, automata of type I generate chaos by computable evolution from uncomputable initial values. But again, it is difficult to obtain satisfactory results about the the classes so generated in terms of local rules.

Finally, there are other classifications whose success, including the foregoing ones, is yet to be judged. They are based, not on the behavior of limiting objects associated with an automaton (such as limit sets, entropies, or fractals), but rather on objects associated with their evolution and long-term behavior (at finite rather than infinite times). For example, the notion of isomorphism in the finite case amounts to equivalence of the connected components (which are simply the basins of attraction) of equilibrium points. The local rule T induces a family of self-maps $\{T^t\}_t$ that act on configuration space and create the various levels of the attraction basins. In fact, one can interpret that way the key idea in Wolfram classification. At least two classification schemes have been suggested to implement this idea on configuration space, with a product measure in which every state occurs with a positive probability.

Definition 8.16 *Let $(X, |*,*|)$ be a compact metric space. A self-map $f : X \to X$ is (respectively, almost) expansive at x iff there exists $\varepsilon > 0$ such that for (almost) every $y \in X$, there exists a time $t \geq 0$ such that $|f^t(x), f^t(y)| \geq \varepsilon$. A family of self-maps $\mathcal{D}$ of X is equicontinuous at a point $x \in X$ if for every $\varepsilon > 0$, there exists a $\delta > 0$ such that for every $f \in \mathcal{D}$, $|f(x), f(y)| < \varepsilon$ whenever $|x, y| < \delta$. The map T is equicontinuous on a subspace $Y \subseteq X$ if the family of iterates*

$\{T^t\}_{t\geq 0}$ *is equicontinuous at every point in* Y, *and it is* nearly equicontinuous *if for every* $\varepsilon > 0$, *there exists a closed subset* $Y \subseteq X$ *of measure* $\mu(Y) \geq \mu(X) - \varepsilon$ *which is* T-*invariant (i.e.,* $T(Y) \subseteq Y$*) and where* T *is equicontinuous.*

In other words, a self-map is expansive if the orbits of any two distinct points eventually get away from one another, and it is equicontinuous if all its iterates act continuously on X in a uniform fashion. The next result asserts that these two notions provide an exhaustive classification of 1D cellular automata. (Recall that a property of a measure space holds *almost everywhere* if the subset of elements where it holds has *full measure*, i.e., the subset of points where it does *not* hold has measure 0). A nearly equicontinuous map is certainly almost everywhere equicontinuous.

Theorem 8.17 *Every 1D cellular automaton is either almost expansive at every point or nearly equicontinuous.*

Another way to classify cellular automata is to interpret Wolfram's idea of self-organization as a tendency to gather dissimilar initial conditions into similar global states under the action of the iteration on configuration space. One can make this idea precise by looking at the the size of the attractors under a suitable notion of attractor (among the several notions available).

Definition 8.18 *Let* $(X, |^{*,*}|)$ *be a compact metric space and* $T : X \to X$ *a continuous self-map. A closed nonempty subset* $A \subset X$ *is an* attractor *of* f *iff there exists an open subset* U *containing* A *such that* $T(\bar{U}) \subseteq U$ *and* $\bigcap_{t\geq 0} T^t(U) = A$. *A set* Q *is a* quasi-attractor *of* T *if* Q *is an intersection* $Q = \bigcap_n A_n$ *of a sequence of attractors* A_n *of* T. *A (respectively, quasi-)attractor is* minimal *if no proper subset is also a (quasi-)attractor. The* basin *of a set* C *consists of all those points* $x \in X$ *whose* ω-*limit sets (i.e., the set of cluster points of their orbits) are contained in* C.

We state, again without proof, a classification based on these notions of attractor and basin of attraction.

Theorem 8.19 *Every cellular automaton* T *falls in exactly one of the three categories characterized by the following conditions:*

HC1: *a unique minimal attractor* A *which has a basin which is open, dense and has full measure;*

HC2: *a unique minimal quasi-attractor which itself is not an attractor but is contained in every one of them, and has a basin which is either of full measure (**HC2a**) or measure 0 (**HC2b**);*

HC3: *at least two disjoint attractors (and hence uncountably many minimal quasi-attractors) both whose basins of attraction have measure 0.*

Wolfram classes do not seem to fit quite nicely into this classification. For example, an automaton with finitely many attractors (such as those in **WC1**) falls in class **HC1**; **WC2** seems to correspond to **HC2** but the latter contains some automata in classes **WC3** and **WC4**.

A modification of this type of classification will be approached in Sect. 11.4.

8.7 Problems

Problems marked * require substantial effort and may need to be looked up.

ABOUT ENTROPY

1. Show that a cellular automaton is nilpotent iff its phase space is a connected digraph.

2. Prove Proposition 8.6. [Use compactness of **C** for the hard part $(1) \Rightarrow (4)$.]

3. Show that the entropy of a self-map T on a compact metric space X is given by

$$ ent(T) \;=\; \lim_{t \to \infty} \frac{\log \#[\mathcal{U} \wedge T^{-1}(\mathcal{U}) \wedge \cdots \wedge T^{-(t-1)}(\mathcal{U})]}{t} , $$

   where $\mathcal{U} \wedge \mathcal{V}$ denotes union of subcovers (consisting of all intersections $U \cap V$ with $U \in \mathcal{U}$, $V \in \mathcal{V}$), f^{-1} denotes the subcover $\{f^{-1}(U)\}_{u \in \mathcal{U}}$, and the $\#$ denotes the minimum number of elements in a finite subcover.

4. Show that the entropy of a 1D cellular automaton T is given by

$$ ent(T) \;=\; \lim_{w \to \infty} \lim_{t \to \infty} \left(\frac{\log R(w,t)}{t} \right) , $$

   where $R(w,t)$ is the number of distinct rectangular windows in all the space-times of T of width w and height (temporal duration) t. [See Problem 3.]

5. Show that the 1D left-shift σ on m states has $ent(\sigma) = \log m$.

6. Show that the entropy of a 1D cellular automaton is always finite. [See Problem 4. For example, the rectangular blocks of the space-time of a 1D automaton of radius r satisfy the inequality $R(w,t) \leq 2r|Q|$, and hence $ent(T) \leq 2r \log |Q|$.]

7. Show that the inequality in Problem 6 is sharp for 1D cellular automata. [Consider the linear rule $T(x)_i := \sum_{j=-r}^{r} x_i \pmod{m}$.]

8. Show that the 1D cellular monomial automaton given by

$$T(x) := x_{i+1}x_{i+2} \pmod{2}$$

has only two temporally periodic configurations x (satisfying $T^t(x) = x$ for some $t \geq 1$), but it has $ent(T) = 2$ (hence it is *not* nilpotent). (Such an automaton shows that the behavior of a cellular automaton on periodic configurations, as commonly done in computer simulations, may miss entirely the core of its full dynamical behavior.) [1 is a quiescent state but 0 is spreading. The limit set $\Lambda(T)$ consists of configurations all-1s **1**, all-0s **0**, and $\ll\bar{0}\bar{1}\gg$, $\ll\bar{1}\bar{0}\gg$, where $\ll\bar{*}$ (respectively, $\bar{*}\gg$) denotes a one-way infinite sequence of the characters *s to the left (to the right). See Problems 4 and 7.]

9. Show that there exists a 1D cellular automaton that has any finite number of temporally periodic configurations x, but which is *not* nilpotent. (Such an automaton shows that, unlike certain families of dynamical systems in the continuum, the topological entropy of cellular automata cannot be expressed as the growth rate of the number of periodic points.) [For example, extend the automaton in Problem 8 to one with 3 states with a modified arithmetic modulo $m = 3$ ($1 \times 1 = 2, 2 \times 2 = 1 \times 2 = 2 \times 1 = 1$) so that 1 is no longer a fixed point but the limit set remains essentially the same.]

10. Show that there exists a 1D cellular automaton T that has only one temporally periodic configuration x, but which is *not* nilpotent (such an automaton is called *aperiodic*). [Extend the automaton in Problem 9 to one with 16 states with an appropriate multiplication so that 1 is no longer fixed but the entropy remains positive.] *

11. Show that there exist aperiodic 1D cellular automata T_n of arbitrarily large entropy $ent(T_n) \geq \log n$ (See Problems 4, 8–9.) [Use the automaton of Problem 10 as a basis for a construction similar to the one used in parts (2)–(4) of the proof of Theorem 8.12.] *

12. Show that there exist cellular automata for which $\overline{\Pi(T)} \neq \Omega(T)$, i.e., the closure of the set of periodic points is different from the set of nonwandering points. (The *closure* $\overline{U}$ of a set U in a metric space X is the smallest closed subset of X containing U. A point x is *nonwandering* if for every $\varepsilon > 0$ there exist points whose orbits start and do return to within distance ε from x in finite time.) [The hint in Problem 10 will do.] *

CLASSES AND ATTRACTORS

13. Show that the rule $T(x)_i := x_i x_{i+1}$ belongs to **HC2a**.

14. Show that the identity rule $T(x)_i := x_i$ belongs to **HC3**.

15. Show that if the basins of two attractors in a compact metric space intersect, then so do the attractors. *

16. Show that every two disjoint quasi-attractors Q_1, Q_2 in a compact metric space can be separated by disjoint attractors A_1, A_2 such that $Q_1 \subseteq A_1$ and $Q_2 \subseteq A_2$.

17. Show that every map on a compact metric space has at most countably many attractors. *

18. Show that no expansive cellular automaton can belong to **CY2**.

19. Show that an expansive cellular automaton belongs to **CY3**.

8.8 Notes

Classifying cellular models can take on a number of meanings, depending on the underlying type of equivalence criterion. The problem of characterizing and classifying problems solvable by *dynamical* systems defined by iteration of local rules of cellular automata and/or neural networks is a difficult and important problem of current interest.

The notion of isomorphism used in Sect. 8.1 for finite networks may be of some use in view of the fact that Las Vegas algorithms that solve the problem with exponentially high probabilty of success have been given in Garzon–Jagota [Ga-Jo].

Although it can be readily formulated for infinite networks, the weak isomorphism problem for infinite spaces is an analog of the well known open problem of isomorphism of symbolic dynamics. The classification of symbolic dynamical systems is by no means complete and much remains to be done (see, for example, Lind–Marcus [L-M]). Classical dynamical systems such as so-called Bernoulli shifts can be classified by a single parameter, their entropy. But for infinite networks, even cellular automata, there are just too many different rules and no natural "parameters" in sight that would lead to a satisfactory classification. Even notions based on more complex structures such as formal languages do not seem capture them in a coherent fashion. For example, a classification based on the Turing complexity of the formal language associated with the blocks in the limit sets was initiated by Hurd [Hu1, Hu2]. However, it was shown by Goles–Maass–Martinez [G-M] that there exist a computation universal cellular automaton whose limit language is regular.

The notion of entropy dates back to the work of L. Boltzmann in the 19th century. The version used here is a more recent dynamical-system approach due to Bowen [B]. The undecidability results about computing cellular automaton entropies are due to Hurd–Kari–Culik [H-K-C], which motivated Problems 1–7 and 10–12. Nilpotent rules are introduced and discussed in Kari [Ka1]. The Rice Theorem 8.13 for cellular automata is established in Kari [Ka2]. The local rule for a solution to Problem 10 can be found in Hurd–Kari–Culik [H-K-C]. A first example originates in Hurd [Hu3]. Since entropies (even measure-theoretic ones) of higher-dimensional euclidean cellular automata may be infinite (see Problem 4), Milnor has generalized it to *higher-dimensional entropies* as a means to quantify information transmission across time and space during cellular automata evolution – see [Mi2, Mi3].

The Wolfram's series of papers in the early 1980s pioneered the use of physical concepts and analogies in research of cellular automata – see his [Wo4]. In particular, Mean Field Theory is used in [Wo2]. Our description follows later work by Gutowitz on mean-field theory [Gut2] and local structure theory [Gut3]. The formalization of Wolfram classes are introduced in Culik–Yu [C-Y], where the results of Sect. 8.3.1 originate. The details of Ishii's formalization in terms of the behavior of almost every configuration can be found in [I]. The decomposition into equicontinuous and expansive maps in Theorem 8.17 according to the action of the family of iterates of the automaton almost everywhere in configuration space originates in Gilman [Gi1, Gi2]. The relationship between this classification and the CY-classification has been addressed by Kůrka [Ku3], where solutions to Problems 18–19 can be found. The classification in Theorem 8.19 in terms of attractor sizes is due to Hurley [Hur], where the details of the proof and solutions to Problems 14–17 can be found. Kůrka has refined both Gilman's alternative in [Ku4] in terms of the relative size of the set of points where it is equicontinuous. Moreover, he extends Hurley's classification to include all continuous dynamical systems on zero-dimensional spaces, not just configuration space. This paper also presents a further approach to a Turing-type classification of cellular automata which will be examined in Chapter 9.

Theorem 8.15 is due to Akira–Kimura[A-K]. Decomposition theory appears to be of little advantage since the complexity of a configuration may quickly increase under mixed composition (which is not commutative). On the other hand, incompressibility and randomness are old ideas in information theory, and the idea of classifying cellular automata rules as information operators is a natural one. However, beyond the work of Svozil [Svo], little progress seems to have been made. One might remark that, by the results in Chapter 7, information processing by infinite discrete neural and automata networks on infinite cellular spaces overflows the classical conceptual framework based on information encoding by *finite* strings. Most infinite configurations carry an infinite amount of information which can be processed in parallel by finitely described programs (local rules) – unlike, for example, ordinary addition and multiplication with ordinary *real* numbers. The import and impact of this type of model is poorly understood (although Chapter 11 presents some preliminary steps in this direction). In particular, the meaning of the notion of 'random configuration' (analogous to that of a Kolmogorov–Chaitin random string [L-V]) when cellular automata are used in place of Turing machines appears to be a novel question (but related to Problem 15 in [Wo1]). That it should be a different notion stems from several observations on how good cellular automata are in generating (pseudo-)random strings – see, for example, Wolfram's [Wo3].

There are other, higher-level criteria to classify neural networks. The semantic classification of functional isomorphism mentioned in Chapter 7, although intriguing from a biological viewpoint, has not been really pursued beyond speculation. Finally, there is the abstract computational approach. The results in Chapter 7 can be regarded as characterizing self-maps of the Cantor set computable in a *single* time step. From a computational point of view, solving the classification of cellular automata and neural networks is analogous to finding a hierarchy of Turing complexity classes (say according to a computational resource, e.g., time or space) in order to determine what problems are solvable by using global dynamics. We will return to this problem in Chapter 11.

References

[A-K]   A. Akira, M. Kimura: Decomposition phenomenon in one-dimensional scope-three tesselation automata with arbitrary number of states. Inf. and Control **34** (1977) 296–313.

[A-K2] A. Akira, M. Kimura: Completeness in tesselation automata. Inf. and Control **35** (1977) 52–86

[B]   R. Bowen: Entropy for group endomorphisms and homogeneous spaces. Trans. Amer. Math. Soc. **153** (1971) 401–414

[C-Y]   K. Culik, S. Yu: Undecidability of CA classification schemes. Complex Systems **2** (1988) 177–190

[F]   C. Fields: Consequences of nonclassical measurement for the algorithmic description of continuous dynamical systems. J. Expt. Theor. Artif. Intell. **1** (1989) 171–178

[G-J]   M.R. Garey, D.S. Johnson: Computers and intractabillity: a guide to the theory of NP-completeness. W.H. Freeman, San Francisco, 1978

[Ga-Jo] M. Garzon, A. Jagota: Efficient neural network isomorphism testing. In: Proc. 2nd Swedish Conference on Connectionism. Erlbaum Publishers, 1994.

[G-Z]   M. Garzon, M. Zhang: Classifying neural networks. R.E. Trahan, Jr. (ed.). Proc. IEEE Southeast Conference, New Orleans, 1990, pp. 567–571

[Gi1]   R. Gilman: Classes of linear automata. Ergodic Theo. and Dynam. Syst. **7** (1987) 108–118

[Gi2]   R. Gilman: Periodic behavior of linear automata. In: Dynamical systems. Lecture Notes in Mathematics, Vol. 1342. Springer-Verlag, Berlin, 1988, pp. 216–219

[G-M]   E. Goles, A. Maass, S. Martinez: On the limit set of some universal cellular automata. Theoret. Comput. Sci. **110**:1 (1993) 53–78

[Gut1] H. Gutowitz (ed.): Cellular automata: theory and applications. Proc. 3rd. Int. Conf. Cellular Automata, Los Alamos, 1991. Physica D **45** (1990).

[Gut2] H. Gutowitz: Mean Field vs. Wolfram Classification. preprint. CNLS, Los Alamos National Lab, NM 1988

[Gut3] H. Gutowitz: A hierarchical classification of cellular automata. In: [Gut1] pp. 136–158

[Hu1]   L. Hurd: Formal language characterizations of cellular automaton limit sets. Complex Systems **1**:1 (1987) 69–80

[Hu2]   L. Hurd: The application of formal language theory to the dynamical behavior of cellular automata. Dissertation, Princeton University, 1988

[Hu3]   L. Hurd: The nonwandering set of a CA map. Complex Systems **2**:5 (1988) 549-554

[H-K-C] L. Hurd, J. Kari, K. Culik: The topological entropy is uncomputable. Ergodic Theory and Dyn. Syst. **12**:2 (1992) 255-265

[Hur]   M. Hurley: Attractors in cellular automata. Ergodic Theo. and Dynam. Syst. **10** (1990) 131-140

[I]   S. Ishii: Measure-theoretic approach to the classification of cellular automata. Disc. Appl. Math **39** (1992) 125-136

[Ka1]   J. Kari, The nilpotency problem of one-dimensional cellular automata. SIAM J. Comput. **21** (1992) 571-586

[Ka2]   J. Kari, Rice's Theorem for the limit sets of cellular automata. Theoret. Comput. Sci. **127**:2 (1994) 229-254

[Ku3]   P. Kůrka: A comparison of finite and cellular automata. In: Math. Foundations of Computer Science MFCS, I. Privara, B. Rovan, eds. Lecture Notes in Computer Science 841, Springer-Verlag, Berlin, 1994, pp. 484-493

[Ku4]   P. Kůrka: Languages, equicontinuity and attractors in linear cellular automata. Preprint, Charles University, Praha, Czech Republic, 1994.

[L]   M. Langton: Artificial life. MIT Press, Cambridge MA, 1989.

[L-P]   W. Li, N. Packard: The structure of elementary cellular automata rule space. Complex Systems **4** (1990) 281-297

[L-V]   M. Li, P. Vitányi: An introduction to Kolmogorov complexity and its applications. Springer-Verlag, New York, 1993

[L-P-L] W. Li, N. Packard, C. Langton: Transition phenomena in cellular automata rule space. In: [Gut1], pp. 77-94

[L-M]   D. Lind, B. Marcus: An introduction to symbolic dynamics. Manuscript, 1993

[Mi2]   J. Milnor: Directional entropies of cellular automaton maps. In: Disordered systems and biological organization, E. Bienenstock et al., eds. Springer-Verlag, New York, 1986, pp. 113-115

[Mi3]   J. Milnor: On the entropy geometry of cellular automata. Complex Systems **2**:3 (1988) 357-386

[Mu]   S. Muroga, I. Toda, S. Takasu: Theory of majority decision elements, J. Franklin Inst. **271** (1961) 376-418

[P-S]   I. Parberry, G. Schnitger: Parallel computation with threshold functions. J. Comput. Syst. Sci. **36** (1988) 278-302

[Svo]   K. Svozil: Constructive chaos by cellular automata and possible sources of an arrow of time. In: Proc. 3rd Int. Conf. in Cellular Automata, 1989. Physica D **45** (1990) 420-427

[Su]   K. Sutner: A note on Culik-Yu classes. Complex Systems **3** (1989) 107-115

[Wo1]   S. Wolfram: Twenty problems in the theory of cellular automata. Physica Scripta T**9** (1985) 170-183

[Wo2]   S. Wolfram: Statistical mechanics of cellular automata. Rev. of Modern Phys. **55**:3 (1983) 601-644

[Wo3]   S. Wolfram: Random sequence generation by cellular automata. Adv. in Applied Math. **7** (1986) 123-169. Reprinted in [Wo4]

[Wo4]   S. Wolfram: Theory and Applications of cellular automata. World Scientific Publishing, Singapore, 1986.

9. Asymptotic Behavior

You can fool all the people some of the time. You can even fool some of the people all the time. But you can't fool all the people all the time.

Abraham Lincoln

The difficulties in classifying cellular automata and networks based on their global behavior have been explained in Chapter 8. The results tend to indicate that a complete classification is, at least in many computational ways, impractical or unfeasible. The study of the specific properties of global behavior of arbitrary automata in terms of their local rules is indeed a most interesting and difficult problem.

Among other reasons for this difficulty is that these problems appear to be of a discrete and combinatorial nature, where classical optimization and analysis are not directly applicable. This chapter deals with positive results that shed light on the long-term behavior of global dynamics. Each technique has only been partially successful and they can be roughly classified as either discrete/combinatorial or analytic. The first section presents some results obtained through combinatorial techniques. Later, several analogies with continuous classical dynamical systems are exploited to gain insight into the nature of the long-term behavior of cellular networks. Despite complex behavior as chaotic as that of maps on the interval, many of them exhibit an interesting property of observability through simulation on computing devices under limitations such as bounded precision, rounding errors, and noise.

9.1 Linear Rules

As usual, the superposition principle affords a number of structural results for linear rules, some of which will be reviewed in this section.

9.1.1 Linear Automata on Tori

A way to understand the behavior of a cellular automaton on an infinite grid is to look at its action on restricted types of configurations. An important type is the family of spatially periodic configurations. On the 1D grid, for example, a local rule must preserve periodicity: T can never increase the period of successor configurations. Although essential features of the global dynamics may be lost in passing to quotient spaces (with limit sets of nilpotent cellular automata, for example), a good deal of information can be gained from looking at projections on finite cellular spaces. We illustrate the technique in this section with 1D euclidean spaces.

Let Γ_n be a cellular space with n cells arranged in a circle, so the cells are indexed with integers modulo n. Configurations are now circular strings, which can be represented as finite words by cutting before the origin. Let p the (temporal) period of a typical configuration. Only certain periods occur for given p and n. Let $\tau_{n,p}$ ($\tau_{n,p}^*$, respectively) be the number of cyclic words of period p (a divisor of p).

It turns out to be somewhat easier to establish a recurrence relation for $\tau_{n,p}$ and $\tau_{n,p}^*$ in terms of the period length than one based on the space size, even in the case of nonlinear cellular automata.

Proposition 9.1 *Let T be a cellular automaton of radius r on a torus Γ_n. If $n \geq 2r + 1$, then the number of fixed points on Γ_n satisfies a linear recurrence relation and is given by*

$$\tau_{n,p} := \sum_{l=1}^{2r} \lambda_l^{n-2r} .$$

for some real numbers λ_l ($1 \leq l \leq 2r$). Likewise for $\tau_{n,p}^*$.

Proof. Since the p-periodic configurations of T are fixed points of T^p, it suffices to look at the fixed points of T. Consider a graph with 2^{2r} nodes representing tuples of site values and adjacencies between two strings x, y wherever

$$\delta(x_{-r}y_{-r+1}\cdots y_{r-1}) = x_0 ,$$

i.e., wherever y's neighborhood can be suffixed to the first symbol of that of x so as to leave the center cell invariant. Let the *invariance matrix* of T be the adjacency matrix $A := [a_{ij}]$ of size $2^{2r} \times 2^{2r}$ defined as follows. Express $i := i_{-r}\cdots i_{r-1}$ and $j := j_{-r}\cdots j_{r-1}$ in binary strings of size $2r + 1$ (pad 0s on the left if necessary). Put $a_{ij} := 0$ unless all $i_l = j_l$ ($-r + 1 \leq l \leq r - 1$) and $\delta(i_{-r}j_{-r+1}\cdots j_{r-1}) = j_0$, in which case $a_{ij} := 1$. The nonzero entries of the powers A^t of the invariance matrix indicate the presence of paths of length t connecting corresponding nodes. The fixed points of the automaton correspond to paths from a node back to itself. Thus $\tau_{n,1}^*$ corresponds to the trace of A^{n-2r}, and so it satisfies the characteristic polynomial of A and hence equals the given sum of the eigenvalues. $\square$

There remains the question of characterizing the attractor structure of the periodic points in the global behavior on tori. As it turns out, the word 'fractional shift' describes well the long-term dynamics on tori of large families of cellular automata.

Definition 9.2 *A map T on Γ_n acts as an s/t-shift on a periodic configuration of period p if*

$$T^t(x) = \sigma^s(x), \tag{9.1}$$

i.e., x is shifted s sites in t iterations. The shift by s/t is fundamental if t is the least positive integer for which (9.1) holds. A configuration on Γ_n is primitive if its least spatial period is n.

Theorem 9.3 *Let s/h be the fundamental shift associated with the limit cycle of a pixel e under the one-sided nearest neighbor rule $T(x)_i := x_i x_{i+1}$. A circular string x appears as a limit cycle for T on a cylinder of size n if and only if it satisfies a recurrence relation*

$$\forall i \; [x_{i-s} = T^h(x_{i \pm h}, \cdots, x_i)] \,,$$

or a similar one obtained by substituting s by $s \pm n$. All limit cycles are generated as solutions of the recurrence on suitable initial conditions, and they have a period dividing $nh/\gcd(n, |s|)$.

Proof. The proof in essentially a case-by-case analysis and it will not be pursued here in detail (see Jen [Je1]). $\qquad\qquad\qquad\qquad\qquad\qquad\qquad\qquad\qquad\qquad$ □

As a corollary one easily obtains the following result for linear rules since all configurations satisfy the relation $x_{i+n} = x_i$.

Corollary 9.4 *Let L be a linear rule on a 1D torus Γ_n. There exist coefficients $b_0, \cdots, b_l$ such that a configuration x appears in a limit cycle of L iff it satisfies the linear relation (all indices are taken modulo n)*

$$\forall i, \; \sum_{j=0}^{l} b_j x_{i+j} = 0 \,.$$

The coefficients b_k can be obtained recursively on n as follows. If B_n is the polynomial with coefficients $B_n(X) := b_l X^l + \cdots + b_1 X^1 + b_0$, then, in fact,

$$B_n(X) = \gcd[a_n(X), X^n - 1] \,,$$

and it can be computed according to

$$B_{p^r m}(x) = [B_m(x)]^{p^r} \,,$$

where p^r is the largest prime power of p dividing n, and $a_n(X)$ is the polynomial corresponding to the kernel of L so that L reduces to multiplication by $a_n(x)$ modulo m. These results can be used to actually generate all limit sequences for linear rules from the primitive configurations, and to solve the inverse problem of identifying all linear rules for which a given sequence of configurations appears as a limit cycle (see Problems 3–5).

9.1.2 Linear Automata on the Line

Spatially periodic configurations of automata must be temporally periodic under arbitrary local dynamics and the previous section deals with the exact nature of the periods. On tori, the converse follows at once for linear rules from the pigeon-hole principle. Perhaps surprisingly, linear rules on the infinite line exhibit a similar property: temporally periodic configurations must also be spatially periodic on the 1D grid. Furthermore, one can set upper bounds that quantify rather closely the spatial period in terms of the temporal period. These results are explored in this section for binary rules.

For a rule T, let the *spread* of T be $\gamma(T) := \max(N) - \min(N)$, where N is the effective neighborhood set of minimum radius of its local rule δ (i.e., $T(x)_i$ does change upon changing one of $x_{i+\min(N)}$ or $x_{i+\max(N)}$). The spread of the identity is 0 by definition and otherwise $\gamma(T) = 0$ iff $T = \mathbf{id} + \sigma^m$ $(m \in \mathbf{Z})$. Let $\alpha(x)$ be the spatial period of a configuration x, and let $\alpha(n)$ be the largest spatial period of all configurations of temporal period n. Recall that τ_p (respectively, τ_p^*) is the number of configurations of temporal period a divisor of (exactly) p.

Theorem 9.5 *Let L be a linear rule with kernel $a \in \mathbf{C}_0$. If L is not the identity, then every temporally periodic configuration of period $\leq p$ has spatial period $\leq 2^{\gamma(L^p)}$. Moreover, if L is not the identity $(a \neq e)$,*

$$\tau_p = 2^{\gamma(L^p)} \quad \text{and} \quad \tau^*(p) = \sum_{d=1}^{p} \mu(d,p)\tau(d) .$$

Here μ denotes the number-theoretic Möbius function defined for positive integers as

$$\mu(p,q) := \begin{cases} 1, & \text{if } p = q; \\ 0, & \text{if } q/p \text{ is divisible by a square of a prime;} \\ (-1)^s, & \text{if } q/p \text{ is the product of } s \text{ distinct primes.} \end{cases}$$

Proof. Since L^p is also linear, a configuration x satisfying $L^p(x) = x$ also satisfies a linear recurrence relation involving at most $\gamma(L^p) + 1$ terms. $\qquad\square$

Theorem 9.5 reveals certain unexpected behavior of linear automata. For example, the rule $T(x)_i := x_i + x_{i+r}$ has $\gamma(L^{2^r}) = 1$ and hence no periodic configurations (other than O) of period 2^{2^r} for any integer $r \geq 1$.

Theorem 9.6 *Let L be a linear rule and p a positive integer. If $\gamma(L^p) \geq 1$ then there exists a matrix A over integers modulo 2 of multiplicative order $\alpha(p)$, called the* companion matrix *of L, such that*

1. *every configuration x of period p is a catenation of blocks bA^t of the form*

$$x \equiv \cdots bA^{-2\gamma(L^p)} \ \ bA^{-\gamma(L^p)} \ \ b \ \ bA^{\gamma(L^p)} \ \ bA^{2\gamma(L^p)} \cdots$$

for some block b of length $\leq \gamma(2^{\gamma(L^p)})$.

2. A is independent of the configuration x of L-period p.

3. $\alpha(n)$ is the least positive integer p satisfying $A^p = Id$.

In particular, if $\gamma(L^p) \geq 1$, for every configuration x of t-period p, $\alpha(x)|\alpha(p)$. Moreover, $\alpha(d)|\alpha(p)$ whenever $d|p$; and $\alpha(2^r d) = 2^r \alpha(d)$ if $d > 1$, but $\alpha(2^r) = 1$.

Proof. Let b be the coefficients of the linear combination that produces $x_{[0,\gamma(p)-1]}$ from the canonical basis of the vector space over $\mathbf{Z}_2$ consisting of all solutions of the linear recurrence of Theorem 9.5. It is easily seen that the companion matrix of this linear transformation satisfies the required conditions.   □

The exact value of the spatial period of a temporally periodic configuration does not seem to admit a simple answer. (However, see Problem 8.)

9.2 Exact Solution

Section 9.1 gave an indication which, in some cases, is in fact a fairly complete description of the limit cycles of a linear automaton. The situation for more general rules is currently unclear. We present in this section an example of what one might term *exact solution* of a cellular automaton by analogy with an exact solution of, say, a differential equation.

A natural idea to understand the behavior of a cellular automaton is to find some relation with known rules. The idea was attempted at the level of local rules and linear automata in Sect. 3.4. The more difficult global level has been encouragingly successful for special 1D nearest neighbor rules. There are only three nontrivial 1D elementary linear rules, rules 60, 90 and 150.

A way to reduce the behavior of a dynamics to a linear rule begins with the observation that the two rules have identical orbits on some configurations. That is the situation with rules $18 = 00010010_2, 128 = 01111110_2$ and $146 = 01011010_2$. Table 9.1 shows the types of configurations on which their evolution agrees with rule 90. The phase space of each of these rules can, in fact, be put into 1-1 correspondence with the phase space of linear rule 90, which will be denoted L throughout this section. The notation $\bar{b}$ indicates a periodic finite configuration of period a block b.

On the other *irregular* blocks, the two actions diverge. However, what saves the reduction to a linear rule is the fact that the portion of a configuration that deviates from linearity collapses uniformly to a single value (0 for rule 18 and 1 for the other two). As long as two irregular blocks do not interact, the linear evolution only differs from the rule's evolution by the extra pixel in the limit configuration of the rule. For example, rule 18 turns 0001101 into 01000. Thus one might insert 0s appropriately to create a barrier that prevents the occurrence of irregular blocks and/or their interaction, apply linear rule 90, and finally retrieve the evolution of the rule by deleting the barriers.

Table 9.1. Configurations with identical evolution under rule 90

Rule	Configurations consisting of
18	O, $\mathbf{1}$, isolated 1s with 0-blocks of odd-length
126	O, $\bar{1}$, 0- and 1-blocks of even length
146	O, isolated 1s with 0-blocks of odd-length

Table 9.2. Inserted blocks for identical evolution under rule 90

rule T	$\kappa :=$ # of *irregular* blocks
18,146	that only contain an even number of 0s or only contain three or more 1s
126	that are of odd length.

Precisely, define the quantity κ and the corresponding transformation for a given configuration x as indicated in Tables 9.2 and 9.3.

Example 9.7 For 1D elementary rule 18,

$$\phi_{18}(\ldots 010001101001011110\ldots) = \ldots 010001010100010101010\ldots,$$

since the irregular blocks are 11, 1001, and 1111. In particular, ϕ_{18} is many-to-one and does not allow reconstruction of x unambiguously.

Proposition 9.8 *For any T of the three nonlinear rules $18, 126, 146$, on an arbitrary configuration x,*

1. *ϕ_T is monotonically non-increasing with time;*

2. *As long as κ is conserved over an interval of time $[t, t + n]$,*

$$\phi_T \circ L(x) = L \circ \phi(x), \tag{9.2}$$

   over the same period of time.

Proof. From the rule tables, verify that a pair of adjacent 1s occurs iff an irregular block has occurred in the previous time step. Hence irregular blocks

Table 9.3. Transformations onto linear rule 90

rule T	$\phi_T(x)$
18	insert an extra 0 in even-length 0-blocks and convert blocks $\bar{1}$ of length ≥ 3 to $1\bar{1}0$
126	insert an extra 0 in odd-length 0-blocks and insert an extra 1 in 1-blocks of odd-length
146	insert an extra 0 in even-length 0-blocks and convert every block $\bar{1}$ of length ≥ 3 to $1\bar{1}0$

cannot be created by time evolution. For the second part, we need only consider irregular blocks. The argument is illustrated with rule 18 and an irregular block $B := 0^l$. The 0 inserted according to ϕ_{18} only influences a light cone of length $l - 2t + 1$ instead of the even length $l - 2t$, and so it is as claimed up to time $t \le \frac{l-1}{2}$. After that time, B evolves into 101, in correspondence with the 11 of T since the local rule of T guarantees that

$$T(\ldots x_i 101 x_{i+4}) \ldots = \ldots 000 \ldots$$

and

$$T(\ldots x_i 11 x_{i+3}) \ldots = \ldots 00 \ldots .$$

From the conservation of κ, mimicking by equation (9.2) follows.   □

Note that condition 2 is satisfied in a limit cycle of T and thus the periodic behavior of all rules T is identical to that of L. However, the correspondence ϕ_T breaks down for transient time when irregular blocks collapse. The problem thus remains of determining the general attractor structure and the transient behavior of the evolution of a configuration as well as of membership in a limit cycle. It is possible, but not pursued here, to refine the transformation ϕ to one Φ_T that gives indication of the limit cycles and the transient behavior.

9.3 Simulation in Continuous Systems

Despite its discrete combinatorial appearance, the long-term behavior of cellular automata bears close analogy to problems in continuous systems. Dynamical systems on continuous spaces, particularly euclidean spaces, have long been objects of study as mathematical systems and some of their behavior is better understood. It is thus tempting to try a direct approach that links cellular automata directly to dynamical systems on the unit interval (herefourth denoted I). Ideally, one would dream of having a so-called *conjugacy* that recodes configurations as (n-tuples of) real numbers in a *continuous bijective* manner, so that the action of the automaton can be simulated by iteration of a dynamical system on the interval. The asymptotic behavior of the automaton would then be transparent from observation of the continuous map. This section explores some known results of this type. Several results for continuous maps of the interval will be freely used below. Proofs as well as background on dynamical systems can be found in Devaney's volume [D]. The reader is also referred to one of several well-known works such as Collett–Eckmann [C-E] or Preston [P] for other more specific results on maps of the interval such as piecewise linear and piecewise monotone maps.

9.3.1 Discrete Computation by Continuous Systems

The fundamental difficulty with the study of automata via continous systems is that configuration space is only a totally disconnected subspace of euclidean spaces (i.e., a Cantor set). As a result, such a tight one-one continuous recoding as described above is only possible into (but never onto) fairly complex subsets of the interval (see Problem 7.9), as will be seen in this section.

Naturally, one should look first at well studied maps of the interval. They include unimodal, analytic, and piecewise linear functions.

Definition 9.9 *A function* $f : I^d \to I^d$ *is piecewise linear if*

- *f is continuous;*

- *there is a sequence* $(P_l)_{1 \leq l \leq n}$ *of convex closed polyhedra (of non-empty interior) such that each restriction* $f_l := f_{|P_l}$ *is affine,* $I^d = \bigcup_{l=1}^{n} P_l$ *and* $\overset{\circ}{P_p} \cap \bigcup_{l \neq p} \overset{\circ}{P_l} = \emptyset$ *for all l.*

PL_d denotes the set of continuous piecewise linear functions on I^d. In the real line, a convex polyhedron P_l is just an interval $I_l = [c_l, c_{l+1}]$ on which the corresponding affine map is of the form $f_l(x) := a_l x + b_l$. $RPL_1 \subset PL_1$ is the set of such functions with all rational coefficients a_l's, b_l's and c_l's. Piecewise analytic and piecewise monotone functions are defined in an analogous way. If f rather has only a countably infinite number of pieces, it is called a countably piecewise linear function, or simply ω-piecewise linear (or -analytic, -monotone, etc.). In this case, some of the I_l's may be reduced to a single point.

The question is to what extent these systems perform any kind of computation. This notion may admit manifold answers and each requires a precise definition. A first notion is real-time (step-for-step) simulation.

Definition 9.10 *Let T be the global map of a machine M (which may be a cellular automaton, a Turing machine, etc.). A map $f : I^d \to I^d$ simulates M if there is an f-stable subset $C \subset I^d$ (such that $f(C) \subseteq C$), and a bijective function $\phi : \mathbf{C} \to C$ such that*

$$T = \phi^{-1} \circ f \circ \phi .$$

The map f simulates M in real *time (respectively, linear time, quadratic time, ...) if a computation of M for t time steps can be performed by iteration of f on the corresponding recoding by ϕ in time t (resp. $O(t)$, $O(t^2)$, ...).*

Since the family of piecewise linear (resp. piecewise analytic, piecewise monotone) functions is closed under composition, linear-time simulation is equivalent to real-time simulation by the same type of map. Intuitively, this means that, in the recoding by ϕ, the iteration of the dynamical system on the subset C mimics (in one or several steps) the computation of the machine. Since C and the interval have their own topologies, the more interesting recodings are those consistent with the topological structure, i.e., are at least continuous (and so injective with a continuous inverse since $\mathbf{C}$ is compact) but can *never* be a homeomorphism.

Theorem 9.11 *There exist infinite families of cellular automata that cannot be simulated by a map of PL_2 with a continuous recoding.*

Proof. Let δ be the local map of a cellular automaton T with two distinguished "stable" states, 1 and 2, such that $\delta(*, q, *) = q$ for any choice of states $*$ and $q = 1, 2$. Let f be a piecewise monotone function simulating T. The set of configurations for which all cells are in a stable state is not countable, therefore one of these configurations is mapped by ϕ to the interior $\overset{\circ}{J}$ of an interval J on which f is monotone. A sequence of iterates of f_J has at most one accumulation point. It suffices to show that a half-line of sites can be simulated by iterating the monotone function f_J. This yields a contradiction for this type of δ because there are one-way infinite configurations which have an infinite number of accumulation points under, say, a left-shift of nonstable states (see Problem 11). $\square$

If we restrict our attention to finite configurations, however, cellular automata can be simulated by Turing machines in time linear in the size of the configuration. To what extent continuous dynamical systems are capable of discrete computation à la Turing has been explored. Although piecewise linear maps on the interval cannot simulate arbitrary Turing machines, they can simulate them on the square, one of several results that we state without proof – see Problem 10.

Theorem 9.12 *Under continuous recodings,*

1. *There exist Turing machines that cannot be simulated by a function of PL_1.*

2. *An arbitrary Turing machine can be simulated in linear time by an ω-piecewise linear function.*

3. *An arbitrary Turing machine can be simulated in linear time by a continuous piecewise monotone function.*

Thus, there is an intrinsic computational difference between piecewise linear euclidean dynamical systems on the line and on the plane. On the other hand, it is worth pointing out that cellular automata can be simulated by many continuous function on I. Indeed, since configuration space $\mathbf{C}$ is homeomorphic to the standard ternary Cantor set $\mathcal{C}$, a cellular automaton is readily conjugate to a continuous function on $\mathcal{C}$. Since the Cantor set is a compact subset of the interval I, this function can be continuously extended to all of I, even in a piecewise linear fashion outside $\mathcal{C}$ (which is a subset of measure 0). Turing and cellular network computation do indeed occur in the 'cracks' of the continuum.

There remains the question of whether actual physical systems (or rather, their traditional models in the continuum) do indeed perform Turing-like or cellular automaton-like computation. Traditionally, these systems are considered to be 'smooth' in the sense that they are differentiable infinitely many times (technically, C^∞). Computation universal smooth maps do exist even in dimension 1, of course over noncompact sets such as the entire real line. They can be conjugated easily into the interval, but at the cost of losing smoothness.

Theorem 9.13 *There exist computation universal maps smooth on the real line* **R**. *There exist computational universal maps of the interval that are smooth except at only one point.*

Finally, if one is willing to settle just for a partial degree of smoothness (differentiability up to some level) *over the entire interval*, the property can be had at the expense of relaxing the requirement of step-for-step simulation to allow simulation of one step by many steps (in an irregular fashion), while preserving the key ingredients in a computation (halting/nonhalting) as dynamical features (periodic/aperiodic points).

Definition 9.14 *Let* $f : X \to X$ *and* $g : Y \to Y$ *be dynamical systems. A* mono-simulation *of* f *in* g *is a continuous injective map* $\phi : X \to Y$ *such that*

$$\forall x \in X, \ \exists m > 0 : \quad \varphi(f(x)) = g^m(\varphi(x))$$

An epi-simulation *of* f *in* g *is a continuous surjective map* $\psi : Y \to X$ *such that*

$$\forall y \in Y, \ \exists m > 0 : \quad f(\psi(y)) = \psi(g^m(y))$$

It can be seen that mono- and epi-simulations preserve the set of periodic points (although not necessarily the length of the periods themselves).

Theorem 9.15 *Every Turing machine can be epi-simulated by an almost everywhere once-differentiable map of the unit interval* I *under a continuous recursive encoding.*

Theorem 9.16 *Every cellular automaton, every boolean random network, and, more generally, every continuous dynamical system on a Cantor set can be simulated by an almost everywhere linear map of the interval.*

The simulation, however, is no longer necessarily mono- or epi-. Nonetheless, it preserves fixed-point and infinitely transient (nonhalting) configurations.

9.3.2 Nonlocal Properties

The previous section shows various possible recodings of computation on discrete models into the dynamics of maps of the interval. The recoding preserves fundamental features of the computation. It is to be expected that facts and properties of dynamical systems in the continuum are correlated with properties of computation by discrete models. Perhaps the dynamical properties of cellular automata may be studied via this embedding. This section presents a special type of recoding that actually makes some automata conjugate to dynamical subsystems of the interval.

Even if generalized to neural and boolean networks, cellular automata can be regarded as continuous maps of the Cantor set C. Since the Cantor set topology is the induced topology from the unit interval in the real line, a *transform* can be defined that maps cellular automata to almost everywhere topologically conjugate dynamical systems on the full unit interval. This transform makes the two systems *semiconjugate* (i.e., the coding ϕ is surjective) when the resulting system on I is continuous. In this case the two dynamical systems share all dynamical properties, for instance they have the same entropy, the Sarkowskii property, and in general, any nonlocal properties (in a precise sense to be defined below). Roughly speaking, the main result is a transfer metatheorem for certain properties of dynamical systems on the unit circle to properties of cellular automata.

9.3.2.1 The ϕ-transform Every element x in the Cantor set C can be expressed by a ternary expansion not containing the digit 1, i.e.,

$$x = \sum_{j \geq 1} x_j t_j \, ; \ \ t_j := \frac{2}{3^j} \, ,$$

where $x_j = 0$ or 1. Let $\phi : C \to I$ be the map given by

$$\phi(x) \ = \ \sum_{j \geq 1} x_j b_j \, , \ \ b_j := \frac{1}{2^j} \, . \tag{9.3}$$

Thus a Cantor set point with finite (periodic) expansion is mapped by ϕ into a dyadic rational number with finite (periodic) binary expansion. For example,

$$\phi(0) \ = \ 0$$
$$\phi(\tfrac{1}{3}) \ = \ \phi(\tfrac{2}{3}) = \frac{1}{2}$$
$$\phi(\tfrac{1}{9}) \ = \ \phi(\tfrac{2}{9}) = \frac{1}{4}$$
$$\phi(\tfrac{7}{9}) \ = \ \phi(\tfrac{8}{9}) = \frac{3}{4} \, , \ \text{etc.}$$

In general, ϕ successively "closes" up the gaps of C in the unit interval by piecewise affinely collapsing each of the middle-third intervals excised in the construction to the midpoint of arising subintervals of I, starting with collapsing $[\tfrac{1}{3}, \tfrac{2}{3}]$ to $\tfrac{1}{2}$. Hence ϕ is continuous and bijective at points with infinite expansion of 0's and 2's, and at most two-to-one on *end-points* of the ternary Cantor set.

Clearly ϕ is not bijective, but it is surjective, and so it has right inverses. There is a way to parametrize the uncountably many right inverses of ϕ as follows. Let $\{I_n\}$ be some enumeration of the intervals successively deleted in the standard construction of the Cantor set. Identify the end-points of I_n with 0 and 1 (say $1 = left$ and $0 = right$). Two end-points of an I_n (equal or distinct) will be called *adjacent* (0 and 1 are considered adjacent as well.) The family of inverses of ϕ are in one-one correspondence with binary expansions x of points in I. ψ_x will denote the inverse of ϕ obtained by choosing x_n as the inverse at the corresponding dyadic point. Every choice of inverse images for finite dyadic rational points gives rise to a *right-inverse*

$$\psi : I \mapsto C$$
$$\psi(\sum_{j\geq 1} x_j b_j) = \sum_{j\geq 1} x_j t_j . \tag{9.4}$$

(Note that ψ is well-defined despite the fact that a real number $x \in I$ may have more than one binary expansion $x = \sum_{j\geq 1} x_j b_j$.)

Definition 9.17 *Let ϕ be a continuous recoding from C to the real unit interval I. The ψ-transform associates to every cellular network $T : C \mapsto C$ a dynamical system $T_\psi : I \to I$ on the interval given by the composition*

$$T_\psi(x) = \phi(T(\psi(x))) . \tag{9.5}$$

Likewise, the ψ-transform into the unit circle S^1 maps a cellular network T satisfying $T(O) = T(1)$ to the quotient map of T_ψ obtained by identifying the end-points of I.

Conversely, the inverse ψ-transform associates to a mapping $f : I \to I$ (or on the unit circle S^1) a map in the Cantor set $f_\psi : C \to C$ given by the composition

$$f_\psi(x) = \psi(T(\phi(x))) . \tag{9.6}$$

It follows at once from Definition 9.17 that the diagram

$$
\begin{array}{ccc}
 & T & \\
C & \longrightarrow & C \\
\phi \downarrow & & \downarrow \phi \\
I & \longrightarrow & I \\
 & T_\psi &
\end{array}
\tag{9.7}
$$

commutes except maybe at those configurations belonging to an adjacent pair. This section deals with the case where ϕ actually gives rise to a full semiconjugation of the two systems T and T_ψ for some choice of inverse ψ. This is not always possible – see Proposition 9.20 below.

Definition 9.18 *Let* $\phi : C \to I$ *be a continuous mapping on a Cantor set* C. *A cellular automaton* $T : C \mapsto C$ *is* ϕ-*continuous if the* ψ-*transform* T_ψ *is continuous for every* ψ.

In order to characterize the nature of ϕ-continuity we introduce the following notions. For the sake of illustration, the encoding of C *into* the middle-third Cantor set defined by the following homeomorphism will be used below. Given a binary configuration $c = (c_i)_{i \in \mathbf{Z}}$, let h_1 be the Cantor set point whose ternary expansion $(x_i)_{i \geq 1}$ is given by

$$x_i := h_1(c)_i \quad := \quad \begin{cases} 2c_{i/2}, & i \text{ even} \\ 2c_{-(i+1)/2} & \text{else} \end{cases} .$$

Thus h_1 conjugates every continuous self-map of C onto a continuous self-map of C and vice versa. For instance, the *1D left-shift (right-shift) of* I is defined as the ψ-transform σ_Ψ, where σ is the corresponding right shift on C, and it will be denoted simply σ_1. It is interesting to note that for every initial point x with a finite expansion, iteration of σ_1 or σ_1^{-1} is eventually given by

$$\sigma_1^t(x) \quad = \quad \frac{1}{4}x$$

These maps are well known in the dynamical systems literature as *Baker maps.*

The foregoing encoding of C to C can be readily modified to an encoding of C to $C \times C$, then to the unit square $I \times I$ through ϕ componentwise. One then obtains a 2D ψ-transform. These alternatives may not be fruitful, however, since not much is known in general about dynamical systems on the square or higher dimensional euclidean spaces.

The following properties of ϕ follow easily from the definitions in Sect. 9.3.2.1.

Proposition 9.19 *The following properties hold:*
(1) ϕ *is continuous.*
(2) *for any choice of* x, ψ_x *is continuous except possibly in some set of dyadic points.*
(3) T_{ψ_x} *is Lebesgue integrable for any choice of* x, *and the value of the integral is independent of* ψ.

Proof. The map ϕ is the restriction of the Cantor–Lebesgue function to the Cantor set, hence is it continuous since the topology in C is the topology induced from the unit interval's. By deleting all pairs of adjacent points of the unit interval, the restriction of ϕ is still continuous, and thus ψ_x is also continuous except possibly at dyadic points. This proves (2). The third statement is an immediate consequence of the second.   □

The question of under what conditions T_ψ is continuous can be answered as follows.

Proposition 9.20 *If $T : C \to C$ is continuous, then the following conditions are equivalent:*

(1)   *T is ϕ-continuous.*
(2)   *T_ψ is continuous for some ψ.*
(3)   *T preserves adjacency (i.e., if x, y are adjacent, so are $T(x), T(y)$).*
(4)   *T_ψ is independent of ψ.*
(5)   *Diagram (9.7) commutes, i.e., $T_\psi \circ \phi = \phi \circ T$, for some (every) ψ.*
(6)   *T and T_ψ are topologically semiconjugate for every ψ.*

In this case, T_ψ will simply be called the ϕ-transform of T, and will be denoted T_ϕ.

Proof. (1) $\Rightarrow$ (2). Obviously.

(2) $\Rightarrow$ (3). Two end-points $a < b$ are adjacent iff $\phi(a) = \phi(b)$. Assume $T(\phi(a)) \neq T(\phi(b))$. In order to prove that T_ψ is discontinuous for every inverse ψ, assume $\psi\phi(a) = a$ (if it is b the argument is analogous). Since $T_\psi = \phi T \psi$, for any ψ, $T_\psi\phi(a) = \phi T(a)$. Let (x^n) be a decreasing sequence in I converging to $\phi(a)$ so that $\psi(x^n)$ converges to b. Since ϕ and T are continuous, $\phi T\psi(x^n)$ converges to $\phi T(b)$. Therefore T_ψ is discontinuous at $\phi(a)$).

(3) $\Rightarrow$ (1). It suffices to show that T_ψ is continuous at dyadic points for every ψ. Let y be a dyadic point with $\phi(a) = \phi(b) = y$ and $a < b$. Now

$$\lim_{x \to y^-} T_\psi(x) = \phi T(a),$$

and likewise

$$\lim_{x \to y^+} T_\psi(x) = \phi T(b).$$

Since T preserves adjacent points, T_ψ is continuous at y, for every y.

(2) $\Rightarrow$ (4) It is enough to verify equality at dyadic points y. Let ψ_1, ψ_2 be two right inverses of ϕ. Since $\psi_1(y)$ and $\psi_2(y)$ are adjacent and T preserves adjacent points,

$$T_{\psi_1}(y) = \phi T\psi_1(y) = \phi T\psi_2(y) = T_{\psi_1}(y),$$

i.e., T_ψ is independent of ψ.

(4) $\Rightarrow$ (5). Diagram (9.7) always commutes at nonend-points. If a is an end-point, and $\psi(\phi(a)) = a$, applying ϕT proves the desired equality. Otherwise, choose ψ' with $\psi'(\phi(a)) := a$. Since T_ψ is independent of ψ, it follows that

$$\begin{aligned} T_\psi\phi(a) &= T_{\psi'}\phi(a)) \\ &= \phi T\psi'\phi(a) = \phi T(a). \end{aligned}$$

Therefore diagram (9.7) commutes.

(5) $\Rightarrow$ (3) If $\phi(a) = \phi(b)$, since the diagram commutes

$$\begin{aligned} \phi T(a) &= T_\psi\phi(a) \\ &= T_\psi\phi(b) = \phi T(b), \end{aligned}$$

i.e.,, T preserves adjacency.

Finally, (5) and (6) are clearly equivalent.   □

One may wonder at this point about the existence of cellular automata satisfying the conditions of Proposition 9.20. The following construction provides a large number of 1D ϕ-continuous cellular automata for which the results below apply. We illustrate with a typical example.

Given two subintervals J, K of I, let $r_{J,K}$ map J affinely onto K after reflecting J about its midpoint (i.e., it maps the left end of J to the right end of K and vice versa). Let $r_J := r_{J,J}$. Let $r_{1,2} : C \to C$ be the self-map of C defined as follows. On $J_1 := [0, \frac{1}{9}]$, r acts as (the restriction of) r_{J_1} (to C). Likewise on $J_3 := [\frac{2}{3}, \frac{7}{9}]$. But on the remaining set, r acts as $r_{[\frac{2}{9}, \frac{1}{3}],[\frac{8}{9},1]}$ $(x < \frac{1}{2})$ or its inverse (otherwise). A picture shows immediately that $r_{1,2}$ preserves adjacent points, and hence its ϕ-transform is continuous. For a ternary expansion of $x \in C$, r flips every digit, except the first two, where 01 and 11 are mapped to 11 and 01, respectively, but 00 and 10 are unchanged. (Thus r is not defined by a local rule *on the standard Cantor set*). Now it is necessary to prove that $r_{1,2}$ is, in fact, a cellular automaton.

Now let T be the 1D cellular automaton defined on 8 symbols $\{0 \equiv 000, 1 \equiv 001, 2 \equiv 010, \ldots, 7 \equiv 111\}$ as follows. Decode each Cantor set point via h_1^{-1} and then replace each bit b with $b11$ if it comes from the first or second digit of the expansion in the Cantor set, $b10$ otherwise (the middle bit is going to indicate which case (flip or not) of the above rule is being applied, the third one which is the case for the two exceptional cells). Let A be the set of symbols of form $*10$ or $*11$. The map T is defined by a *local* rule δ transforming each symbol in A exactly the same way $r_{1,2}$ does (modulo the encoding), and mapping every other state to 0 otherwise. Since T is defined by a local rule mapping O to O, T is a cellular automaton on 8 symbols. By construction, T can be conjugated onto $r_{1,2}$ on C and therefore $r_{1,2}$ is indeed a cellular automaton. With little additional effort, it can be converted into a binary 1D cellular automaton with a neighborhood of radius 6.

9.3.2.2 Sarkovskii's Theorem

The continuity of T_ψ allows one to transfer properties of real-valued dynamical systems to certain cellular automata. As a typical example consider Sarkovskii's theorem. The *Sarkovskii order* is the decreasing ordering of the natural numbers given by the sequence

$$3 \quad \triangleright \quad 5 \quad \triangleright \quad 7 \quad \triangleright \quad \ldots \quad 2.3 \quad \triangleright \quad 2 \cdot 5 \quad \triangleright \quad 2 \cdot 7 \quad \triangleright \quad \ldots$$
$$2^2 \cdot 3 \quad \triangleright \quad 2^2 \cdot 5 \quad \triangleright \quad 2^2 \cdot 7 \quad \triangleright \quad \ldots \qquad \qquad \cdots$$
$$2^n \cdot 3 \quad \triangleright \quad 2^n \cdot 5 \quad \triangleright \quad 2^n \cdot 7 \quad \triangleright \quad \ldots \quad \cdots \quad \triangleright \quad \cdots 2^3 \quad \triangleright \quad 2^2 \quad \triangleright \quad 2 \, .$$

This ordering is closely related to the behavior of periodic orbits of maps on the real line. A celebrated result, Sarkovskii's theorem asserts that if a continuous dynamics $T : \mathbf{R} \mapsto \mathbf{R}$ has a point of period p_0 then it has a point of every period $p \trianglerighteq p_0$ in Sarkovskii's order.

On the other hand, it has been shown in Sect. 9.1.2 that even *linear* cellular automata on the line may have an infinite number of forbidden periods, e.g., the linear map sum of 2 pixels $T = \sigma^0 + \sigma^m$ has no periodic points of period a power of 2. This means that Sarkovskii's statement does not hold for cellular automata in general. However, the property holds for ϕ-continuous cellular automata.

Theorem 9.21 *If a cellular automaton is ϕ-continuous and it has a configuration of period $p_0 \trianglerighteq 3$ in Sarkowskii's order, then it has points of every period $p \trianglerighteq 2p_0$, except possibly 2 or 4.*

In particular, the linear cellular automaton given by the pointwise sum $T(x)_i := x_i + x_r$ mentioned in Sect. 9.1.2 is not ϕ-continuous.

Theorem 9.21 follows from the next two lemmas.

Lemma 9.22 *Let T be a ϕ-continuous cellular automaton.*
  (1) If T has a periodic point of period t then T_ϕ has a periodic point of period t or $t/2$.
  (2) If T_ϕ has a periodic point of period t then T has a periodic point of period t or $2t$.

Proof. If $T^t(x) = x$, then $T_\phi^t \phi(x) = \phi T^t \psi \phi(x) = \phi(x)$. Hence the period k of $\phi(x)$ divides t, say $t = qk$. Since $T^k(x)$ and x are adjacent, so are $T^{2k}(x)$ and $T^k(x)$. Hence $T^{2k}(x) = x$ or $T^{2k}(x) = T^k(x)$. In the first case, $k = t$ or $k = t/2$. In the second,
$$x = T^{(q-2)k} T^{2k}(x) = T^{(q-1)k}(x),$$
which is impossible since x has period t and $(q-1)k < t$. The proof of (2) is similar and will be omitted. □

The *degree* of a map f of the unit circle can be roughly described as the number of times the range of f goes around the circle (positive if clockwise, negative if counterclockwise) as its argument increases around the circle once. The proof of the following result can be found in [B-G-M-Y].

Lemma 9.23 *Let $f : S^1 \to S^1$ be a continuous self-map of the unit circle.*
  (1) if $|\deg(f)| > 1$ then f has periodic points of all periods, except only period 2 in case $\deg(f) = -2$;
  (2) if f has a fixed-point and a periodic point of period n_0, then it has periodic points of all periods $n \triangleright n_0$.

The ψ-transform also preserves a number of dynamically significant properties. For instance,

Theorem 9.24 *The ϕ-transform preserves topological entropy.*

Proof. First prove that $h(T) \geq h(T_\psi)$. Every (n, ε)-spanning set $E_{(n,\varepsilon)}$ for T of minimal cardinality yields a spanning (n, ε') set $\phi(E_{(n,\varepsilon)})$ for T_ψ, where
$$|x - y| < \varepsilon \Rightarrow |\phi(x) - \phi(y)| < \varepsilon'.$$

In fact, given $x \in I$ there exists $a \in E_{(n,\varepsilon)}$ such that

$$|T^j \psi(x) - T^j(a)| < \varepsilon , j = 0, 1, \ldots, n - 1 .$$

Therefore, for each $j = 0, 1, \ldots, n - 1$,

$$|\phi T_\psi^j \psi(x) - \phi T_\psi^j(a)| \quad = \quad |T_\psi^j \phi(x) - T^j \phi(a)| < \varepsilon'$$

Thus any minimal (n, ε')-spanning set $F_{(n,\varepsilon')}$ for T_ψ satisfies $T(n_{\varepsilon'}) \le (n_\varepsilon)$. It follows that $ent(T_\psi) \le h(T)$.

The converse follows from an application of the following known result due to Bowen [Bo1, Theorem 17].

Theorem 9.25 *Let $(X, d), (Y, d')$ be compact metric spaces and $T : X \to X$, $s : Y \to Y$, and onto $\pi : X \to Y$ be continuous maps. If $\pi \circ T = s \circ \pi$, then*

$$ent_d(T) \le ent_{d'}(s) + \sup_{y \in Y} ent_d(T, \pi^{-1}(y)) .$$

If moreover X and Y are compact, then $ent_d(T) = ent(T)$ and $ent_{d'}(s) = ent(s)$.

Taking $X := \mathcal{C}, Y := I$ and $\pi := \phi$, ϕ^{-1} has at most two points, so $ent_d(T, \phi^{-1}(y)) = 0$ for all $y \in I$ and therefore $ent(T) \le ent(T_\Psi)$. $\qquad\square$

Theorems 9.21 and 9.24 naturally raise the question of what other properties can be imported to dynamical systems on Cantor sets. Toward this end, let Ω be a dynamical property, i.e., a property left invariant under homeomorphisms. Say that a property Ω of a compact perfect space X is *localizable* if X has the property Ω but there exists a countable dense subset E such that the subspace $X - E$ does not have Ω. Otherwise Ω is said to be *nonlocalizable*. The following properties are nonlocalizable: sensitive dependence on initial conditions [D], chaoticity (in the sense of Li & Yorke's [L-Y]), chain recurrence, topological mixing, etc.

Theorem 9.26 *Every nonlocalizable property Ω is invariant under ϕ-transforms, i.e., if a cellular automaton T is ϕ-continuous, then T has Ω if and only if T_ϕ does.*

The proof requires use of the fact that the Cantor set is *countable-dense homogeneous*, i.e., given any two countable dense subsets X_1, X_2 of $\mathcal{C}$, there exists a homeomorphism φ of the Cantor set such that $\varphi(X_1) = X_2$. (A proof of this fact can be found in [Be].)

Proof. Let E be a countable dense subset of I, so $\phi^{-1}(E)$ is also a countable dense subset X of $\mathcal{C}$. Since $\mathcal{C}$ is countable-dense homogeneous, assume without loss of generality that X is the set of endpoints. Since Ω is nonlocalizable, $T|_{\mathcal{C}-X}$ satisfies Ω. Therefore, $T_\phi|_{I-X}$ also satisfies Ω. $\qquad\square$

9.4 Observability

The evolution of cellular automata and networks is usually observed in computer simulations on parallel and sequential machines of various types. Even in special-purpose cellular automata type machines, an infinite number of cells is out of the question. Even on a large finite number of cells, the corresponding iteration on actual implementations might introduce at each step negligible errors that necessarily accumulate under iteration, perhaps propagating out of control. Thus, it is quite clear that, in general, a computer simulation of an orbit of a given dynamical system (referred to in the sequel as a *pseudo-orbit*) is, in fact, far from the true orbit. Hence it is important to know when a numerical process is actually approximated by a trajectory of the real phenomenon.

In the classical theory of dynamical systems, this problem motivated the study of the so-called shadowing property. Roughly speaking, systems satisfying the shadowing property are *globally observable* in the sense that their pseudo-orbits on real computers are uniformly approximated by actual orbits so that the long-term orbital behavior is to some extent captured by pseudo-orbits. The shadowing property thus seems to be a desirable long-term property of dynamical systems.

This section presents several results on this other aspect (see the discussion in Chapter 6) of the general problem of simulation and implementation. First, the basic question of whether the identity map has the property. Unlike the situation with real spaces, the identity does, and, as a matter of fact, it characterizes totally disconnected compact spaces. Now, if 1D global dynamics are toggle or linear, they have the shadowing property. It follows that corresponding maps induced by neural networks over activation sets which form a group have the property. However, not all types of even cellular automata have the property.

Some notation and precise definitions follow that aim at including the case of the continuum as well. All results in this section hold for distance and a map of a compact metric space X (such as the unit interval of the real line or configuration space) unless more specifically stated. Also, the distinction between x_i, the state of site i, and the nth element x^n of a sequence will reside in the nature of the subindex (i, j, k for sites; l, m, n, t for positive integers). Recall that a sequence $\{x^n\}_{n \geq 0}$ is an orbit if and only if $f(x^n) = x^{n+1}$, for $n \geq 0$.

Definition 9.27 *Let $f : X \to X$ be a continuous map of a compact metric space with metric $|*, *|$. Given a number $\delta > 0$, a δ-pseudo-orbit is a sequence $\{x^n\}$ so that the distances $|f(x^n), x^{n+1}| < \delta$ for all $n \geq 0$. The map f has the shadowing property, or is (globally) observable, if and only if for every $\varepsilon > 0$ there exists a $\delta > 0$ so that every δ-pseudo-orbit $\{x^n\}$ is ε-approximated by the orbit, under f, of some point $z \in X$, i.e.,*

$$\forall n \geq 0, \ |x^n, f^n(z)| < \varepsilon.$$

9.4.1 Observability of the Identity

The simplest question concerns the shadowing property of the identity function. Surprisingly, the identity map defined on a real closed interval $[a, b]$ does not have the shadowing property. In fact, just consider $\varepsilon := |b - a|/4$. For no $\delta > 0$ is there an orbit that ε-approximates the pseudo-orbit $\{x_n\}$ defined by $x_n := a + n\min\{\varepsilon/2, |b - a|/4\}$ if $a + n\min\{(\varepsilon/2, |b - a|/4\} \leq b$, and b otherwise.

Theorem 9.28 *The identity map* id *of a compact metric space X is observable iff X is totally disconnected.*

Proof. Assume X is not totally disconnected. Let Ω be a connected component with positive diameter of a point $a \in X$. Since the closure of a connected set is connected, Ω is closed. Choose two distinct points $x, y \in \Omega$ and put $\varepsilon := |x, y|/4$. For any $\delta > 0$ and any two points $\alpha, \beta \in \Omega$, there exists a finite chain of open balls $B_1, \cdots, B_k$ of radii $\delta_0 > 0$, where $\delta_0 := \min\{\varepsilon/4, \delta/2\}$, such that $\alpha \in B_1$, $\beta \in B_k$ and $B_i \cap B_{i+1} \neq \Phi$ for any two consecutive balls B_i, B_{i+1}. Let $z_i \in B_i \cap B_{i+1}$. Obviously $x, z_1, z_2, \cdots, z_{k-1}, y, y, \cdots$ is a pseudo-orbit which is not ε-traceable.

For the sufficiency, let $\varepsilon > 0$ and $\{B(x, \varepsilon/2) : x \in X\}$ be a covering of X. For each x there exists a closed and open subset C_x such that $x \in C_x \subset B(x, \varepsilon/2)$ since $B(x, \varepsilon/2)$ is disconnected. Since $\{C_x : x \in X\}$ is a covering of X, there exists a finite subcover $C_{x^1}, \cdots, C_{x^k}$ of closed and open subsets. Now define

$$U_1 := C_{x^1}, U_2 := C_{x^2} - C_{x^1}, \quad \cdots \quad U_k := C_{x^k} - \left(\bigcup_{j<k} C_{x^j}\right).$$

The sets $U_1, \cdots, U_k$ are *clopen* (i.e., open and closed), disjoint, and cover X. Let $0 < \delta := \min\{\varepsilon, |U_i, U_j| : i \neq j\}$ (the distance between two sets A and B is the infimum of distances $|x, y|$ between pairs $x \in A$ and $y \in B$). Every δ-pseudo-orbit $\{x^n\}$ of **id** lies entirely in one (and only one) C_{x^j}. Since their diameter

$$diam(C_{x^j}) < \varepsilon,$$

for any point $y \in C_x$, $|x^n, y| < \varepsilon$, as required. $\square$

Remark. Note that X being totally disconnected is, moreover, necessary and sufficient for **id** to have the asymptotic shadowing property (where pseudo-orbits are traceable only asymptotically). The pseudo-orbit

$$x, z_1, z_2, \cdots, z_{k-1}, y, z_{k-1}, \cdots, z_1, x, z_1, z_2 \cdots, \cdots$$

is not asymptotically traceable (with x at the appropriate position depending on the threshold N).

9.4.2 Toggle Rules and the Extension Property

It will be shown in Sect. 9.4.3 that cellular automata in higher dimensions can be analyzed as (nonhomogeneous) neural networks *on the line* with neighborhoods uniformly bounded in size, although they are not necessarily local and/or homogeneous. Consequently, this type of network is explored first.

Definition 9.29 *Let T be a global map of configuration space in the line. A site j T-influences site i if there exist $x, y \in C$ such that $x_l = y_l$ for all $l \neq j$ but $T(x)_i \neq T(y)_i$. The effective neighborhood N_i of a site i is the set of indices*

$$N_i(T) := \{j : j \ T\text{-influences} \ i\}.$$

(Reference to T will be dropped if T is clear from context.)

It is only necessary to consider locally finite maps T, for which all the N_i's are finite. Note that a rule is *strictly local* per the definition in Sect. 7.4.1 iff $N_i \subseteq \{i\}$ for every site i (see Problem 7.18).

Lemma 9.30 *Every strictly local CA has the shadowing property.*

Proof. Since there is no interaction between or across sites, every ϵ-pseudo-orbit can be ϵ-traced by the orbit of the first element. □

Definition 9.31 *The i-range $R(T)_i$ of a global map T is the projection of the range $R(T)$ of T to site i. The product range of T is the cartesian product $\prod_i R(T)_i$ of all i-ranges of T. In this subspace with the induced metric, a closed ball will be denoted $PB[x, \epsilon]$. A map T is i-toggle at position $j \in N_i$ iff the restriction of T to $R(t)_j$ is surjective onto $R(T)_i$ for arbitrarily fixed assignments of the site values outside of site j. The map T is toggle iff for each i, T is i-toggle at some site from each of the sets*

$$N_i^* - \bigcup_{j<i} N_j \quad \text{and} \quad N_i^* - \bigcup_{j>i} N_j, \tag{9.8}$$

where N_i° denotes $N_i - \{i\}$. The neighborhoods N_i are said to be left-threaded (respectively right-threaded, threaded) if the first (second, both) neighborhood condition(s) (9.8) hold(s) for every site i.

Example 9.32 The XOR rule on any number of neighbors in binary is i-toggle at any position. The left-shift is i-toggle at position $i + 1$ on the 1D integer grid, but not at any other position. In fact, every binary linear rule is toggle at positions on which the next-state of the center cell effectively depends. Later, we will use the fact that the cartesian product of two toggle rules is toggle, as can be easily checked.

The following result is used to prove that toggle maps have the shadowing property.

Lemma 9.33 *Let X be a compact metric space and T a continuous map on X. If for every $\epsilon, \delta > 0$ the map T satisfies*

$$\forall x \in R(T), \; PB[T(x), \epsilon + \delta] \; \subseteq \; T(PB[x, \epsilon]), \tag{9.9}$$

then every δ-pseudo-orbit can be ϵ-traced by an orbit of T. In particular, if for every ϵ there exists such a δ, T has the shadowing property.

A *pseudoblock* B is the restriction of a 1D configuration x to a set of sites (called its *support* and denoted $\underline{B}$). It is a *block* if the sites are finite in number and contiguous. There is an obvious operation of concatenation between pseudoblocks B, B' with disjoint supports, simply denoted BB', whose support is the union of the supports of B and B'. The action of T restricts to pseudoblocks B as follows. The image $T(B)$ is a pseudoblock with support $\{i : N_i \subseteq \underline{B}\}$ so that $T(B)_i := T(x)_i$, where x is any configuration such that x agrees with B on $\underline{B}$. In this case, the pseudoblock $T(B)$ is said to be *in the range of T*.

Definition 9.34 *A global map T satisfies the* extension property *(EP) if for every pair of pseudoblocks B, D with the supports of $T(B)$ and D disjoint, there exists a pseudoblock B' whose support is disjoint from $\underline{B}$ such that*

$$T(BB') = T(B)D.$$

The EP is said to hold on blocks *(EPB) if this property holds only wherever $T(B), D$ and $T(B)D$ are blocks.*

Proposition 9.35 *Every toggle map has the extension property on blocks.*

Proof. By induction on the cardinality of $\underline{D}$. If $\underline{D} = \{k\}$ and k is an extra cell to the right of $\underline{T(B)}$, by toggleness, T is k-toggle at some cell $i \in N_k^\circ - \bigcup_{j<k} N_j$. Therefore there exists an extension of the pseudoblock B as desired. The inductive step follows likewise. $\square$

Proposition 9.36 *Every toggle map such that either $i \in N_i$ or $i < j$ (respectively, $i > j$) for all $j \in N_i$ is observable.*

Proof. It suffices to verify condition (9.9). We establish the statement under the hypothesis $\imath \in N_i$ since the other one follows analogously. For given $\epsilon > 0$, the condition is verified with $\delta := \epsilon$ for the distance given by $\frac{1}{2^n}$, where n is the largest positive integer such that $x_{[-n,n]} = y_{[-n,n]}$. Observe that this distance decreases exponentially as x, y agree on larger site blocks. Let $u \in PB[T(x), 2\epsilon]$ agree with $T(x)$ on a maximal block B with support $[-n, n]$. Let $\underline{B}^-$ denote the pseudoblock obtained by adding to $\underline{B}$ all sites which influence a site in B. Let $D := u_{[-n-1, n+1]}$. By toggleness, there exists a pseudoblock B' that coincides with x in the whole interval $[-n-1, n+1]$ whose image is D. The extension property now guarantees a configuration y that coincides with x on $[-n-1, n+1]$ and such that $T(y) = u$. Clearly $|x, y| \leq \epsilon$, as desired. $\square$

9.4.3 Observability of Linear Cellular Automata

Recall from Chapter 3 that a generalized superposition principle can be defined where the set Q of states is endowed with a binary operation. (For instance, an abelian group structure makes it a modulo over the ring of integers.) The operation can be extended componentwise to the entire configuration space, herein indicated by simple concatenation. A self-map of configuration space is *linear* if it satisfies the superposition principle under this operation, i.e., if

$$\forall x, y \in \mathbf{C}, \ \ T(xy) = T(x)T(y).$$

With an identity element, the superposition principle holds iff the operation is associative and commutative according to Proposition 3.13. In this section we prove that linear maps in a much broader class than cellular automata satisfy the shadowing property using condition (9.9).

In order to simplify a general treatment, higher dimensional grids are mapped to the 1D grid in any of many known ways. Of necessity, the encoding does not preserve locality, but it does preserve linearity. This mapping destroys locality in the sense that the next-state of a site in the line depends on sites that are arbitrarily far away but correspond to neighboring sites in the plane. However, they can be still called neighbors in the sense of Definition 9.29.

Theorem 9.37 *Every linear cellular automaton over an abelian group of states is observable.*

Note that there are linear maps without shadowing (of course, they are not continuous, see Problem 14).

Theorem 9.37 follows from the following intermediate results. First, we prove that

Proposition 9.38 *Every linear continuous map with uniform neighborhoods including the center cell over an abelian group whose order is square-free satisfies condition (9.9).*

Second, continuity of linear maps is characterized as follows.

Proposition 9.39 *A linear map T is continuous iff every N_i is finite.*

The shadowing property is in general not preserved under uniform limits. Still, this statement is true for the maps involved in the proof of the following result.

Proposition 9.40 *Every locally finite (respectively, linear) map is a uniform limit of (linear) maps with uniformly bounded neighborhoods.*

Proof (of Proposition 9.38). First, note that all the neighborhoods are threaded. Second, it suffices to prove the result when the number of states is prime since every finite abelian group is the direct sum of abelian groups of prime order. A linear map T thus splits into a cartesian product $T' \times T''$ of linear maps which is obviously toggle if each T' and T'' are toggle.

We prove the result for a prime number of states by contradiction. Assume T is not i-toggle and let $k \in N_i$ and a pseudoblock B with support $N_i^k := N_i - k$ be such that $T(B)_{ki} : R(T)_k \to R(T)_i$ is not surjective for some choice of states in N_i^k, i.e. $T(B)_{ki}(r) = T(B)_i(s)$ with $s \neq r$. By superposition, $T(O)_i(r-s) = 0$. Since $r - s$ is a generator of S, $T(B)_{ki}$ is constantly 0 when N_i^k is all 0. This contradicts that k influences i, and hence, T is toggle. Therefore, the result follows from Proposition 9.36. □

Proof (of Proposition 9.39). Recall that T is continuous iff whenever a sequence $\{x^n\}$ converges to x, the sequence $T(x^n)$ also converges to $T(x)$. If N_i contained an infinite sequence of sites $\{j_n\}$, the sequence given by $x^n := \sum_{j_k \leq j_n} x_{j_k} e^{j_k}$ (where e^j denotes the pixel configuration given by $e_i^j := \delta_{ij}$) would converge to some configuration x, but the sequence $\{T(x^n)_i\}$ would not have a limit since each $T(e^{j_k})_i$ is nonzero. Therefore N_i must be finite. The converse is an obvious consequence of the nature of the product topology. □

Proof (of Proposition 9.40.) For the given map T and $n \geq 1$, consider the rules T_n with neighborhood $N_i(T_n) := [i - n, i + n]$ given by $T_n(x)_i := T(x_{[i-n,i+n]})_i$, where $x_{[i-n,i+n]}$ is obtained from x by replacing the values of x with 0 outside of $[i - n, i + n]$. Note that $|N_i(T_n)| \leq 2n + 1$. Because T is locally finite, eventually $N_i(T) = N_i(T_n)$ for all i. Therefore, $|T(x), T_n(x)|$ becomes arbitrarily small uniformly on x, as $n \to \infty$. Under these conditions, it is clear that the T_ns are linear if and only if T is linear. □

Proof (of Theorem 9.37). Every abelian finite group is a direct sum of abelian groups of prime power order, so the result will follow from Proposition 9.38 and the case where the number of states is the power of a prime. In this case, it is known that some composition power of T is either identically 0 or all its nonzero coefficients are units (see the proof of Theorem 3.10), which implies that T is toggle. In both cases, T is observable (see Proposition 9.36). □

Theorem 9.37 is, in a sense, best possible since linear automata over semi-groups (even the *multiplicative* integers modulo 2) may not be observable, as the following Example 9.41 shows.

Example 9.41 (See Problem 8.8). The one-way 1D binary cellular automaton over $\mathbf{Z}_2$ given by

$$T(x)_i := x_{i+1} x_{i+2}$$

is *not* observable. It is easy to see that if an initial configuration x^0 has a first value $x_k = 0$ to the right of the origin, then the cell at the origin will see a cone

```
11111111111011111111111111011111111
11111111100111111111111001111111
111111100011111111111000111111
1111100001111111111000011111
11100000111111111000001111
10000001111111111000000111
00000011111111000000011
00000111111000000001
000011111000000000
00011110000000000
0011100000000000
011000000000000
10000000000000
0000000000000
```

Fig. 9.1. The space–time of a nonobservable cellular automaton

of at least $2t - 1$ zeros before seeing any 1s again, after *the first* 0 arrives at the origin $\lceil \frac{k}{2} \rceil$ steps later, as illustrated in Fig. 9.1. By adding noise periodically outside the window about the origin determined by δ, a δ-pseudo-orbit will force a periodic temporal sequence of states at the origin. Thus for $\varepsilon := 1$, for every $\delta > 0$ it is possible to find a pseudo-orbit which cannot be ε-traced by T. Since an infinite block of 1s to the right travels left, a tracing point must contain 0 arbitrarily far from the origin. Hence no initial condition z will have an orbit with successive states arrive at the origin cell in the right sequence so as to match those generated by the noise added since for large enough time, the window of 0s will be too large to agree with the given pseudo-orbit.

9.4.4 Observability in Neural Networks

We are finally ready to state and prove the main result of this section. For the proof we need the following result which is a slight variation of Theorem 7.17.

Theorem 9.42 *A continuous self-map of configuration space is induced by a locally finite neural network if and only if*

1. *T is continuous;*

2. *The support of every pixel's image $T(e^j)$ is finite;*

3. *T is the composition $T = F \circ L$, where F is strictly local and L is linear.*

Theorem 9.43 *Every neural network with uniform neighborhoods including the center cell over an abelian group of states whose order is square-free is observable.*

Proof. Let F and L be maps as in Theorem 9.42 and let L also satisfy the hypotheses in Proposition 9.38 for a given neural network T. Since F is strictly local

$$PB[F(x), \epsilon] \subseteq F(PB[x, \epsilon]).$$

Since L satisfies condition (9.9) it follows that,

$$PB[F \circ L(x), 2\epsilon] \subseteq F \circ L(PB[x, \epsilon]).$$

Now it is easy to see that condition (9.9), and hence the shadowing property, holds for $F \circ L$, as well.   □

Since cellular automata are particular cases of neural networks, it follows in particular that the same type of cellular automata on Cayley graphs are observable.

Theorem 9.44 *Every linear cellular automaton whose state set is an abelian group is observable, i.e., has the shadowing property.*

The class of (even linear) observable cellular automata and neural networks does not appear simple to describe. Only a necessary condition is known, which we state without proof. Let $W(\varepsilon, t)$ be rectangular windows in the space-times of T of width $w := \lceil \log(1/\varepsilon) \rceil$ and height (temporal duration) t. These windows can be regarded as words of length t over an alphabet of $|Q|^w$ symbols, the *scenes* that an observer at the origin will see as the evolution of arbitrary initial configuration collapses on her space-time window. [See Problem 8.4.]

Theorem 9.45 *Let T be a cellular automaton over a state set Q. If T is observable, then for every $\varepsilon > 0$ the words over the alphabet Q^w form a regular language.*

Finally, it is possible to construct an infinite number of topologically inequivalent continuous self-maps of a Cantor set which are not observable. Let $I_n := (a_n, b_n)$ $(n \geq 1)$ be a sequence of intervals successively deleted in the construction of the standard ternary Cantor set C with $0 < a_n < b_n < a_{n+1} < b_{n+1}$ for all $n \geq 1$ so that the sequences $\{a_n\}, \{b_n\}$ converge to 1. Let $C_n := [b_n, a_{n+1}] \cap C$ $(n \geq 0)$ be the trace of the Cantor set in the given intervals, where $b_0 := 0$. Since every two totally disconnected, compact, perfect and Hausdorff spaces are homeomorphic (see Problem 7.9), let h be the map of C mapping C_0 onto $C_0 \cup C_1$, C_n onto C_{n+1}, and leaving 1 as a fixed point. The map f_1, obtained by juxtaposing a copy of h on the first half of the ternary Cantor set and a copy of h^{-1} on the second half, is a homeomorphism of C with one fixed point which is an attractor on the left and a repellor on the right, hence without the shadowing property. Similar maps f_n with exactly n fixed points can be constructed for every $n \geq 1$.

9.5 Problems

The problems marked * may need to be looked up.

LINEAR MAPS ON TORI

1. Prove that the rule $T(x)_i := x_{i-2} + x_{i-1} + x_i \pmod 2$ has a 4/2-fundamental shift. Conclude that a configuration x belongs to limit cycle on Γ_8 iff it satisfies

$$2x_i + x_{i+1} = 2x_{i+2} + 2x_{i+3} + 2x_{i+5} = 0.$$

[The corresponding polynomial $x^5 + 2x^3 + 2x^2 + x + 2$ divides $x^8 - 1$ exactly.]

2. List all the limit cycles of the rule in Problem 1.

3. Show that a configuration c on Γ_n is a fixed point of the rule T of a given radius r induced by c as follows. δ leaves invariant center cells i whose neighborhoods satisfy $x_j = c_{i+j}$ for each j, and otherwise assigns an arbitrary fixed value $\neq x_i$ to other neighborhoods (indices are, naturally, integers modulo n).

4. Show that the rule of radius r induced by a primitive finite configuration c in Problem 3 is the unique fixed point on *all* cylinders Γ_n iff $r \geq \lceil \frac{m+1}{2} \rceil$, where m is the length of the longest self-match of c (a *self-match* is a pair of identical substrings of consecutive characters at different locations).

5. Generalize Problem 3 to an arbitrary finite number of primitive strings. [Let n be the least common multiple of all lengths, c be the longest substring common to strings formed by concatenations of copies of each primitive string and m the corresponding radius as in Problem 4.]

LINEAR MAPS ON THE LINE

6. Verify the results in Table 9.4 for the 1D euclidean rule $T(x)_i := x_i + x_{i+1} \pmod 2$.

7. Show that a configuration x has temporal period p under rule 90

$$T(x)_i := x_{i-1} + x_{i+1} \pmod 2$$

iff it satisfies the spatial recurrence relation

Table 9.4. Some properties of temporal periods for 1D XOR

p	τ_p (period p)	τ_p^* (smallest period p)
1	1	1
2	1	0
3	4	3
4	1	0
5	16	15
6	16	12
7	64	63
8	1	0

$$x_i = \sum_{j=0}^{p} \binom{p}{i} x_{p+i-2j} \, .$$

Conclude that the companion matrix of the rule is given by

$$\begin{bmatrix} 0 & 0 & 0 & 0 & 0 & 1 \\ 1 & 0 & 0 & 0 & 0 & 0 \\ 0 & 1 & 0 & 0 & 0 & 1 \\ 0 & 0 & 1 & 0 & 0 & 1 \\ 0 & 0 & 0 & 1 & 0 & 1 \\ 0 & 0 & 0 & 0 & 1 & 0 \end{bmatrix} \, .$$

8. Let L be a 1D linear rule with $\gamma(L) \geq 1$. Show that the spatial period of a t-periodic configuration x of period p is the smallest integer $n(\bar{x})$ such that the polynomial $Q(X) := \sum_{l=0}^{n'(\bar{x})} \lambda_l X^l$ divides $X^{n(\bar{x})} + 1$, where the λ_l are the coefficients is the shortest linear dependence among the boolean vectors $\bar{x}, \bar{x}A_p, \bar{x}A_p^2, \cdots, \bar{x}A_p^{n'(x)}$. [$\bar{x}$ is the initial condition that generates x for the recurrence relation of Problem 7.]

SIMULATION IN CONTINUOUS SYSTEMS

9. Show that the left-shift is a topologically transitive cellular automaton (i.e., there exists a configuration ω whose orbit is dense in configuration space, i.e., it comes within arbitrary small distance from any other configuration). [Consider a length-increasing enumeration $w_1, w_2, \cdots$ of all finite words and let $\omega := w_1 w_2 \cdots$.]

10. Show that a pushdown stack can be simulated by a family of piecewise linear map of the unit square.

11. Show that there exist Turing machines that cannot be simulated by a map of PL_1 on the interval. [A piecewise linear map cannot have infinitely many k-cycles for infinitely many k – see [C-G-K].]

12. Show that linear rule 90 given by $T(x)_i := x_{i-1} + x_{i+1} \pmod 2$ cannot be simulated by a piecewise linear map in the unit square.

13. Show that a countably piecewise linear function can simulate two-stack pushdown automata, and hence arbitrary Turing machines. [The stacks contents and internal state can be coded as fractions $1/2^h 3^{s_1} 5^{s_2}$ in a sequence of turning points converging to $(0,0)$ in the plane – see [C-G-K].]

OBSERVABILITY

14. Show that there exist linear maps over a group state set which are not observable in configuration space. [Consider a basis modulo 2 of configuration space containing the all-ones configuration **1**, put $T(\mathbf{1}) := \mathbf{1}$, $T(x) := O$ for all basis vectors $x \neq \mathbf{1}$, and extend linearly to **C**. The pseudo-orbit $\mathbf{1}, \mathbf{1}, \mathbf{1}', O, O, \cdots$ cannot be $\frac{1}{16}$-traced if the $'$ indicates an appropriately small perturbation of **1**.]

15. Show that a linear map in euclidean space is observable iff it is hyperbolic (i.e., all its eigenvalues λ satisfy $|\lambda| \neq 1$). *

16. Show that there exist two neuron networks with real-valued activations that are not observable. [Take for activation functions the piecewise linear-threshold map (valued 0 for $x < 0$, 1 for $x > 1$ and the identity in between) and for linear part, a map that rotates the euclidean plane 45° clockwise followed by a stretch by a factor of $\sqrt{2}$. The global dynamics leaves a segment of positive length pointwise invariant.]

17. Show that if a linear map in the euclidean plane has complex eigenvalues, then its composition with the piecewise linear-threshold transfer function is observable iff it does not leave a segment of positive length invariant. *

18. Show that linear neural networks can produce scenes at the origin which form a nonrecursive set. [See the construction in the proof of Theorem 6.5.]

9.6 Notes

The study of long-term behavioral properties of cellular automata via combinatorial analysis of modular polynomials originates in a well known paper of Martin–Odlyzko–Wolfram [M-O-W]. Jen has further exploited the technique to describe in detail the combinatorial relations between spatial and temporal periods on tori. Most of the results in Sect. 9.1.1 appear in Jen [Je1, Je2, Je3], although preliminary headways appear in Grassberger [Gra] and Ito [I]. Those in Sect. 9.1.2 are due to Cordovil–Dilão–DaCosta [C-D-DC]. Wuensche has produced an atlas of phase spaces of 1D cellular automata [W-L, Wu2] and random networks [Wu1, Wu3], as well as a software environment to generate them.

Study of cellular automata via dynamical systems is hardly a new idea and can be traced back to Wolfram [Wo1] and Grassberger [Gra]. Chris Moore establishes an early result for generalized shifts, as well as the existence of computation universal smooth maps in the plane [Mo1] and the interval [Mo2, Mo3]. The algorithmic complexity of decision problems for simple dynamical systems on the continuum seems to be a fairly unexplored area. For example, it is an open problem if it is decidable whether a given rational number $x \in I$ eventually becomes a fixed point under the iteration of a given piecewise linear map with rational coefficients. This problem and pertaining results in Sect. 9.3 appear in Cosnard–Garzon–Koiran [C-G-K], where it is conjectured that no piecewise linear map in a finite dimension can simulate arbitrary cellular automata. The observation in Problem 12 is due to R. Bartlett [B]. Other notions of simulation, mainly from a symbolic dynamics approach using generalized shifts, have been studied by C. Moore [Mo1]. The notion of mono- and epi-simulation originates in Kůrka [Ku1, Ku2]. Theorems 9.15 and 9.16 are due to Bartlett–Garzon [B-Ga1]. Questions of this type are of interest in trying to find milder notions and stronger results on the simulation of cellular automata by dynamical systems on the continuum than those in Definitions 9.10 and 9.14.

The question of simulating Turing machines on analog neural networks with finitely many activations has been explored in a series of papers by Siegelman and Sontag [Si-So1, Si-So2], who first established that such a simulation is possible with finitely many neurons and the saturated linear function. The number of neurons was reduced to 2 and proven best possible by Koiran in [C-G-K].

Chaotic behavior of cellular automata as dynamical systems, diffusion and analogies to strange attractors are observed by Grassberger in [Gra]. In fact, he uses a 2D encoding to prove that *discrete* deterministic systems are capable of chaotic behavior. The ϕ-transform and derived properties in Sect. 9.3 appear in Botelho–Garzon [Bo-Ga1]. The shadowing property is a classical notion in the theory of continuous dynamical systems that can be traced back to G. Birkhoff and the classical work of Anosov–Sinai, although the terminology used herein is due to Bowen [Bo2]. It has also been referred to as a property of *stochastic stability* by Morimoto [Mor]. The sufficient condition for observability contained in Lemma 9.33 is closely related (in fact, it follows by a similar proof) to a sufficient condition for shadowing due to Coven–Kan–Yorke [C-Y-K, Lemmas 2.3-2.4]. Toggle rules appear in a particular form in the celebrated paper of Hedlund [H] under the name *permutive*. The exploration of the observability of cellular automata and neural nets in terms of shadowing originates in Botelho–Garzon [Bo-Ga2], which falsely claims that all discrete neural networks are observable. The main result in Sect. 9.4 is established in the corrigendum to that paper. Theorem 9.45 is due to Kůrka [Ku3]. Counterexample 9.41 is also due to Kůrka and easily generalizes to the family of one-way monomial cellular automata, as studied in Bartlett–Garzon [B-Ga2]. It appears difficult to give a characterization of exactly what maps on configuration space or on the interval possess the property (see also, e.g., [Bo2, C-Y-K]) although some positive results on observability of analog neural networks are shown in Garzon–Botelho [Ga-Bo], where a solution to Problems 16–17 can be found. To give a characterization of exactly which maps of configuration space possess this property is indeed an interesting open problem.

References

[B]   R. Bartlett: Discrete computation in the continuum. Doctoral disserta-
            tion, Department of Mathematical Sciences, The University of Memphis,
            1994

[B-Ga1]   R. Bartlett, M. Garzon: Computation universality of monotonic maps of
            the interval. Preprint.

[B-Ga2]   R. Bartlett, M. Garzon: Monomial cellular automata. Complex Systems
            7:5 (1993) 367–388

[Be]   R. Bennet: Countable dense homogenous spaces, Fund. Math. **74** (1971)
            189–194

[Bo1]   R. Bowen: Entropy for group endomorphisms and homogeneous spaces.
            Trans. Amer. Math. Soc. **153** (1971) 401–414

[Bo2]   R. Bowen: On axiom A diffeomorphisms. In: CBMS Regional Conference
            Series in Math. 35, American Mathematical Society, 1978

[B-S-S]   L. Blum, M. Shub, S. Smale: On a theory of computation over the real
            numbers; **NP**-completeness, recursive functions and universal machines.
            Bull. AMS **21**(1989) 1–46

[B-G-M-Y]   L. Block, J. Guckenheimer, M. Misiurewicz, L.S. Young: Periodic points
            and topological entropy of one-dimensional maps. In: Nitecki and C.
            Robinson (eds.): Global Theory of Dynamical Systems. Lecture Notes
            in Mathematics 819. Springer-Verlag, New York, 1980, pp. 18–34

[Bo-Ga1]   F. Botelho, M. Garzon: On dynamical properties of neural networks.
            Complex Systems **5**:4 (1991) 401–413

[Bo-Ga2]   F. Botelho, M. Garzon: Boolean neural networks are observable. Theoret.
            Comput. Sci **134** (1994) 51–61. Corrigendum, ibid., forthcoming

[C-E]   P. Collet, J.P. Eckmann. Iterated maps on the interval as dynamical
            systems. In: Progress in Physics, vol. 1. Birkhäuser, Boston, 1980

[C-D-DC]   R. Cordovil, R. Dilão, A. Noronha Da Costa: Periodic orbits of additive
            cellular automata. Discrete Comput. Geometry (1986) 277–288

[C-G-K]   M. Cosnard, M. Garzon, P. Koiran: Computability properties of low-
            dimensional dynamical systems, Ext. Abs. in STACS'93. Lecture Notes
            in Computer Science, Vol. 665. Springer-Verlag, New York, 1993. Full
            version in Theoret. Comput. Sci. (1994), in press

[C-Y-K]   E. Coven, I. Kan, J. Yorke: Pseudo-orbit shadowing in the family of tent
            maps. Trans. AMS **308** (1988) 227–241

[Cu]   P. Cull: Dynamics of neural nets. Trends in Biolog. Cybernetics **1** (1991)

[D]   R.L. Devaney: An introduction to chaotic dynamical systems. Addison-
            Wesley Publishing, Reading MA, 1986

[Ga-Bo]   M. Garzon, F. Botelho: Observability of neural network behavior. Ext.
            Abs. in: Proc. 6th Neural Information Processing Systems Conference, J.
            Cowan et al. (eds.). Morgan-Kaufmann CA, 1993 pp 455-462

[Gi]   R. Gilman: Classes of linear automata. Ergodic Th. Dynam. Syst. **7**
            (1987) 108–118

[Gra]   P. Grassberger: Chaos and diffusion in deterministic cellular automata. Physica D **10** (1984) 52–58

[G-Y]   P. Guan, Y. He: Exact results for deterministic cellular automata with additive rules. J. Statistical Physics **43**:3/4 (1978) 445–455

[Gut6]   H.A. Gutowitz: Transients, cycles and complexity in cellular automata. Phys. Review A (1991)

[H]   G.A. Hedlund: Endomorphisms and automorphisms of the shift dynamical System. Math. Syst. Theory **3** (1969) 320–375

[Ho]   J.G. Hocking, G.S. Young: Topology. Addison-Wesley, Boston MA, 1969

[I]   H. Ito: Intriguing properties of global structure in some class of finite cellular automata. Physica D **31** (1988) 318–338

[Je1]   E. Jen: Cylindric cellular automata. Comm. Math. Physics **118** (1988) 569–590

[Je2]   E. Jen: Linear cellular automata and recurrence systems in finite fields. Comm. Math. Physics **119** (1988) 13–28

[Je3]   E. Jen: Exact solvability and quasi-periodicity of one-dimensional cellular automata. Nonlinearity **4** (1991) 251–276

[Ku1]   P. Kůrka: Universal computation in dynamical systems. Preprint, Charles University, Praha, Czech Republic, 1993

[Ku2]   P. Kůrka: Simulation in dynamical systems and Turing machines. Preprint, Charles University, Praha, Czech Republic, 1993

[Ku3]   P. Kůrka: Languages, equicontinuity and attractors in linear cellular automata. Preprint, Charles University, Praha, Czech Republic, 1993

[L-Y]   T. Li, J. Yorke: Period three implies chaos. Amer. Math. Monthly **82** (1975) 985–992

[M-O-W]   O. Martin, A.M.Odlyzko, S. Wolfram: Algebraic properties of cellular automata. Comm. Math. Phys. **93**(1984) 219–258.

[Mo1]   C. Moore: Generalized shifts: unpredictability and undecidability in dynamical systems. Nonlinearity **4** (1991) 199–230.

[Mo2]   C. Moore: Generalized one-sided shifts and maps of the interval. Nonlinearity **4** (1991) 727–745

[Mo3]   C. Moore: Smooth maps of the interval and the real line capable of universal computation. Preprint, 1993

[Mor]   A. Morimoto: Some stabilities of group automorphisms. Kyoto Daigaku Sur. Kokyuroku **313** (1977) 148–164

[P]   C. Preston: Iterates of piecewise monotone mappings on an interval. In: Lecture Notes in Mathematics, Vol. 1347. Springer-Verlag, New York, 1988

[Si-So1]   H. T. Siegelman, E. D. Sontag: Turing computation with neural nets. Appl. Math. Lett. **4**:6 (1991) 77-80

[Si-So2]   H. T. Siegelman, E. D. Sontag: On the computational power of neural nets. In : Proc. Fifth ACM Workshop on Computational Learning Theory(COLT). Morgan-Kaufmann, San Mateo CA, 1992

[Wa]   P. Walters: On the pseudo-orbit tracing property and its relationship to stability. Lecture Notes in Mathematics, Vol. 668. Springer-Verlag, New York, 1978, pp 231–244

[Wo1]   S. Wolfram: Computation theory of cellular automata, Comm. Math. Physics **96** (1984) 15–57

[W-L]   A. Wuensche and M.J. Lesser: The global dynamics of cellular automata: an atlas of basins of attraction fields of one-dimensional cellular automata. In: Santa Fe Institute Studies in the Science of Complexity, Vol. 1. Addison-Wesley, Reading MA, 1992

[Wu1]   A. Wuensche: Basins of attraction in disordered networks, In: I. Alexander, J.Taylor (eds.). Artificial Neural Networks. Elsevier, Amsterdam, 1992

[Wu2]   A. Wuensche: Complexity in one-dimensional cellular automata: gliders, basins of attraction and the Z-parameter. Cognitive Science Research paper, University of Sussex, 1993

[Wu3]   A. Wuensche: Memory far from equilibrium: basins of attraction of random boolean networks. In: Proc. Conf. on Artifical Life, Université Libre de Bruxelles, 1993

10. Some Inverse Problems

One picture is worth a thousand words.
Old folk saying

As explained in Chapter 2, conventional computing requires encoding of information into strings of symbols over a given alphabet. Classical computation thus deals with recognition and generation problems of formal languages consisting of words (strings of symbols). Classical computational theories originated in attempts to understand calculation as performed by humans. All the resultant models (Turing machines, Church's λ-calculus, Chomsky grammars, Markov algorithms, etc.) are based on the seemingly sequential nature of *conscious* human calculation. They are inherently sequential.

Simultaneously, in the course of the last four decades, our notion of computation has changed radically as a result of a paradigm shift in both physical and cognitive science. That many forms of *perception*, and, in general, *cognition* are essentially a form of computation is now the working hypothesis of many contemporary cognitive scientists. In this new sense, it is becoming clear that classical computation theories, based on one-dimensional representations known as *words*, can only process indirectly, through a transduction into strings, much of the semantics of the information contained in the original 'data structures'. In cases not directly concerning language, this reduction becomes more of a disadvantage, e.g., when encoding some real-life scene (say, a piece of art like the Mona Lisa; or a picture of Abraham Lincoln) into a word description. All words, however expanded, belabored, or compounded, seem to fall short of conveying the whole information content of the picture itself.

This chapter discusses some basic results on tools for problem solving with cellular models. First, the basic problem of synchronization. Second, the problem of information transmission. Third, the various models of language recognition. Fourth, the much scarcer results about higher dimensional generalized words, which should probably be taken as primitive objects if we were to build up on the old paradigm of string processing in the context of parallel processing and distributed representation. The last two are considered because a transition to parallel computation cannot afford to ignore the achievements of sequential computing, which will necessarily have to be incorporated one way or other. (It must be said, however, that it is not clear whether words, languages and/or automata, the key actors in sequential computation theory, will remain the basic elements of a theory of parallel computation.)

10.1 Signals and Synchronization

One of the fundamental problems with parallel processing is the management of communication between parallel processors and their coordination. Due to locality of interaction, global communication must take place as an emerging property. The problem has been exemplified in the so-called *firing-squad synchronization problem*, FSP. Its simplest version on the line consists of endowing, with a common program, a row of soldiers (initially on a quiescent state) who are lined up next to a special site (a general) so that at some time upon his initial order '*Fire!* ', all the soldiers will *simultaneously* enter a common designated *firing* state. The number of soldiers in the squad can be, in principle, arbitrarily large, and the general and soldiers can only deliver messages to neighboring sites who can hear their voices. The FSP prototypes the general problem of analyzing communication between sites in parallel network computation.

Solutions to the problem are presented in this section. The general and soldiers are represented at first by cells in a 1D euclidean space, with the general located at the origin. The notion of *signal* has been suggested as the right tool for the understanding of information transmission in cellular automata computation. This section shows that, indeed, it is a key ingredient in solutions to the FSP, and, furthermore, in several results on language recognition, as seen in the following sections.

10.1.1 Synchronization of a Line

A solution to the FSP is a common local rule (which is, naturally, independent of the number of soldiers) that each and every soldier can execute so as to accomplish simultaneous firing upon an initial external excitation of the general's site (assumed below to be the origin) from the quiescent state. At least three states are required: *quiescent, fire!* for the general, and *firing!* for the soldiers. The quiescent soldier on the far right is supposed to be distinguishable from the next empty cell in an initial configuration and it is not usually counted. The goodness of a solution can be measured in terms of time, or in terms of the numbers of states of the automaton. Every solution requires time $\geq 2n - 1$ since the soldier on the other end from the general at least has to hear the order and send back an answer (however, see Problem 1).

On the other hand, it is relatively easy to obtain solutions with many states. One of the earliest has the general send two waves, one at the speed of light (moving one cell to the right per time step) and another slow wave at a third of light speed. The right end soldier reflects the fast wave so it meets the slow wave at the midsoldier, who now acts as a general for the two halves. The problem is thus reduced recursively to synchronizing each of the halves (which may share a soldier if there's an odd number of them). Finally, a cell fires when it and one of its neighbors are both in the 'fire' state. The space-time of this solution is shown in Fig. 10.1 (waves indicated by lines, as though they were in a continuum). It

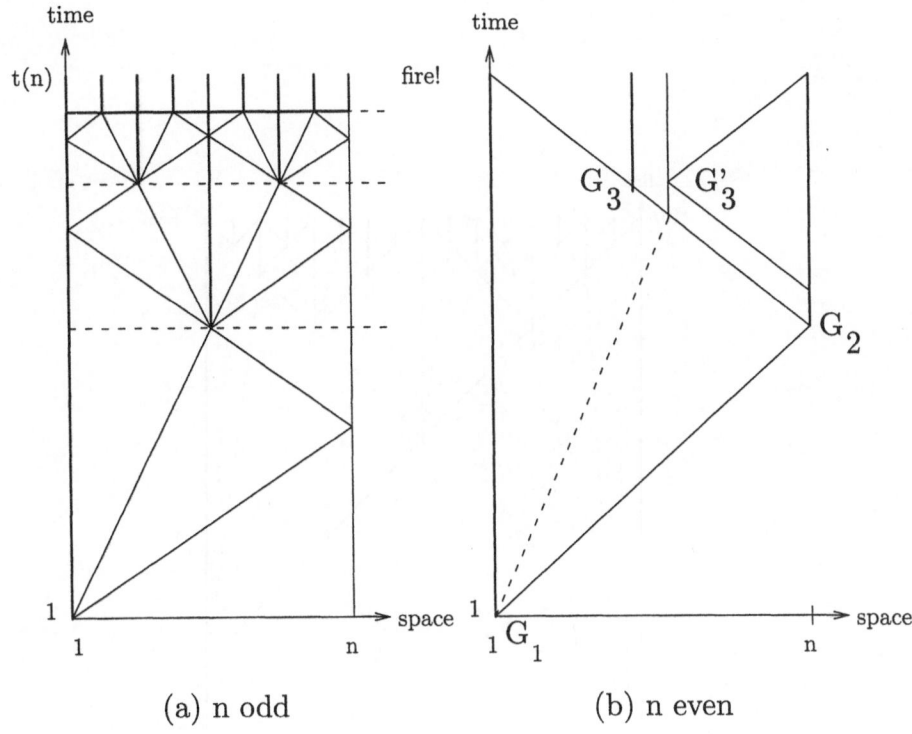

Fig. 10.1. Geometry of MacCarthy-Minsky's solution to FSP

requires time $O(3n - 1)$ (improvements are possible but without achieving the optimal), but it also requires a large number of states.

A time optimal solution proceeds similarly until the fast wave reaches the right-end soldier and makes it act as general (with the squad to its left of size $\frac{n}{2}$). However, in this solution the general has continued to send a family of successively slower waves that intersect the returning fast wave at positions $\lceil \frac{n}{2^l} \rceil$ ($l = 1, \cdots, \lfloor \log n \rfloor - 1$) and make the corresponding midline soldiers act also as right-end generals for the halves to their left. The idealized waves and intersections of the space-time are shown in a continuum-like diagram (cells shrunk to points) in Fig. 10.2. (There are actually two cases depending on the parity of n, but they can be easily specified per this general strategy.) This solution is also time minimal and takes only 8 states.

These two solutions might appear efficient because the partitions are symmetric at midpoints. However, generals need to alternate left- and right-end positions along successive halves, which, in fact, requires superfluous states. A third solution using 6 states makes the recursive partition of the firing squad line $\frac{2}{3}$ of the way (as in the construction of the Cantor set). The fast wave trav-

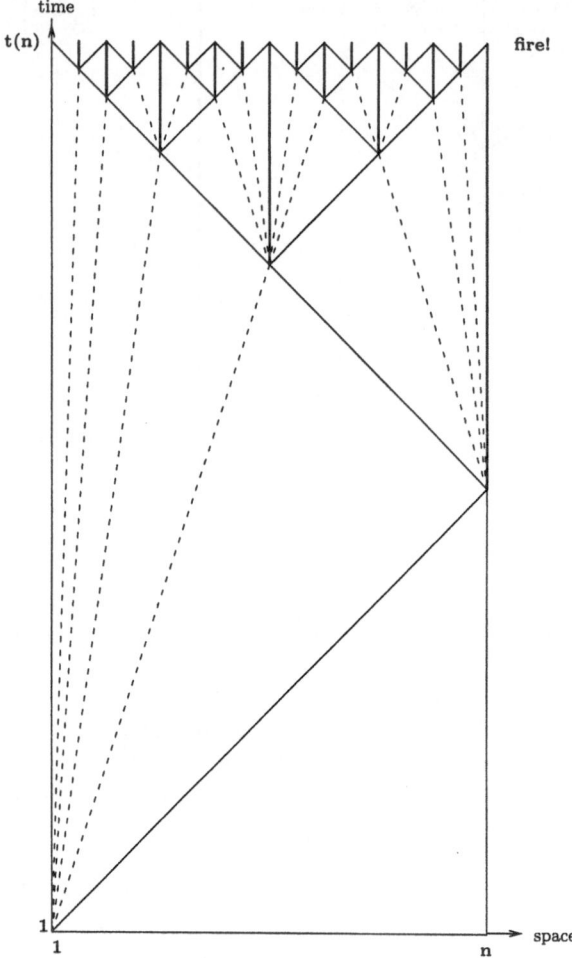

Fig. 10.2. Geometry of Waksman-Balzer's solution to FSP

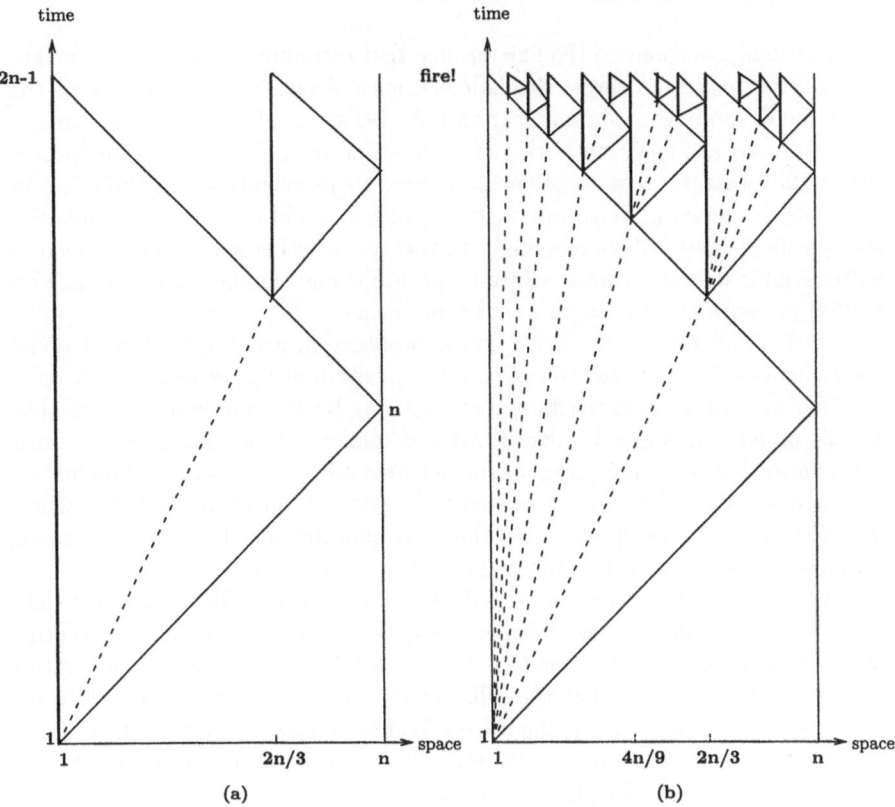

Fig. 10.3. Geometry of Mazoyer's solution to FSP

els at the appropriate speed and, after reflection, it intersects the slow wave at the $\frac{2}{3}$-soldier, who then becomes a general and handles the synchronization of the last third of the cells. The fast wave continues its return toward the general and recursively intersects slower waves at positions $(\frac{2}{3})^l n$ $(l = 1, 2, \cdots)$, spawning generals who handle the corresponding third of the remaining line (strictly speaking, thus described, this solution really requires 8 states, but two of them can be eliminated). The sketch of the space–time of this solution is shown in Fig. 10.3.

A state optimal solution is unknown. Computers have been programmed to verify that no solution exists with 4 states, but the existence of a 5 state solution has neither been proved nor disproved. The existence of such a solution is quite possible in view of the fact that there are so many solutions to FSP that the solution set is not even recursively enumerable.

10.1.2 Synchronization of a Network

The foregoing solutions to the FSP assume that communication between neighbors occurs without delays. Realistically, such delays should be part of the model, and, moreover, communication links between soldiers may need be described by an arbitrary network (not just a line or a path). One can assume that each message takes some constant time to propagate along the edges of an arbitrary network. With any luck, a solution would still be independent of the specific delay(s). If so, one might further ask whether such a solution exists with variable delays, or even arbitrary positions and number of generals. This section presents answers to all of these problems.

Let FSP(D) denote the firing squad problem on a digraph D on n nodes (one might add subindices to indicate the position of the general(s)). A basic solution in time $2n$ constructs a spanning tree by any method (for example, Prim's or Kruskal's greedy algorithm) and implements a solution on a linear array formed with all the nodes of the network using a 'snaking' technique (see Problem 4). This idea can be refined to obtain a solution in time $4R$, where R is the radius of the network, i.e. the maximum distance between the general and any of the soldiers, which may be better in some cases.

A similar solution works for FSPU(D)$_\tau$, the problem with uniform constant delay τ for a message to travel between any two nodes of a digraph D. Another solution of the problem in time $(\tau+1)^2 + (\tau+1)(2n+1)$ consists in generating a signal in the array so that the cells change their states synchronously every $\tau+1$ steps according to, say, the Balzer–Waksman strategy. However, it is easy to see that this strategy or its variants will not work for the problem FSPN(D) with variable unbounded delays since a linear array can no longer be broken into two equal parts with equal delays. A better strategy breaks up the problem into pieces with approximately the same delay radius, i.e., the maximum delay Δ between the general and any other cell. With this strategy, one obtains a solution in time

$$f(\Delta) = O(106\Delta^3 + 2\Delta + 2\tau_{max} + 5),\tag{10.1}$$

where τ_{max} is the maximum delay between any two cells in the network D. The solution of FSPN(D)$_\tau$ proceeds in two stages: (a) contruct a minimum delay rooted spanning tree Π; and (b) make the tree fire. Part (a) is easily accomplished with a single-source shortest path algorithm in time $O(2\Delta + 2\tau_{max} + 2)$. Part ($b$) is accomplished by dividing the tree Π into subtrees Π_l ($1 \le l \le n$), each of delay radius less than Δ, and running the same algorithm recursively from the root of each subtree in time $f(\Delta) - f(\Delta_l)$. Solving this recurrence yields (10.1).

10.1.3 Signals in Dimension 1

The fundamental ingredient in the solutions of the FSP and its variations is the notion of a 'wave' traveling across the cellular space. An analysis of the information processing capabilities of cellular automata may consist in a systematic study of the communication complexity between sites. Although this approach has not been fully explored in the literature, the notion of a *signal* has been suggested to be a fundamental concept.

Intuitively, a *signal* is a wave of information propagating in the cellular space according to the 'physical laws' (laws of motion) of the space. In particular, when the neighborhood N of a cell consists of nearest neighbors, a signal would be seen as the trace of a particle-like object traveling along a connected line and bouncing back and forth between sites of the automaton (which either, reflect, receive, or pass it along, possibly modified). For instance, a pixel signal traveling at the speed of light (one cell per unit time) away from the origin in 1D space appears as a 45° horizon line in the space–time of the automaton. In fact, the various solutions of the FSP consist of the description of the appropriate signals to carry the message "Fire when you and both your neighbors are ready to" from site to site. Thus one way to understand the capabilities of cellular automata for information transmission and processing consists of describing the types of signals that they can generate. Although this is still a largely unexplored area (even the notion of a signal does not seem to have a well accepted formalization), some initial results are available in euclidean dimension 1 and they will be described in this section.

Since time is irreversible, a signal can be regarded as a discrete path (or curve) in space–time, that is, a union of segments joining neighboring sites in consecutive time instants. In particular, signals can be represented as connected curves. Such curves will be called *constructible* if at any given instant during the evolution of some cellular automaton from a pixel initial configuration, a site i is on the signal s if and only if its state s_i belongs to a given subset of distinguished *signal states* S. A signal is clearly invariant under space translations, so we will assume the source pixel located at the origin of the space. A signal is entirely contained in the light cone with vertex at the origin and sides at a 45° angle. A clockwise rotation by the same angle reduces the signal to an ordinary curve in the first quadrant of a cartesian plane (nonnegative integer coordinates). A signal can thus be formalized as the graph of a corresponding function $f : \mathbf{N} \to \mathbf{N}$, which will be referred to as the *graph of the signal*. (As usual when dealing with integer-valued functions, we will describe them by real-valued functions whenever possible, on the assumption that at points of discontinuity, the segments in the graph of the function will be connected by vertical segments.) Equivalently, one can represent such a function by a 2D configuration x_{i_x,i_y}, or by a *Moore machine* (i.e., finite-state machine with an additional output function $\Delta : Q \to Q$ of the states), which on input each of the columns $x_{i_x,1}, x_{i_x,2}, x_{i_x,3}, \cdots$ (lower characters first) gives as output the corresponding states of the next column $x_{i_1+1,1}, x_{i_2+1,2}, x_{i_3+1,3}, \cdots$.

Example 10.1 The signal LOG$_2$ is constructible as the discretization of the function $\log_2(n + 1)$. The key observation is that the 'area' under this curve consists of binary integers written out successively, least significant digit on the

x-axis, most significant on the curve. Two transformations are necessary in order to convert this configuration into the space–time of a cellular automaton. First, the successive columns w^l of this 2D array can be obtained recursively by adding 1 to the previous column in binary starting with the word $w^0 := 10000\cdots$, as shown in Fig. 10.4(a). Let M' be the finite-state machine that on input the successive characters in a binary integer w (least significant first) outputs the corresponding digits of the binary integer $w + 1$ (see Fig. 10.4(b), where the outputs are a function of the states as indicated inside their circles).

Second, by adding states 1 and $0'$, M' can be transformed into a Moore machine M whose output alphabet is identical to the state set and which realizes on a cellular automaton with one neighbor the graph of the desired signal, as shown in Fig. 10.4(c). In a similar way one can verify that the signals $n + \log_m(n)$ are constructible for arbitrary $m > 1$, as well as the identity signal – see Problem 8.

In the language of signals, the solution of inverse problems by cellular automata would reduce to applications of a signal calculus. In order to develop it, signals need to be characterized by observable macroscopic attributes. Two of them are the ratio and the speed. The *ratio* is the time (number of iterations) ρ_n required by the signal to reach cell n *for the first time* from the origin. The *speed* of the signal is n/ρ_n. A signal of eventually constant maximum speed 1 will be referred to as a *photon*. Among the few steps taken toward mapping out the geography of constructible signals, one of the known results bears a distinct resemblance to Borodin's gap theorem in recursive function theory.

Theorem 10.2 *A constructible signal is either eventually a photon or lower bounded by the signal of ratio $n + \log_2(n)$.*

Proof. It suffices to show that the corresponding graph is either constant or lower bounded by $\log_2(n)$. Consider the words w^l consisting of successive states of the cells l in a signal of ratio ρ_n, or, equivalently, the lth column of space–time, for $1 \leq l \leq n$. If $\rho_{n_0} - n_0 < \log_2 n_0$ for some n_0, the pigeon-hole principle guarantees the existence of two identical columns $w^{n_0} = w^{n_0+\tau}$ since there only exist n distinct words of length $\log_m n$. Since the cellular automaton is homogeneous, the repetition is propagated periodically, i.e., $\rho_{n+\tau t} = \rho_n + \tau t$. and hence $\rho_n - n = \rho_{n_0} - n_0$ for infinitely many $n's$. Since by definition $\rho(n) - n$ is increasing, the value $\rho_n - n$ is necessarily constant. $\square$

10.1.4 Clocks

In addition to the problem of information transmission, an important aspect of the applications of cellular networks to particular problems is the type of mechanisms available for permanent storage of information. No systematic development exists to handle this problem. In this section we present a type of mechanism in 1D spaces, a so-called *clock*, that can be used by a source cell

11111111111111111111111111111111
11111111111111110000000000000000
11111111100000000111111100000000
11110000111100001111000011110000
11001100110011001100110011001100
10101010101010101010101010101010

(a) The area under the signal

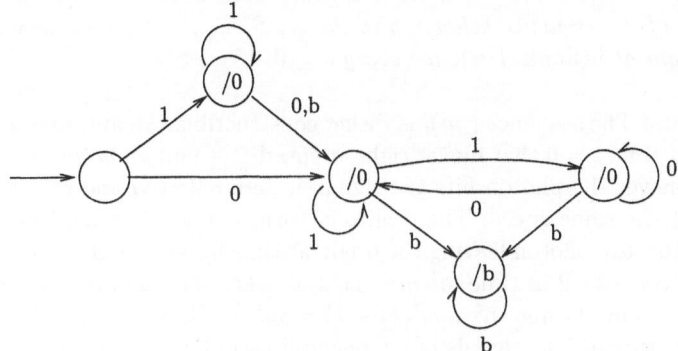

(b) The Mealey machine adding 1 to its inputs

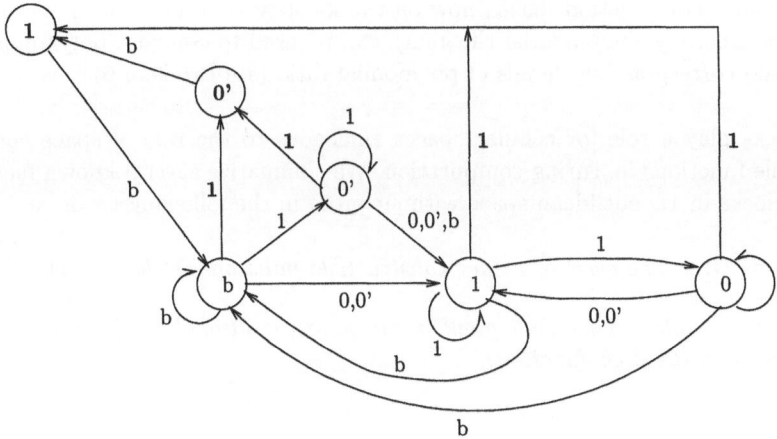

(c) The transition diagram for the signal

Fig. 10.4. Constructibility of a LOG₂ signal

to store a sequence *in the cellular space* and retrieve it dynamically using the laws of motion of the automaton. The sequence is distributed over the cells in the space, in principle arbitrarily far from the source, as opposed to a specific localization of information in a static way in the states of particular cells. The information hits the source back periodically at certain later times, whence the name.

Definition 10.3 *A one-way infinite word* $w := (w_t)_{t\geq 0}$ *over an alphabet* S *is* ca-contructible *if there exists a cellular automaton* T *with state set* Q *containing* S *so that the origin cell of a 1D cellular automaton enters state* f_t *at time* t *for every* $t \geq 0$, *if started on pixel* $f_0 e^0$ *a time* $t = 0$. *An increasing function* $f : \mathbf{N} \to \mathbf{N}$ *is* Fisher constructible *(or simply, a clock) if there exists a cellular automaton* T *as before that ca-constructs some word* w *such that* $w_{f(n)}$ *is the* nth *element of the sequence belonging to the set* S, *i.e.,* T *ticks in elements of* S *at the origin at instants* $f(n)$, *for every* $n \geq 0$.

Example 10.4 The sequence $(m^t)_t$ is Fisher constructible. An automaton emits a signal ν_0 at time $t = 0$ that moves right at speed $\frac{m-1}{m+1}$ and a photon ν_1 at time $t = m$. Whenever the photon hits the signal, it gets reflected and sent back to the origin at the same speed. The origin, in turn, bounces it back toward the signal ν_0. Thus, the photon leaving the origin at time m^t will catch up the signal ν_0 at site $m^t(m-1)/2$ in time $m^t(m-1)/2$; it takes the same time to bounce back to the origin at time $m^t + m^t(m-1) = m^{t+1}$. These exponential clocks can be easily turned into signals of exponential ratio by shifting the mirror at the origin one site to the right every time the photon bounces (see Problem 7).

A similar construction, based now on the identity $(n+1)^2 = n^2 + 2n + 1$ (or more generally the binomial theorem), can be used to contruct polynomial clocks and corresponding signals of polynomial ratio (see Problem 6).

Clocks play a role for cellular spaces analogous to the role of space constructible functions in Turing computation. We summarize several known facts about clocks in 1D euclidean space without proof in the following theorem.

Theorem 10.5 *The class of Fisher constructible functions is closed under*

1. *linear combinations with diophantine positive rational coefficients (followed by the floor function);*

2. *composition;*

3. *taking minima and maxima.*

In fact, the following result raises the (as yet unanswered) question whether ca-constructibility on the line is just an analog of or is rather identical to the notion for Turing machines. Recall that a function f on $\mathbf{N}$ is Turing constructible if some Turing machine M uses exactly $f(n)$ tape cells for some input of length n, for every $n \in \mathbf{N}$.

Proposition 10.6 *Every increasing Fisher constructible function is Turing space constructible.*

Proof. Let f be Fisher contructible on state set S by a 1D local rule δ (which is assumed, without loss of generality, to have radius 1) starting with a pixel at the origin. Construct a Turing machine M that simulates the action of δ starting with the pixel as the first symbol on the tape so that at all times t, M has effectively used only the first $t + 1$ cells. (It may be necessary to add an extra symbol for M to mark cells just in case δ does change some quiescent states as the wave of activation travels away from the origin.) In addition, M keeps a counter of how many times the origin cell has entered a distinguished state from S. M halts when the counter reaches n, the length of the input. By Fisher constructibility, M must then be at site $f(n) + 1$ and so will have used $f(n)$ cells on its tape. □

10.2 Formal Language Recognition

Classical theories treat words in two apparently different ways: automata (recognizing devices) and grammars (generating devices). Much of the classical treatment of languages goes into proving that these two approaches are essentially equivalent (viz., Kleene's Theorem, Chomsky's Theorems, etc.). With this hindsight, cellular automata can be construed as an interesting synergy of these two concepts. This aspect of cellular automata computation is explored in this section for formal language recognition.

10.2.1 Models

Language recognition appears as an inverse problem to cellular models. It is necessary to specify precisely how the input strings are to be given, how the model is to process them, and finally, how it is to communicate answers about membership in the target language. This section describes the most common models: one-way and two-way automata, and iterative arrays. These devices are one-dimensional, and the following notation will be used to describe 1D configurations, in addition to the notation introduced in Chapter 2. Superposition of strings indicates concatenation whenever possible. $\bar{\omega}^{\gg}$ denotes a periodic word infinite to the right with period ω. Likewise to the left. $^{\ll}\bar{\omega}^{\gg}$ is a two-way infinite periodic string of period ω. Recall that an underlined state $\underline{*}$ is located at the origin of the cellular space.

Definition 10.7 *A cellular automaton recognizer (in short, car) M consists of a 1D cellular automaton δ on a state set Q, distinguished disjoint subsets*

Σ, Q_a, Q_r *(alphabet, accepting and rejecting states, respectively) and a special border state iota* ι *such that* $\delta(\iota, *) = \iota$ *for every character* $*$. *The string language* L *over the alphabet* Σ *recognized by* M *in time* $t(n)$ *consists of all words* $\omega := \omega_1 \cdots \omega_n \in \Sigma^*$ *such that,*

$$T^{t(|\omega|)}(\lll \bar{\iota} \underline{\omega_1} \cdots \omega_n \bar{0} \ggg)_0 \in Q_a$$

i.e., words that take the origin cell to an accepting state at time $t(|\omega|)$ *when giving the automaton the word* ω *as an initial configuration sandwiched between border states on left and quiescent states on the right.*

Thus a *car* can utilize the quiescent portion of the cellular space on the right of the input word as auxiliary storage for intermediate computation, but will never alter the portion of the tape to the left of the origin.

The entire computation can be visualized in a space–time diagram where a grandmother cell (like a general at the origin) attempts to collect information about the given string given on the right, performs some auxiliary computation, and finally makes a decision about the nature of the string by entering the appropriate final state.

It was shown in Theorem 4.7 that one-way (say, right) cellular automata are as powerful as two-way automata on the euclidean line. According to definition 10.7, information flows left to right initially and then is collected back to the origin. It may save some communication time to have the last input cell play the role of a grandmother cell that will decide acceptance or rejection.

Definition 10.8 *A* one-way cellular recognizer *(for short, ocar), is a one-way car in which acceptance (or rejection, respectively) of a word* $\omega := \omega_1 \cdots \omega_n$ *determined by the condition*

$$T^{t(|\omega|)}(\lll \bar{\iota} \underline{\omega_1} \cdots \omega_n \bar{0} \ggg)_n \in Q_a \text{ (or } Q_r). \tag{10.2}$$

The input to both *cars* and *ocars* are fed all at once as an initial configuration. The local rule has to take care of whatever communication the grandmother needs in order to gather enough information to recognize the input. This may appear difficult to accomplish for a local rule, particularly since an *ocar* cell i will never get to see the activity of cells $j > i$ (or $j < i$).

A third recognition mode allows the leftmost cell to be the only input channel. The input symbols of ω are fed *one after another* to the cell, and a special symbol $\$$ is clamped thereafter to permanently signal the end of the input. The information flows into quiescent cells on the right as the automaton is iterated and, as with *ocars*, a grandmother nth cell eventually reaches a decision.

Definition 10.9 *A* one-way iterative array *(for short, oia), is a one-way cellular automaton* M *with an input window at cell 0, through which the t-th symbol of the input string is entered at time* $t-1$ $(1 \leq t \leq n)$, *followed by a constant symbol*

$. *The language accepted by M in time $t(n)$ consists of all strings $\underline{\omega_1}\cdots\omega_n$ that, when fed as the configuration $\underline{\omega_1}\cdots\omega_n\bar{\$}^{\gg})_n$, lead cell n to an accepting state in Q_a at some time $t \leq t(|\omega|)$.*

In this case, all workcells have access, in principle, to the entire input string and, moreover, the automaton has direct knowledge of the length of the string n due to the clamping of the \$s after the last symbol of ω. Hence *oias* are much easier to understand and design as sequential programs since they compute in 'waves' going from the input cell to the grandmother cell.

Example 10.10 Several well known fairly complex languages can be recognized by *cars*, *ocars* and *oias*.

1. Dyck languages (i.e., well-formed parenthetical expressions made on a set of parentheses $\{a_1, b_1, \cdots a_r, b_r\}$) are recognized by *oias*. For example, for $r = 1$, a word over the alphabet $\{(,)\}$ can be recognized as follows. If the left boundary cell contains a ')', it sends forward to the grandmother cell a reject signal. Otherwise, the leftmost cell is marked with a special state. Each cell keeps a copy of its input symbol. Left parentheses propagate to the right at light speed. When one meets a right parenthesis, the latter changes state to an extra symbol c_1 and stops the signal, if it matches; otherwise, it sends a signal for grandmother to reject. When, and if, the leftmost parenthesis is replaced by a c, it sends forward an acceptance signal. Grandmother accepts if it is told to do so, else it rejects. Note that acceptance actually takes place in n steps, i.e., in *real time*.

2. More generally, context-free languages can be recognized by *oias*.

3. The language QBF (quantified boolean formulas), which is **PSPACE** complete, can be recognized by an *oia*.

There is a class of languages corresponding to each of the recognition models, which will be denoted by prefixing with $\mathcal{L}$ the name of the model. One has $\mathcal{L}CAR(t(n))$, $\mathcal{L}OCAR(t(n))$ and $\mathcal{L}OIAR(t(n))$, and the corresponding unions $\mathcal{L}CAR$, $\mathcal{L}OCAR$ and $\mathcal{L}OIAR$ for all $t(n) \geq 0$.

Theorem 10.11 *$\mathcal{L}CAR$ is closed under intersection, complementation and reversal.*

Proof. Union and intersection can be done using a *car* whose state set and local rule is the cartesian product of the corresponding given elements as in the solution to Problems 2.8–9. Likewise for complementation. For reversal, it suffices to show that there is a cellular automaton that reverses a given word, for then one can serially compose with a *car* that recognizes the original language (see Problem 13). Reversal of a given word can be accomplished by the same technique just described. A *car* M can simply place an additional marker $\diamond$ at the first quiescent cell $n + 1$. Whenever a cell has a right neighbor in state $\diamond$, it replaces the symbol a_i by the $\diamond$ and propagates it to the right up to the first quiescent cell. When the grandmother cell sees a $\diamond$, all of the input string has been reversed over the cells $[n + 1, 2n]$, so she can initiate a synchronization of the nonquiescent unmarked cells. At 'firing!', the unmarked cells proceeds to simulate the *car* M accepting the original language. Upon acceptance or rejection, cell $n + 1$ (acting as a grandmother for M') sends a signal to the grandmother at the origin, which accepts or rejects accordingly.   □

It is immediate from the definitions that

$$\mathcal{LOCAR}(t(n)) \subseteq \mathcal{LCAR}(n + t(n))$$

because, upon acceptance by simulation of an *ocar*, cell n can send cell 0 an OK signal so it accepts as well. Also,

$$\mathcal{LOCAR}(t(n)) \subseteq \mathcal{LOIAR}(3n - 1 + t(n))$$

since an *oia* M can simply shift each input symbol to the right up to the first quiescent cell, where it is held until all the input is fed and simulation of the given *ocar* M' commences. The appearance of the signals \$s triggers off a synchronization process that occupies M for the next $2n - 1$ steps, at the end of which it simulates the witness *ocar* M' for the given language, and finally accepts or rejects accordingly. (Some regrouping of the cells may be necessary so that the *oia* M only uses the first n cells of the cellular space.)

As a matter of fact, the apparent strenghts and weaknesses of these models balance each other out overall.

Theorem 10.12 *The three models are equivalent in recognition power, i.e.,*

$$\mathcal{LCAR} = \mathcal{LOCAR} = \mathcal{LOIAR}.$$

Proof. The first equality follows from the results in Chapter 4. There only remains to show that $\mathcal{LOIAR} \subseteq \mathcal{LOCAR}$. Let a language L be recognized by an *oia* M'. An *ocar* M accepts L by shifting its input symbols to the right and feeding them to cell $n + 1$, which acts as an input cell for the segment $[n + 1, 2n]$, where a *car* can be simulated to accept or reject the original word. (Again, some regrouping is necessary to comply with the requirement that only the first n cells are used.) Thus the reversal L^R, and hence L, belongs to $\mathcal{LCAR} = \mathcal{LOCAR}$.
        □

10.2.2 A Speedup Theorem

Turing machines have the property of speedup, i.e., wherever one of them recognizes a language in time $t(n)$, it can always be redesigned to recognize it in time $\frac{1}{\kappa}t(n)$, for any constant $\kappa > 0$, at the price of additional states (grouping together several cells of the original machine into one). For cellular automata, one can get an analogous speedup in communication among cells by enlarging the scope of vision (i.e., the neighborhood) of the center cell. The problem with the first approach (the *weak form* of the speedup) is that the new automaton does not recognize the original language but an encoding of it in a larger alphabet. A *strong form* which does not require even a larger scope is more difficult, although possible, to establish.

Theorem 10.13 *For every fixed neighborhood radius, time $t(n) \geq n$ and integer $\kappa \geq 1$,*

$$\mathcal{L}CAR(t(n)) = \mathcal{L}CAR(t(n)/\kappa).$$

Proof. The key idea in avoiding the larger scope is to have cells collect information from their neighbors and then synchronize their update, using a solution to a local firing squad problem for a neighborhood large enough for the speedup to be performed as desired. (The details of the proof are somewhat cumbersome and will be omitted.)   □

Even the characterization of the two smallest time complexity classes, namely the cases $t(n) = n$ (real time) and $t(n) = O(n)$ (linear time) offers unexpected difficulties, as there has been conflicting evidence for equality.

10.3 Picture Languages

In the previous section, cellular automata have been mostly applied to solving string problems using one-dimensional representations. Communication, a key ingredient in solving problem on a massively parallel computer, is severely restricted in 1D spaces due to the fact that every cell is a bottleneck since it disconnects the space in two halves. Communication is much less congested, and therefore harder to analyze, in higher dimensions because of higher connectivity (i.e., any two cells are connected by many more paths). On the other hand, as argued in Sect 10.3.1 below, pictures are richer and more natural ways to represent information. Therefore, it is quite possible that more efficient computation can be achieved by cellular automata in higher dimensions and other cellular spaces. The purpose of this section is to show some exploratory results in this direction.

The classical theory of computation also addresses the question of characterizing functions that can be calculated by automatizable procedures. When information is encoded in strings, the question amounts to computation in terms of expansions. Because of parallelism and pictorial representation, cellular automata present other possibilities that are also explored in this section.

10.3.1 The Issue of Representation

The basic problem in higher dimensional cellular spaces is the lack of standard well known representations. Perhaps paradoxically, from the point of view of naturally occurring information processing (e.g., cognitive processes), distributed information representation is a natural object. Known facts about anatomy and physiology in animals and humans readily confirm the assertion that known *brains* manipulate information in terms of visual and mental 'images' rather than the ordinary words that appear necessary for *description* of the process at higher levels of abstraction or human communication (e.g., written language). Artificial neural networks are the current mechanisms postulated as natural ways of modeling this kind of processing. Similar developments have taken place in the study of phenomena hitherto considered the exclusive domain of natural sciences such as physics, chemistry, etc. using cellular automata, and their generalizations (see Chapter 12). More generally, the physical universe itself appears to be as massively parallel. Thus, it is desirable to develop a theory of information representation and processing on structures more expressive than ordinary strings.

Any notion of *representation* requires two basic ingredients: first, a *writing medium* to hold a physical embodiment of the representation, and, second, a way to make *marks* on this medium in various ways to produce different representations. The nature of the marks may be taken to be continuous or discrete. Our main interest lies in marks of a discrete character, although the number of such marks may be finite or countably infinite. The main aspect to consider in a generalization of words as representations to the context of parallel processing is the fact that the most important elements that make up a representation (such as a picture) are *not* its pixels (e.g., gray-levels) themselves, but rather it is certain features of their *contiguity and relative position* to one another that seem to make up most of the information content of the picture itself. Other realistic assumptions are imposed upon this kind of model. For instance, it is physically impossible for anybody (including current computing devices) to draw, one by one, an infinite number of discrete characters or symbols, or in fact to distinguish contiguity relations between a pixel and an infinite number of many others. Thus it seems it is impossible to assume infinitely many neighbors for any given cell. There is also the issue of anyone's ability to discriminate a continuum of differences in a continuous mark and *control* them *individually* for the purpose of communication. Hence if one wants to consider a realistic generalization, it seems to be necessary to assume only marks of a discrete character, independently of their discrete or continuous nature.

This intuition is made rigorous by choosing a cellular space as the medium of expression. Pictures and images are much more naturally represented as configurations in 2D and 3D or even less familiar spaces than as one-dimensional encodings. It was shown in Chapter 6 that one can assume without loss of generality that the underlying cellular spaces have the structure of Cayley graphs. Configurations of finite support can be regarded as generalized words holding an information content. One can then regard local dynamics of cellular automata and neural networks as some sort of *rewriting* rule applied to the symbols (states) at the cells in terms of their context (symbols at neighboring cells). It would be natural then to distinguish between terminal and nonterminal symbols and arrive at the notion of a π-*grammar*. Once a notion of word is in place, a *language* of generalized words is defined in the usual way as a subset of configuration of finite support in a cellular space.

The purpose of this section is to survey the sparse known literature about generalized languages of this sort. In this context, configurations of finite support will be referred to as π-*words*. The presentation is made in increasing cellular space complexity as measured by a natural parameter: the order ot growth as introduced in Section 2.5. Spaces of linear growth are spaces of finite bandwidth. Spaces of polynomial growth essentially correspond to spaces on nilpotent groups (more precisely, nilpotent-by-finite). At the other end of the spectrum are spaces of exponential growth, such as the 4-ary tree of the Cayley graph of the standard free group. Finally, the new spaces of intermediate growth.

In view of the results in Chapter 6, one must raise the important question of how π-computation behaves intrinsically, beyond the horizon of Turing computation. In particular, one must address the question of how fruitful it is to consider π-words directly instead of their linear encodings. The results of Sect. 10.3.4 show hard evidence that, semantic blurring of everyday meaning of images by string encodings notwithstanding, linear encodings have very little to offer beyond very difficult problems. In this regard, the results of this section are intended to open many more questions than they actually answer.

10.3.2 π-Languages in Spaces of Linear Growth

The simplest type of a nonlinear writing medium π is a cycle or ring of cells. First consider *cyclic automata* (only one cell fires at a time, and the active cell is indicated by a read/write head). The simplest type only assumes a very restricted writing ability: it has only a finite number of (physical) pebbles that it can use to mark the cells of the graph. In particular, it will not be able to write the same symbol again until its previous occurrence has been erased.

As it turns out, 1-pebble automata behave essentially like finite and 0-pebble automata: *they recognize regular unary languages of π-words; they are equivalent to their nondeterministic counterparts, have solvable decision problems* (such as emptiness, finiteness and so on), etc. There is a sharp contrast with the situation for cyclic 2-pebble automata, however. Even in this restricted case, *2-pebble cyclic automata have most of the same decision problems recursively unsolvable.* In fact, 2-pebble automata are essentially equivalent to *logspace*-bounded Turing machines. Thus decision problems about cyclic π-languages have an infinite complexity by classical standards. Nonetheless, each recognizable π-language is still recognizable in polynomial time by Turing machines.

The processing power of full versions of cellular automata in 1D spaces has been, perhaps inevitably, measured against the backdrop of Turing complexity classes. We present several results that give a parallel implementation of some classical sequential automata.

10.3.3 2D-Euclidean Languages

Next in the hierarchy are groups of polynomial growth. The structure of these groups is already known. They turn out to be the nilpotent-by-finite groups. Thus, they are linear groups, and it is known that their word problem can be mapped onto 1D data structures in polynomial time. This, in principle, would solve the problem. However, as discussed before, the semantics of 2D representation is entirely lost in the process. One- and higher-dimensional languages have been studied in the areas of pattern recognition and image processing, and as such, offer a number of results impossible to survey here. We single out some results that give a parallel implementation of some classical sequential automata in the euclidean line, and some interesting languages in the 2D euclidean plane.

Definition 10.14 *A nondeterministic cellular automaton is called* state-partitioned *if its neighborhood is the von Neumann neighborhood, the state set Q is a cartesian product $Q = C \times N \times E \times W \times S$, and the local rule is a function only of the central component of the center cell, and the nearest components of the states of its neighbors. The automaton is* locally reversible *if the local rule is injective (or equivalently, bijective).*

For example, in a partitioned 1D cellular automaton, each state x_i has three components $x_i := x_i^w x_i^c x_i^e$ and the local rule is of the form $\delta(x_{i+w}^e x_i^c x_{i+e}^e)$. In a partitioned 2D cellular automaton each state x_i has five components $x_i := x_i^c x_i^n x_i^e x_i^w x_i^s$ corresponding to the five neighbors in the 2D von Neumann neighborhood *news*. The local rule is of the form $\delta(x_i^c x_{i+n}^s x_{i+e}^w x_{i+w}^e x_{i+s}^n)$.

Lemma 10.15 *The global dynamics of a partitioned cellular automaton is injective (respectively, surjective, reversible) if and only if so is its local rule.*

Proof. We illustrate the argument with injectivity on 1D automata. If the global dynamics T is not reversible, i.e., $T(x) = T(y)$ for a pair of distinct configurations x and y, then

$$\delta(x_{i+w}^e x_i^c x_{i+e}^w) = \delta(y_{i+w}^e y_i^c y_{i+e}^w),$$

for some cell i where $x_i \neq y_i$ (i.e., where some components must differ as well), which contradicts the injectivity of δ. Conversely, assume that the local dynamics δ is not reversible but the global dynamics T is. Let $x_{-1}x_0x_1$ and $y_{-1}y_0y_1$ be two different triples mapped to a common next state by δ. Extend these neighborhoods of the origin to whole configurations x, y in an arbitrary way but so that $x_{i-1}^w = y_{i-1}^w$, $x_i^c = y_i^c$ and $x_{i+1}^w = y_{i+1}^w$ for all i with $|i| > 1$. It is easy to check these conditions imply that $T(x) = T(y)$. Obviously $x \neq y$, i.e., the global dynamics T will not be injective either.   □

Theorem 10.16 *Every 1D cellular automaton can be simulated by a 1D partitioned cellular automaton. In particular, Turing computation can be performed with partitioned reversible cellular automata.*

Proof. Without loss of generality assume that the neighborhood of the given cellular automaton M is von Neumann's. Let the state set be Q' and the local rule δ'. A partitioned automaton δ with state set $Q := Q' \times Q' \times Q'$ and local rule

$$\delta(p^w p^c p^r) := (\delta'(p^w p^c p^r)\delta'(p^w p^c p^r)\delta'(p^w p^c p^r))$$

simulates M if each original configurations x' is coded by the configuration x given by $x_i := x_i' x_i' x_i'$, for all i, on the new state set.

The other statement follows from the results in Chapter 5 that 1D cellular automata can simulate Turing machines is we establish that Turing machines can be simulated by reversible Turing machines. In principle, a Turing machine can record all the previous history of its computation (e.g., by never rewriting other than the blank symbol and recopying the entire nonblank tape to the right appropriately modified to reflect the next move of a Turing machine) so that its moves are reversible.   □

It also follows from Theorem 10.16 that there exists computation universal 1D reversible cellular automata.

Many attempts to generalize the Chomsky hierarchy (regular, context-free, context-sensitive, recursively enumerable) of string languages to the euclidean plane have resulted in a great variety of generalizations. The following are particularly relevant to cellular automata because they are defined in terms of a local recognizing procedure consisting of sliding a window of fixed size around the π-words and verifying that only certain patterns have occurred. In this section the word *array* will refer to an ordinary array of support a finite rectangle of cells. The rectangular subblocks of size $r \times s$ of an array a will be denoted $B_{r \times s}(a)$.

Definition 10.17 *A 2D array π-language is* locally testable *if it is closed under the equivalence relation $\equiv_{r,s}$ among configurations of having identical subblocks of size $r \times s$, for some $r, s \geq 1$. A π-language $L \subseteq C_0$ is* local *if there exists a finite number of arrays $\mathcal{Q}$ such that*

$$L := \{x \in C_0 : B_{2 \times 2}(x) \subseteq \mathcal{Q}\}.$$

A 2D π-language R over an alphabet A is recognizable *if there exists a local language L over some alphabet Σ and a morphism $\varphi : \Sigma \rightarrow A$ such that $R = \varphi(L)$. A Blum–Hewitt automaton is a sequential 2D finite-state machine* *that traverses input 2D arrays from the most bottom left cells (making NEWS moves) and accepts by entering a final state somewhere within the array. A one-way array cellular acceptor (or 2D ocar) is a 2D version of an ocar with a grandmother cells at the top rightmost cell.*

No definitive characterizations are known for π-languages in the plane, not even in the simple case of array languages. We give the following results for illustration without proof.

Theorem 10.18

1. *Every locally testable array language is accepted by a deterministic Blum–Hewitt automaton (but not conversely).*

2. *Every array language accepted by a nondeterministic Blum–Hewitt automaton is recognizable.*

3. *The class of recognizable array languages is identical to the class of 2D languages $\mathcal{LOCAR}_{2D}$.*

10.3.4 Recognition over Spaces of Exponential Growth

By contrast, groups of superpolynomial growth cannot be mapped efficiently into 1D structures. Therefore, their study offers additional theoretical interest. Spaces of exponential growth (free groups) are well known groups whose Cayley graphs are trees. Studies of *sequential* automata on trees have yielded important results on the decidability of monadic second-order theories of successor, as discussed in Sect. 5.4. However, no systematic study of analogous picture languages recognizable by cellular automata on these groups exists. Of the few facts known about cellular automata on infinite trees, the following theorem can be regarded as a proof of the parallel computation thesis mentioned in Chapter 1 for tree cellular spaces. Some definitions are required.

Recall from Chapter 2 that D_n denotes the group presented by

$$D_n := \langle\, A_1, \cdots, A_\nu | A_1^2 = \cdots = A_\nu^2 = 1 \,\rangle.$$

If $\nu = 1$ one obtains a cyclic group of order 2 so assume $\nu \geq 2$. Since each generator is its own inverse, its Cayley graph is a tree in which every node i has n neighbors $iA_1, \cdots, iA_\nu$. One can naturally extend the notion of *cars* and *ocars* of Sect. 10.2 as long as he specifies two things: how the input is to be given to the automaton and the criterion for acceptance. The input/output tape is the content of the subpath given by nodes expressible as a combination of the first two generators A_1, A_2. The automaton is given a word over the alphabet S and it is to determine membership in a given string language L by stabilizing in an appropriate configuration x_{yes} or x_{no}, as specified in Chapter 6. For a given complexity class of functions $\mathbf{D}$, let $\mathbf{D}(\Gamma)$ be the class of string languages thus accepted by cellular automata within some time function $t(n) \in \mathbf{D}$ over space Γ.

Theorem 10.19 *The following equalities hold:*

(a) $\mathbf{P}(D_3) = \mathbf{PSPACE}$

(b) $\mathbf{P}(D_\nu) = \mathbf{P}(F_\nu) = \mathbf{P}(F_2)$ *for all* $\nu \geq 4$.

Proof. It is clear that if Γ' is a subgraph of Γ containing the input/output tape, then $\mathbf{P}(\Gamma') \subseteq \mathbf{P}(\Gamma)$ by an obvious extension of the local rule. Now the free group on two generators F_2 contains free subgroups on any number of generators and a copy of the tree D_3. This remark establishes all but the first inclusion in part (b). To prove that $\mathbf{P}(F_\nu) = \mathbf{P}(D_3)$, observe that D_3 contains a subgroup isomorphic to F_2 (see Problem 2.3) so that a simulation of an iteration of a cellular automaton over F_2 can be easily accomplished mimicking the arguments in Theorem 6.4.

To prove the inclusion $\mathbf{P}(D_3) \supseteq \mathbf{PSPACE}$ in part (a), just observe that polynomial time Turing reductions (and, more general programs) can be simulated by a cellular automaton. So, it suffices to check that some **PSPACE**-complete problem can be recognized in polynomial time by some cellular automaton on D_3. Quantified boolean formula QBF in prenex normal form is one such. An instance has the form $\amalg_1 v_1 \cdots \amalg_l v_l \psi$, where each $\amalg_*$ is either $\forall$ or $\exists$ and ψ is a quantifier-free boolean formula. A cellular automaton T can verify the validity of such a formula in six stages as follows. First, it marks the rightmost symbol of the input and marks it as the root of a tree (level 0). Second, T labels the successors of the root with quantifiers the various quantifiers $\amalg_*$ (level 1); T also marks the 2^l sites at the lth level *read* (they will each represent a particular assignment of truth values to $v_1, \cdots, v_l$ by interpreting, starting from the root, a left turn as 0 and a right turn as 1). Third, at each cell i marked *read*, T makes a copy of ψ at each cell at the lth level, disjoint of all the others. Fourth, T evaluates the formula ψ for the truth values assigned by an i, simultaneously for all *read* nodes *is* at the lth level. Fifth, based on these values, T evaluates the quantifiers in the intermediate nodes between the *read* cells toward the root according to their label (universal or existential) assigned in stage 2. Finally, T cleans up the workspace and writes out the appropriate answer in the input/output tape. It is evident that each of these stages can be carried out in polynomial time by a Turing machine and hence T answers in polynomial time.

The inclusion $\mathbf{P}(D_3) \subseteq \mathbf{PSPACE}$ in part (a) is easier. Although a local rule δ accepting in polynomial time $p(n)$ might use $O(2^{p(n)})$ cells in workspace (which prevents a brute force simulation by a Turing machine), it is easy to reduce acceptance of an input of size n to evaluation of an exponential number $c^{p(n)}$ of certain terms, which can be done in space $O(p(n))$ as described in the next paragraph. First, reduce the problem to evaluation of the state of one cell at time $p(n)$. Second, associate to each cell i at time $t \leq p(n)$ a term of the form $\tau_{i,t+1} := \delta(\tau_{i,t}, \tau_{iA_1,t}, \cdots, \tau_{iA_\nu,t})$; the term $\tau_{i,0}$ is the state of cell i in the initial configuration on the given string of length n. The value of the term $\tau_{i,t}$ is the state of the cell i at time t.

These terms are constructed from boolean variables and cell states as atomic terms (of depth 0 and size 1, respectively) through iterated application of δ. The *depth* (respectively, the *size*) of a term $\delta(\tau_1, \cdots, \tau_\nu)$ is $1 + \max\{depth(\tau_1, \cdots, \tau_\nu)\}$, the longest nested chain of applications of δ (respectively, $1 + size(\tau_1) + \cdots + size(\tau_\nu)$). A term of size $O(\nu^d)$ and depth d can be evaluated by chunking it into pieces of size d (which require only space of the same size) using a postorder traversal of the tree that represents its construction.   □

Again, little else is known about picture π-languages on cellular spaces of exponential growth.

10.3.5 Recognition over Spaces of Subexponential Growth

Now we turn our attention to the innermost and least known layer in the growth hierarchy. The layer of π-languages on Cayley graphs of subexponential growth have most decision problems unsolvable and a very complex structure, even for sequential recognizers of the type looked at in Sect. 10.3.2.

The first known family of groups of exponential growth can be described as a parametrized family of groups of isometries of the infinite complete binary tree. Some definitions are required to state the results of this section. Recall that if X generates a group G, every element of G can be expressed as a word $\omega \in (X \cup X^{-1})^*$. This expression is never unique (for instance, insertion and/or deletion of words of type AA^{-1}, $A \in X^{\pm 1}$, does not change the group element). Recall the following algorithmic problem.

Definition 10.20 *The* word problem *of the group G with respect to the generating set X is the formal language*

$$\mathsf{WP}(G, X) := \{\omega \in (X \cup X^{-1})^* : \omega = 1_G\}$$

consisting of words that multiply to the identity of G.

The Turing time and space complexity of the set $\mathsf{WP}(G, X)$ are group-theoretic properties independent of X, so X can be omitted. Also, if the elements of X are involutions (i.e., satisfy $A^2 = 1$), the alphabet of $\mathsf{WP}(G)$ does not have to be symmetrized with X^{-1}.

One can construct groups of subexponential growth as groups G_χ of isometric permutations of the infinite complete binary tree defined by a parameter χ, a one-way infinite ternary string. The Turing complexity of such a string can be precisely defined as follows. (As usual, expressions such as $f(log \, log \, n)$, etc. will be understood to mean $f(\lceil log_2 log_2 \, n \rceil)$, etc.)

Definition 10.21 *Let $f : \mathbf{N} \to \mathbf{N}$ be a function on the set of natural numbers $\mathbf{N}$. An infinite ternary sequence χ is said to belong to a complexity class* **DSPACE**(f) *if there exists a Turing machine that computes the jth digit χ_j of χ in space bounded above by $f(n)$ given the $n = \lceil \log j \rceil + 1$ digits representing its input j in binary. Likewise for* **DTIME** *and the corresponding nondeterministic complexity classes (i.e., in computing χ_j nondeterministic moves are allowed and halting only occurs when χ_j has been found).*

The construction can be succinctly described as follows. Let Υ be the set of one-way infinite *paths* from the root of the complete infinite binary tree (which we will identify with the tree itself). Each element γ of Υ can be regarded as an infinite binary sequence $(\gamma_i)_{i \geq 1}$ of edges, each of which will be referred to as a *left* or *right turn* and denoted 0 or 1, respectively, according to its orientation from its parent. The *complement* of a left (respectively, right) turn is the corresponding right (left) turn. The group G_χ is a group of permutations of Υ generated by four bijections $a, b_\chi, c_\chi, d_\chi$, which act as follows on a path γ in Υ. The generator a complements the first turn γ_1 of γ and leaves all the other turns γ_i ($i \geq 2$) invariant. Pictorially, a swaps the two halves of the infinite tree Υ. The action of b_χ, c_χ, and d_χ is determined by the respective infinite rows $U := u_1 u_2 \cdots u_n \cdots$, $V := v_1 v_2 \cdots v_n \cdots$, and $W := w_1 w_2 \cdots w_n \cdots$ over $\{I, S\}$ (meaning I=identity, S=swap) obtained from χ by substitution of its digits according to the following table:

$$0 := \begin{cases} S \\ S \\ I \end{cases}, \quad 1 := \begin{cases} S \\ I \\ S \end{cases}, \quad 2 := \begin{cases} I \\ S \\ S \end{cases}.$$

The generator b_χ, for each $i \geq 1$, leaves invariant all turns $\gamma_1, \ldots, \gamma_i$ up to and including the first left turn γ_i of γ, and complements the next turn γ_{i+1} *if* the corresponding i^{th} entry u_i in the first row U of b_χ is S. Otherwise, b_χ leaves γ invariant. The generators c_χ, d_χ are defined in an analogous way except that their action now depends on the entries of the second and third rows V and W of χ, respectively. These generators are involutions, i.e., they are their own inverses. Further, the following relations hold:

$$bc = cb = d, \quad cd = dc = b, \quad bd = db = c \tag{10.3}$$

(for notational simplicity the subscript χ will be left out when it is clear from the context). Therefore, every element of G can transformed to a product of the generators a, b, c, d in which no two of b, c, d appear in consecutive positions, and all generators appear without consecutive repetitions.

To illustrate this construction put $\chi := 012\ 012\ 012\ \cdots$, so that the generators b_χ, c_χ, d_χ correspond to the rows of χ given by:

$$
\begin{aligned}
U &= SSI\ SSI\ SSI\ \cdots \\
V &= SIS\ SIS\ SIS\ \cdots \\
W &= ISS\ ISS\ ISS\ \cdots.
\end{aligned}
$$

Hence b_χ maps a path γ to a different path if and only if the first left turn γ_i of γ occurs when i is not congruent to 0 modulo 3; c_χ maps a path γ to a different path if and only if the first left turn γ_i of γ occurs when i is not congruent to 2 modulo 3; and d_χ maps a path γ to a different path if and only if the first left turn γ_i of γ occurs when i is not congruent to 1 modulo 3. Since the first left turns in the path γ are not complemented by any of these generators, a second application of the same generator will complement again the same turns and return the path to its original shape. This means that a, b, c, d are all involutions.

These groups, knwon as the *Grigorchuk groups* are infinite, residually finite, and have the following properties:

(1) if all three symbols 0, 1, and 2 repeat infinitely often in χ, then G_χ is a 2-group;

(2) if exactly two of 0, 1, and 2 repeat infinitely often in χ, then G_χ contains elements of infinite order but contains no free (sub)semigroup on two generators;

(3) if at least two of 0,1, and 2 repeat infinitely often in χ, then G_χ is *not* finitely presentable and has subexponential growth.

In this notation one can gauge the difficulty of π-word problems in terms of Turing complexity by the following three results. Let's say that a language L *separates* the inclusion of two space complexity classes **DSPACE**$(f) \subseteq$ **NSPACE**(f) iff $L \in$ **NSPACE**(f)–**DSPACE**(f). Likewise for time complexity classes.

Theorem 10.22 *If $f : \mathbf{N} \rightarrow \mathbf{N}$ is a nondecreasing function such that $f(n) \geq 2^n$ and if* **DSPACE**$(f) \neq$ **NSPACE**(f), *then some word problem* WP(G_{χ_L}) *of a Grigorchuk group separates the inclusion*

$$\mathbf{DSPACE}(f(\log \log n)) \subseteq \mathbf{NSPACE}(f(\log \log n)).$$

Theorem 10.23 *If $f : \mathbf{N} \rightarrow \mathbf{N}$ is a nondecreasing function such that $f(n) \geq 2^{2^{n+1}}$ and if* **DTIME**$(f) \neq$ **NTIME**(f) *then some word problem* WP(G_χ) *of a Grigorchuk group separates all the inclusions*

$$\mathbf{DTIME}(\frac{f(\log \log n - 2)}{\log^{1+\varepsilon} n}) \subseteq \mathbf{NTIME}(f(\log \log n - 1)\log n + n^2),$$

provided that $\varepsilon > 0$. In particular, if **DTIME**$(2^{2^{n+3}}) \neq$ **NTIME**$(2^{2^{n+3}})$, *then a Grigorchuk group word problem separates the inclusion*

$$\mathbf{DTIME}(n^2/\log^2 n) \subseteq \mathbf{NTIME}(n^4 \log n).$$

Corollary 10.24 *If $g(n) \geq \log n$ and $g(n)$ is space constructible, then there exists a complete word problem of a Grigorchuk group in* **NSPACE**(g) *with* *respect to logspace reductions. Likewise for deterministic and time complexity classes with $g(n) \geq n^2$ and linear-time reductions.*

In particular, there exist word problems of groups of subexponential growth of any degree of Turing unsolvability.

On the other hand, observe that ordinary word problems can be posed as simple π-languages by entering every input as a nonquiescent configuration in unary described by following the corresponding generators on the cellular space. As such, the problem admits a trivial solution, even by a sequential 1-pebble automaton which initially drops its pebble to mark an end of the path as an origin, then walks along the nonquiescent path and decides at the end of the word whether it has returned to the origin or not.

Thus, the study of π-languages on arbitrary cellular spaces remains a puzzle closely related to old standing questions in classical complexity theory.

10.4 Problems

SIGNALS AND SYNCHRONIZATION

1. Show the following is a solution to the FSP in real time n. The general at the origin is given a counter up to n. Each cell in a quiescent state initializes its counter with the counter value at its left minus 1. Each soldier in a nonquiescent state decreases its counter by 1. A cell fires when it reaches state 0. Give at least two reasons why this solution is undesirable.

2. Find a solution of FSP on the line in time $2n - 1 - k$ if the general occupies the site at position k instead of position 0, where $1 \leq k \leq n/2$.

3. Modify the Waksman–Balzer's solution to solve FSP on a torus of n cells in time $n - 1$ with bidirectional flow of information and in time $2n - 2$ with one-way flow.

4. Solve FSP(D) for a tree D on n vertices. [Reduce to the problem on a line by a suitable traversal.]

SIGNALS AND CLOCKS

5. Show that the nonquiescent wave front along the light cone of a pixel's evolution in a cellular automaton is eventually periodic, even if the pixel is immersed in a uniform configuration consisting of a nonquiescent state.

6. Construct a quadratic clock. Use it as a basis of an induction to construct polynomial clocks. [Use the binomial theorem.]

7. Construct a signal of exponential ratio. [Use the exponential clocks constructed in Sect. 10.1.4.]

8. Show that $n + \log_m n$ and $n!$ are Fisher constructible.

9. Show that the ratio $f(n)$ of every ca-constructible signal is Fisher constructible.

10. Give an example of a Fisher constructible function which is not the ratio of a signal. [Consider the complement $f'(n) := n + |\{l : f_l - l < n\}|$ of a superexponential constructible function f_l, such as the factorial, which must be sublogarithmic.]

11. Show that if f is a Fisher constructible signal and m is a positive integer, then

    (a) mf is also Fisher constructible; [Make a signal f_n emit a photon that bounces back from a signal of slope $\frac{m+1}{m-1}$ to arrive at the origin at time mf_n.]

    (b) $\frac{1}{m}f$ is also Fisher constructible.

    Therefore Fisher constructible signals are closed under multiplication by a rational number.

12. Show that Fisher constructible signals are closed under addition. [If $f_n \leq g_n$, $f_n + g_n = 2f_n + (g_n - f_n)$.]

LANGUAGE RECOGNITION

13. Show how to perform serial composition of two cellular automata T_1, T_2 on finite configurations, the first of which halts by stabilizing all its non-quiescent cells in a certain 'Fire!' state.

14. Show that the following sets over any finite alphabet Σ are recognizable in real-time by oias: (a) $\{a^n b^n : n \geq 0\}$ (b) the set of palindromes $\{ww^R : w \in \Sigma^*\}$; (c) $\{ww : w \in \Sigma^*\}$. (R is the reversal operator for strings.)

15. Show that the language $\{0^n 1^{n+m} 0^m : n, m \geq 0\}$ can be recognized in real-time by an oca.

PICTURE LANGUAGES

16. Show that the emptiness problem for local π-languages is unsolvable. [See the proof of unsolvability of 2D PARENT SEARCH.]

17. Show that the class of recognizable π-languages is closed under row and column 2D-catenation (obtained by lining up arrays of the same size, horizontally or vertically respectively, and catenating them in the obvious way).

18. Show that the language of squares over a one-letter alphabet is recognizable but not locally testable.

19. Show that recognizable languages properly include languages accepted by Blum–Hewitt automata. [The latter are not closed under row or column catenation.]

20. Show that the class of languages accepted by deterministic Blum–Hewitt automata are closed under boolean operations.

21. Show that for every group Γ, $\mathbf{P}(\Gamma) = co\mathbf{P}(\Gamma)$.

10.5 Notes

The firing squad synchronization problem was first posed informally by J. Myhill in 1957, who wanted to synchronize all the parts of a self-reproducing machine to turn them on simultaneously. It was stated formally in Moore [Moo] and given a number of solutions by MacCarthy–Minsky (as cited in [Mi, p. 28-9]) in time $3n - 1$, and minimal time $2n - 1$ by Waksman [Wa], Balzer [B], and Mazoyer [Ma1, Ma2] with 16, 8, and 6 states respectively. These sources provide also the actual local rule of the cellular automaton solution. The question remains whether there is a minimal time solution with 5 states. Several variations in Problems 2–3 appear in Moore–Langdon [M-L] and Culik [Cu]. Mazoyer [Ma2] has also shown that FSP can be solved with only one bit of information exchanged. The firing squad problem on the plane is considered by Kobayashi [K]. The solutions presented here for the generalized FSP to arbitrary finite graphs are due to Jiang [J]. One can further solve $\text{FSP}_G(D)$ with several generals at the vertices of a subset G, some of which may be initially activated, some of which (but not all) may be initially idle and may be excited by the external world at different times. A solution is easily obtained if one of the generals could be elected as a leader by treating the remaining generals as soldiers, but it may become difficult (sometimes impossible) to have the generals break possible ties. In fact, Jiang [J] gives a solution for FSPU with a bounded number of generals in time $O(\Xi)$, where Ξ is the diamater of the graph (maximum distance between any two vertices), as well as a solution for FSPN in time $O((\Phi + \tau_{max})^3)$, where Φ is the delay diameter (the maximum transmission delay) of the network. He further shows that solutions do not exist for an unbounded number of generals.

Signals were used by von Neumann in his design of a computation universal constructor. His use in solutions to specific problems was continued by Fisher, who designed a cellular automaton to generate the sequence of primes in real-time [F]. Recently, it has been further developed by J. Mazoyer and his students. The results in Section 10.1.4 originate in their survey [M-T]. Signals and clocks formalize an important aspect of the fundamental problem of the transmission and storage of information in cellular automata processing. Despite the importance of this and the problem of information storage in specific applications of cellular automata, as pointed out in [L], no systematic study has been initiated in the literature beyond the few results presented here for the 1D case. The study of computation by cellular automata can be regarded as part of the field of communication complexity for massively parallel computers.

Language recognition was one of the first natural applications of cellular automata in the blooming of formal languages and grammars in the 1950–1970s. Early works appear in Fisher [F], Cole [C], and A.R. Smith III [Sm1, Sm2, Sm3]. Generalizations to higher dimensional languages (also called *array languages*) is also an old idea, as witnessed by the early papers of Blum–Hewitt [B-H], Selkow [Se], A.R. Smith III [Sm3], and Rosenfeld [Ros] (where one can find a survey of early work up to 1979). One-way iterative arrays are considered in Hennie [H] and Dyer [D]. The notion of one-way automata has been generalized to arbitrary Cayley graphs in Róka [Rok]. 2D array languages have always been subject of active study as witnessed by the extensive literature in journals on image processing and pattern recognition – see, for example, the survey by Inoue–Takanami [I-T] and Saoudi et al. [N-S-D, S-R-D]. The exposition in Sect. 10.3.3 follows the works of Morita et al. [Mo1, M-U] and Giammarresi–Restivo [G-R]. A witness to the strict inclusion of locally testable languages in the class recognized by Blum–Hewitt automata in Theorem 10.18 can be found in Inoue–Takanami–Nakamura [I-T-N]. Computational problems for circular π-words recognized by *sequential* automata have been studied by Garzon [Ga1, Ga2]. Groups of polynomial growth have been recently characterized precisely as the nilpotent-by-finite groups by deep results of Gromov [Gro]. Thus they are linear groups, and their word problems can be solved in logspace, hence in polynomial time, by a result of Lipton–Zalcstein [L-Z]. Theorem 10.19 is due to Mycielski–Niwiński [M-N]. Groups of subexponential growth were discovered by Grigorchuk [Gri] as counterexamples to a conjecture of Milnor [M]. The more combinatorial presentation of his construction given in Sect 10.3.4 appears in Garzon–Zalcstein [G-Z].

Recent developments in a number of applications – see, e.g., [Wol, Gut1, R, Gut5, Gut6] – show the approach of Sect. 10.3.3 may turn out to be fruitful in a systematic development of symbolic-like processing via cellular automata. However, more work is needed to develop enough of a treatment comparable to the theory of computation for string languages in order to decide whether this approach will be the most natural one in the context of fine-grained parallel processing.

References

[B]   R. Balzer: An 8-state minimal solution to the firing-squad synchronization problem. Inform. and Control **10**(1967) 22–42

[B-H]   M. Blum and C. Hewitt: Automata on a 2-dimensional tape. Proc. 6th IEEE Annual Symp. on Switching and Automata theory (now FOCS) (1967) 179–190

[C]   S.N. Cole: Real computation by n-dimensional iterative arrays. IEEE Trans. on Computers **C-18** (1969) 349–365

[Cu]   K. Culik: Variations on the firing squad problem and applications. Inf. Proc. Letters **30** (1989) 153–157

[D]   C. Dyer: One-way bounded cellular automata. Inf. and Control **44** (1980) 261–281

[F]   P.C. Fisher: Generation of primes by a one-dimensional iterative array. J. Assoc. Comput. Mach. **12** (1965) 388–394

[Ga1]   M. Garzon: Cyclic Automata. Theoret. Computer Sci. **53** (1987) 307–317

[Ga2]   M. Garzon: Cayley Automata Theoret. Computer Sci. A **108** (1993) 83–102

[Ga3]   M. Garzon: Graphical Words and Languages. In: Proc. Int. Conf. on Words, Languages and Combinatorics, Kyoto, 1990. M. Ito, ed., World Scientific Publishing, Singapore, pp. 160–178

[G-Z]   M. Garzon, Y. Zalcstein: The complexity of Grigorchuk groups with application to cryptography. Theoret. Computer Sci. **88**:1 (1991) 83–98

[G-R]   D. Giammarresi, A. Restivo: Recognizable picture languages. Int. J. of Pattern Recognition and Artif. Intel. **6**:2/3 (1992)

[Gri]   R. Grigorchuk: Degrees of growth of finitely generated groups and the theory of invariant means. Math. USSR Izv. **25** (1985) 259–300

[Gro]   M. Gromov: Groups of polynomial growth and expanding maps. Inst. Hautes Etudes Scientifiques Publ. Math. **53** (1981) 53–78

[Gut1]   H. Gutowitz (editor): Cellular automata: theory and applications. Proc. 3rd. Int. Conf. Cellular Automata, Los Alamos, 1991. Physica D **45** (1990) 431–440. Also issued as a separate book by MIT Press, 1992

[Gut5]   H.A. Gutowitz: Statistical Properties of Cellular Automata in the Context of Learning and Recognition, Part I: Introduction. In: Learning and Recognition – A Modern Approach, K.H. Zhao. (ed.) World Scientific Publishing, Singapore (1989), pp. 233–255

[Gut6]   H.A. Gutowitz: Statistical Properties of Cellular Automata in the Context of Learning and Recognition, Part II: Inverting Local Structure Theory Equations to Find Cellular Automata With Specified Properties. In: Learning and Recognition – A Modern Approach, K.H. Zhao. (ed.) World Scientific Publishing, Singapore (1989), pp. 256–280

[H]   F.C. Hennie: Iterative arrays of logical circuits. MIT Press, Cambridge MA, 1961

[I-T-N]   K. Inoue, I. Takanami, A. Nakamura: A note on two-dimensional finite automata. Inf. Process. Lett. **7** (1978) 49–52

[I-T]   K. Inoue, I. Takanami: A survey of two-dimensional finite automata. In: Proc. 5th Int. Meeting of Young Computer Scientists, J. Dassow and J. Kelemen (eds.). Lecture Notes in Computer Science, Vol. 381. Springer-Verlag, Berlin, 1989, pp. 72–91

[J]   T. Jiang: The synchronization of nonuniform networks of finite automata. Proc. IEEE Symp. on Foundations of Computer Science FOCS **30** (1989) 376–381

[K]   K. Kobayashi: The firing squad synchronization problem for two-dimensional arrays. Inf. and Control **34** (1977) 177–197

[L]   C.G. Langton: Computation at the edge of chaos: phase transitions and emergent computation. Physica D **42** (1990) 12–37

[L-Z]   J. Lipton, Y. Zalcstein: Word problems solvable in logspace. J. Assoc. for Comput. Mach. **24** (1977) 522–526

[Ma1]   J. Mazoyer: A six-state minimal time solution to the firing squad synchronization problem. Theoret. Comput. Sci. **50**(1987) 183–238

[Ma2]   J. Mazoyer: A minimal time solution to the firing squad synchronization problem with only one bit of information exchanged. Rapport de Recherche 89-03, LIP-École Normale de Lyon, France

[M-T]   J. Mazoyer, V. Terrier: Signals in one dimensional cellular automata. Rapport de recherche, LIP École Normale Supérieure de Lyon, France, 1993

[M]   J. Milnor: Advanced Problem 3603, Amer. Math. Monthly **75** (1968) 685–686

[Mi]   M. Minsky: Finite and Infinite Machines. Prentice-Hall, Englewood Cliffs NJ, 1967

[Moo]   E. Moore: Sequential machines, selected papers. Addison-Wesley, Reading MA, 1964

[M-L]   F.R. Moore, G.G. Langdon: A generalized firing squad problem. Inf. and Control **12** (1968) 212–220

[Mo1]   K. Morita, Y. Yamamoto, K. Sugata: Two-dimensional three-way array grammars and their acceptors. Int. J. of Pattern Recognition and Artificial Intelligence. **3**:3/4 (1989) 353–376

[M-U]   K. Morita, S. Ueno: Parallel generation and parsing of array languages using reversible cellular automata. Preprint

[M-N]   J. Mycielski, D. Niwiński: Cellular automata on trees, a models for parallel computation. Fund. Informaticae **XV** (1991) 139-144

[N-S-D] M. Nivat, A. Saudi, V.R. Dare: Parallel generation of finite images. Int. J. of Pattern Recognition and Artif. Intel. **3**:3/4 (1989) 279–294

[Rok]   Z. Róka: One-way cellular automata on Cayley graphs. Preprint. LIP, École Normale Supérieure de Lyon, 1993

[Ros]   A. Rosenfeld: Picture languages (formal models for picture recognition). Academic press NY, 1979

[R]   P. Rujàn, Cellular Automata and Statistical Mechanical models. J. Statistical Physics **49** (1987) 139–232

[S-R-D] M. Saoudi, K. Rangarajan, V.R. Dare: Finite images generated by GL-systems. Int. J. of Pattern Recognition and Artif. Intel. **3** (1989) 459–467

[Se]   S.M. Selkow: One-pass complexity of digital picture properties. J. Assoc. Comput. Mach, **19** (1972) 283–295

[S-S]   H. Siegelman and E. Sontag: On the computational power of neural nets. Proc. 5th Comput. Learning Theory Conf. COLT (1992) 440–449

[Sm1]   A.R. Smith III: Cellular automata and formal languages. Proc. 11th IEEE Symp. on Switching and Automata Theory (now FOCS) (1970) 216–224

[Sm2]   A.R. Smith III: Two-dimensional formal languages and pattern recognition by cellular automata. Proc. 12th IEEE Annual Symp. on Switching and Automata Theory (now FOCS) (1971) 144–152

[Sm3]   A.R. Smith III: Cellular automata complexity tradeoffs. Inform. and Control **18** (1971) 466–482

[Sm3]   A.R. Smith III: Real-time language recognition by one-dimensional cellular automata. J. Assoc. Comput. Mach. **6** (1972) 233–253

[T]   J. Tits: Free subgroups in linear groups. J. of Algebra **20** (1972) 250–270

[Ya1]   Y. Yamamoto, K. Morita, K. Sugata: Context-sensitivity of two-dimensional regular array grammars. Int. J. of Pattern Recognition and Artificial Intelligence. **3**:3/4 (1989) 259–319

[Wa]   A. Waksman: An optimal solution to the firing squad synchronization problem. Information and Control **8**(1966) 66–78

[Wol]   S. Wolfram: Theory and Applications of Cellular Automata. World Scientific, Singapore, 1986

11. Real Computation

Mais les vrais voyageurs sont ceux-là qui partent
pour partir ··· Et sans savoir pourquoi, disent toujours: Allons!
Charles P. Baudelaire

Like formal language recognition, computation of real-valued functions is a well-defined area of classical computation. The concept was the motivation behind the celebrated Turing paper on computable numbers and has been studied ever since. Given the fact that sequential computing devices (Turing machines and the like) are countable in number and process information coded as finite strings of symbols over a bounded alphabet, the various notions of computable function are necessarily of an approximative nature. Now, generalizations of computability have been made on various types of assumptions.

In classical analysis, one takes a certain class of primitive functions (e.g., polynomials, sines, and cosines) and operations (additions, multiplications, and convergent series) and obtains the well-known complexity hierarchies of continuous functions (Taylor series) and periodic functions (Fourier series). From a more computational point of view, the notion of BSS-complexity has proved of interest in providing a framework for the study of sequential complexity of real functions vis à vis dynamical systems. Information and complexity arguments made in Chapter 6, however, make these approaches unfeasible from a practical perspective. While it is desirable to have solutions that are precision independent, it is physically impossible to carry out measurements to infinite precision. Yet, it is necessary to cope with the facts that a real number generally contains an infinite amount of information, that there are uncountably many of them, and that physical objects seem to be able to handle representations of this type in very short times in a highly parallel fashion. What is needed are representations and procedures capable of penetrating the infinitary nature of real numbers through the finiteness of feasible hardware.

From the point of view of parallel computation, analog approaches can be taken based on *finite* sets of real-valued functions such as the saturated linear and sigmoids functions. One obtains at least the class of continous functions within relatively low complexity. However, there remain three somewhat unsatisfactory issues. First, one still assumes that the basic functions can be calculated to unbounded precision in *unit time*. This assumption is somewhat unrealistic in terms of digital hardware implementation and communication time required to perform these operations even on parallel machines. Second, despite infinite precision, the errors involved in the approximations run out of control under iteration of the function. Third, very few (only countably many) functions are computable despite the fact that the notion of computation is, except in very few cases, still based on approximations that come arbitrarily close to the values of the function being computed, but are approximations nonetheless (even under the ideal conditions of Turing machines).

On the other hand, cellular automata and boolean networks offer the advantages of local parallel algorithms. In addition, they afford notions of *exact* computability that are model independent, robust under iteration, uniformly approximable (for all functions, by considering finite portions of the cellular space), hence easily implementable in hardware. To some extent, the problem is dual of the simulation of cellular networks by continuous systems dealt with in Sect. 8.2. In particular, the fundamental difficulty to overcome is the dual problem of representation of real values. Representations in use since the beginning of time, expansions and the like, are essentially one-dimensional. The nature of parallel processing seems naturally to impose a distributed representation as the best prospect to fully exploit the computational power of cellular models. Yet, such representations have hardly been considered in the literature, and the few that have been, only very recently. This chapter explores the capabilities of cellular networks as models for parallel calculation of real-valued functions.

Due to massive parallelism, one can expect to be able to compute functions on cellular models faster that on sequential counterparts. In fact, certain *real-valued* functions are now *exactly* computable. Moreover, they are computable in *constant time*. Exact computation is explored in the next section. For one-dimensional encodings, functions computable in constant time admit a sequential characterization as functions computable in *on-line* mode. This definition is generalized to variable time under a natural halting condition and more general representations. Since only restricted classes of functions are exactly computable, Sect. 11.3 explores two natural alternatives for approximate computation that exploit the observability property of neural networks.

11.1 Representation and Primitives

The fundamental issues to resolve in dealing with real-valued computation concern (i) how to represent real values; and (ii) what type of procedures can be used to perform on them desired operations.

The issue of representation admits two type of solutions. The first and oldest type of representation, an infinite expansion in a fixed base m, is a recording of an idealized measurement process with respect to successively smaller units (elements in a geometric sequence $\{m^{-n}\}_n$) Even more efficient and flexible systems, such as signed-digit arithmetic, Avizienis' redundant systems, etc., are still *inherently* sequential. Further, all these systems aim at a type of representation appropriate for linear memories and sequential processes, hence still essentially one-dimensional. Usually this alternative also requires the ability to perform basic operations on these representations in a single time step, regardless of the precision of the numbers involved. This is the classical mathematical solution, in which polynomials (addition–multiplication combinations) play a central role of desirable computational primitives.

However, exact evaluation of real-valued functions on the continuum is un-thinkable in general on conventional computer arithmetic, where only approximations to real values (such as truncations of infinite discrete expansions) can be represented and manipulated (even on idealized models such as Turing machines). More symbolic and/or compact representations (such as polynomials for algebraic numbers, fast Fourier transforms, etc.) require at any rate basic representations of real coefficients, and they hence still present the same basic problem. This option leads to the analog BSS models discussed earlier.

Now, it appears very unlikely that, for parallel processing in the wild (by nature) this is the way quantities are represented, even assuming analog encodings (to infinite precision and/or irrational bases). It is not hard to imagine a second type of representation by state patterns in a cellular space. In fact, this is the type used in cellular automata simulations closely resembling several physical phenomena. Distributed representations have also been key to the success of certain applications of neural networks. The most general of this kind of representation is one in which real numbers are represented by initial input configurations of a cellular network. Even minor deviations of arithmetic expansions yield interesting results. Yet, these representations can be very distributed, perhaps higher-dimensional or even more complex, depending on the connectivity of the automaton's grid (although many results below are representation independent).

The following definition allows for both types of representation.

Definition 11.1 *A representation of real numbers* $\mathbf{R}$ *is a partition of configuration space* $\mathbf{C}$ *given by a function* $\phi : \mathbf{C} \to \mathbf{R}$, *the configurations in* $\phi^{-1}(x)$ *representing a real number* $x \in \mathbf{R}$. *Equivalently, one can describe* ϕ *by a multivalued function*

$$\psi : \mathbf{R} \to \mathbf{C}$$

that encodes x *in several ways as a configuration in* $\psi(x)$. *Likewise one can define representation for subsets of reals by replacing* $\mathbf{R}$ *with a subset* X.

Example 11.2 In the standard m-ary representation, $\phi^{-1}(x)$ consists of at most two expansions, more often one. There are variations thereof. Recall that the bandwidth of a (possibly infinite) grid is a measure of the thinness of the grid (See Definition 6.3). A one-dimensional representation can be more generally regarded as any distributed representations on a digraph of finite bandwidth.

The iterative nature of arithmetic calculation makes it desirable to adopt a few encodings uniform for various operations. The ψ-representation may be more desirable when the domain of ϕ is not all of $\mathbf{C}$. Thus we shall also refer to ϕ as a *decoding* function. But since the encodings are themselves the object of research, in principle, no restriction is placed on the type of representations (e.g., bijective, recursive with respect to oracles for the expansions of the given number x, continuous, etc.)

The second fundamental issue, that of basic operations allowable on representations, is a more contextual issue that determines the type of computation. Various alternatives are discussed in the following sections.

Without loss of generality, we only consider functions defined on the unit interval $I \subset \mathbf{R}$. The extensions to unbounded functions and several variables are straightforward once the one-variable primitives are known.

11.2 Exact Computation

In this section cellular automata are taken as the basic allowable operations. Thus, for a given function f, the basic problem is how to operate on an arbitrary configuration, representing a real number x, by full cellular automata transformations so as to eventually obtain some representation in $\psi(f(x))$.

11.2.1 Constant-time Computation

Because of massive parallelism, cellular automata can perform complex transformations in a single time step. Thus, it is necessary to begin as follows.

Definition 11.3 *A function $f : I \to I$ is π-computable in constant* time *in representation ϕ, if there exists a neural network T and a positive integer K such that on input any ϕ-representation of a real number x as initial condition, the iteration of T produces at time K a ϕ-representation of $f(x)$.*

In particular, a statement of exact computability requires proof that the equivalence relation defined by the representation ϕ is preserved by the action of the automaton, i.e., that

$$\phi(a) = \phi(b) \quad \Rightarrow \quad \phi(T^K(a)) = \phi(T^K(b)) \,. \tag{11.1}$$

Example 11.4

1. Obvious examples of constant-time computable functions are multiplication (division) by a fixed integer m in radix m, corresponding to the (left-) right-shift $T := \sigma$ given by

$$\sigma(x)_i := x_{i+1} \,.$$

2. In case ϕ happens to be a semiconjugacy of a cellular automaton onto a real-valued function $f : I \to I$, i.e., when $\phi T = f\phi$, $\phi(a) = \phi(b)$ implies

$$\phi T(a) = f\phi(a) = f\phi(b) = \phi T(b) \,.$$

Thus T π-computes f in representation ϕ in one step. Proposition 11.5 below shows there is an infinitude of continuous functions on the unit circle thus π- computable.

3. Likewise, every cellular automaton T with m states gives rise to a real-valued function of the interval π-computable in one step in any representation coarser than the partition induced by T, i.e., for which

$$\phi(a) = \phi(b) \Rightarrow T(a) = T(b).$$

In fact, one can then define $f(x)$ for $x \in I$ by putting

$$f(x) := \phi T(a),$$

where a is an arbitrary element in $\phi^{-1}(x)$. The condition above implies that f is well-defined.

The equivalences shown in Chapter 5 show that for 1D (finite-bandwidth) representations Definition 11.3 below is *robust* in the sense that the same class of functions can be computed if one uses cellular automata, neural networks or automata networks.

The question of interest is whether there are any interesting π-computable functions of the interval.

Proposition 11.5 *Let T be an arbitrary 1D cellular automaton. If there exists a countable dense subset $E \subseteq \mathbf{C}$ that can be partitioned into two-element subsets stable under T, then T computes a continuous, integrable real-valued function of the unit circle with the same entropy and which is completely determined by the value of its integral. In particular, there exist infinite families of continuous functions on the unit circle exactly π-computable in constant time.*

Proof. As noted in Problem 7.9, configuration space is homeomorphic to the ternary Cantor set in the unit interval by a homeomorphism mapping any dense subset of $\mathbf{C}$ onto the set of end-points (with finite ternary expansions). After appropiate recodings, one is thus reduced to a continuous self-map T of the Cantor set that preserves end-points. The statements now follow from Proposition 8.3.2 and Theorem 8.3.4. In fact, the proof establishes that these functions are computable by 1D cellular automata.   □

Since cellular automata are closed under composition, π-computable functions in constant time are actually computable in unit time, allowing arbitrary encodings. The Hedlund–Richardson theorem doesn't settle, however, the question of characterization of the set of functions π-computable in constant time for at least two reasons. First, under the product topology, configuration space does not represent the unit interval, but only its Cantor subsets. Second, encodings may not be continuous. In fact, with the same encoding defined in Sect. 8.3.3, which is continuous, one can obtain functions on I computable in constant time with a countable number of discontinuities. The fact that discontinuous functions are π-computable in constant time shows, incidentally, that even constant-time π- computation is a genuinely different concept since Turing computable functions must be continuous in the ordinary metric of the interval. The essential ingredient that brings about this departure is not the change to more general representations (the representation used above is recursive if an oracle is available for the function's argument) but rather massive parallelism.

Nonetheless, we can give a partial result for the case of representations of finite bandwidth. In this case, constant-time π-computation turns out to be equivalent to *on-line* computation by a finite-state machine. On-line computation is a mode that exploits pipe-lining of sequential operations in order to speed up results to (quasi)real-time. There, the basic computing element is a 'black-box' (of various types, maybe including a fixed number of registers, bounded memory, look-up tables and the like). The input x is represented as a string (possibly infinite) whose digits are fed one at a time, usually most-significant first. After a certain constant delay d, the box outputs the corresponding digits in the representation of the image $y = f(x)$.

Definition 11.6 *A function $f : I \to I$ is* computable on-line *by a finite-state machine M with delay d if for some (maybe multi-valued) one-way infinite representation ϕ of finite bandwidth, on being given as input the successive digits x_t of a ϕ-representation of x at each time $t \geq 0$, M outputs the successive digits $\psi(f(x))_{t-d}$ of some ϕ-representation of $f(x)$ for all $t \geq 0$, for every real value $x \in I$.*

Theorem 11.7 *A real function is π-computable in constant time with respect to one-way infinite representations of finite bandwidth, if and only if it is computable on-line by a finite-state machine.*

Proof. (Sketch – see Problem 2). By Theorem 3.2.6 further assume that the cellular automaton is one-way by a slight modification of the representation. Thus the theorem asserts the equivalence between an infinitely long evaluation by a finite-state machine, and a simultaneous evaluation by an infinite row of copies of the same machine in d parallel steps.   □

Expectedly, constant time evaluation appears very stringent. In fact, it is known that, with respect to redundant signed binary representation (in which addition can be performed in constant time regardless of the size of the operands), only piecewise linear functions with rational break points and rational slopes are actually π-computable, among sufficiently smooth functions (continuous and with first derivative continuous except at finitely many points). The full potential of cellular automata for real function evaluation has clearly not been realized this way.

We can enlarge this class of functions in at least two ways: extend Definition 11.3 or relax exactness (i.e., go for approximations).

11.2.2 Variable-time Computation

In order to generalize Definition 11.3, it is necessary to establish a *halting criterion* to decide *when* to read off the value of the function. In general, this is a controversial issue due to massive parallelism and the absence of a centralized control and halting states occurring in Turing machines. In the previous case this problem does not arise since one simply keeps a clock up to constant time that signals when to read the result. Now, for variable time a natural halting criterion is *stability*, i.e., the automaton transforms the initial configuration until no more changes are effected and the automaton has arrived at a stable configuration. This stable configuration is then read off as the corresponding encoding of the value of the function. This is the approach taken in this section.

Definition 11.8 *Let* $X \subseteq \mathbf{R}$ *be a set of real numbers. A function* $f : X \to \mathbf{R}$ *is* π-computable *if there exist encoding and decoding maps*

$$\psi : X \to \mathbf{C} \ \text{and} \ \phi : \mathbf{C} \to \mathbf{R}$$

as well as a neural network with global dynamics

$$T : \mathbf{C} \to \mathbf{C}$$

such that for each $x \in X$ *there exists an integer* $K := K(x) \geq 0$ *such that* T *will eventually stabilize on any input in* $\psi(x)$ *and this fixed point is an encoding of the value* $f(x)$, *i.e.,*

$$\text{Stability:} \quad T^{K+1}(x) = T^K(x), \ \text{and}$$
$$\text{Soundness:} \quad f(x) = \phi[T^K(\psi(x))].$$

Three remarks are in order about this notion. First, the nature of the encoding and decoding functions have been left deliberately open. They must be also *computable* in some sense, lest everything becomes π-computable. Second, the requirement that the automaton enter an equilibrium state may seem too stringent. Its purpose, however, is to ensure that the automaton 'knows' when it has arrived at the value of the functions, as opposed to just accidentally passing through the appropriate value, leaving to the user the burden of picking it out at the appropriate time (which defeats altogether the purpose of the device). Third, the behavior of the automaton T at configurations not encoding values in X is left unspecified. At values x for which f is not defined, T may stabilize with the wrong value or it may never stabilize.

As mentioned above, by attaching a clock that counts up to a constant and then forces stability, it is easy to see that this is a generalization of constant time computability. Furthermore, this is a generalization of ordinary computability, by the construction in Chapter 6. (In this construction, T never enters an equilibrium state where f is undefined.) More generally, Theorem 5.3.2 can be rephrased as follows.

Proposition 11.9 *The characteristic function of every countable set of natural numbers, including recursive functions and recursive encodings of the* HALTING PROBLEM, *is π-computable on recursive encodings.*

For finite bandwidth there are two additional results. First, calculations can be performed on a common universal network described in Theorem 5.3.4. Secondly, the global stability condition is consequently equivalent to a distinguished "grandmother" cell which signals when the value of $f(x)$ is available by entering some special state.

The last two results raise the question whether every real-valued function is computable in variable time. The usual cardinality argument is not applicable since the same cellular automaton can compute many functions as the encodings varies. But indeed, there exist π-uncomputable functions in finite bandwidth. The first example of such a function follows easily from the unsolvability of STABILITY for finite bandwidth by automata within the same class given in Theorem 5.3.7.

In order to gain some insight into the nature of functions of the interval computable in variable time, we restrict our attention to continuous bijective representations.

Definition 11.10 *Let X be I or $\mathbf{C}$. A function $f : X \to X$ is reducible to a function $g : I \to I$ if there exists a Cantor subset $C \subseteq I$ and a bijection $\phi : X \to C$ mapping f onto $g_{|C}$, i.e., such that for all $x \in X$,*

$$f(x) = \phi^{-1} \circ g \circ \phi(x).$$

If $f \equiv T$ is the global dynamics of a cellular network and ϕ is a conjugacy, g is said to simulate *the network in real time. A class of functions $\mathcal{D}$ is π-complete if every π-computable function f is reducible to some $g \in \mathcal{D}$.*

Cellular automata can be seen as continous maps on the Cantor set and thus can be easily extended to appropriate continuous functions on I (even piecewise linear almost everywhere – see also Problem 3). Coupled with the computation universal cellular automata of Chapter 6 one easily obtains complete maps.

Proposition 11.11 *There exist continuous maps of the interval which are complete in the class of π-computable functions. Likewise for the subclasses of functions π-computable in each euclidean dimension.*

Thus, although in general *exact* π-computable functions may be discontinuous, they remain some sort of reduct of continuous functions to Cantor subsets of their domain. On the other hand, there is no evidence that there exist representations for which addition and multiplication of *real numbers* (the building blocks of conventional mathematics and computer arithmetic) are exactly π-computable in general. Nevertheless, the notion remains of interest, at least for the well established fact that the dynamical behavior of many physical (respectively, cognitive) phenomena can be approximated by cellular automata (neural networks).

A possible line of further development presented next will conclude this chapter. A more satisfactory characterization of π-computable functions in variable time, either exactly or approximately, remains an open question.

11.3 Approximate Computation by Neural Nets

Many times exact computation of a real-valued function is not required or desirable. Rather, information on the overall qualitative behavior of the map may be enough. One of the difficulties with approximation in general is that upon iteration (a common occurrence during function evaluation) the errors inherent in the approximation propagate in an unknown and uncontrollable way, even for a low number of iterations. It is desirable to approximate arbitrary continuous dynamical systems to any degree of accuracy through an unbounded number of iterations. The main idea in this section is to capture the dynamical behavior of a map f through observation of the dynamical behavior of a set of primitives which are, in some sense, more accessible (e.g., they are biologically motivated, are directly observable on parallel computer simulations, or have a relatively simple dynamical behavior).

Every continuous system in configuration space can indeed be approximated through all iterations by infinite boolean networks, when one requires approximation of arbitrary exact *orbits* of the given map. However, this result no longer holds when the orbital behavior of approximant neural networks is not observable exactly due to the presence of random noise. In that case, not all continuous functions can be approximated by neural nets, but they can approximate large classes of discontinuous maps. The results for finite analog nets are not as definitive. Neural nets can nonetheless approximate large families of both continuous (including chaotic maps) and discontinuous maps (including baker maps and maps with dense periodic points). No precise characterization is known of the type of functions that can be reached this way via analog neural networks.

11.3.1 Relative Shadowing

Observability of neural net global dynamics has been studied via the notion of shadowing in Chapter 9. The action of a map amounts under this purview to a good approximation of the orbits of the map by pseudo-orbits of the same map f (obtained, for example, by using discrete and finitary character approximations on currently available computing machinery). In practice, however, f is generally unknown and one is not necessarily allowed computation of f. In fact, actual computer systems impose a very limited number of maps $\{g_\alpha\}_\alpha$ that are actually available for the study of f because they are the only maps actually programmable and observable on a computer in *exact* detail. The second notion of π-computability proposed in this section follows this pattern, taking the global dynamics of cellular networks to be basic maps implementable on a parallel computer (either exactly or approximately).

The question of approximating arbitrary continuous dynamical systems to any degree of accuracy through an unbounded number of iterations can be formalized in two essentially different ways. Let $\mathcal{F}(X)$ denote the set of all self-maps of a metric space $(X, |*, *|)$ with the uniform metric

$$|f, g|_u := \sup_{x \in X} |f(x), g(x)|.$$

Let $C(X)$ denote the subset of continuous maps of X (or just C if X is the unit interval I of the real line $\mathbf{R}$). Recall that a sequence $x_0, x_1, \cdots$ is an orbit of g if for all $n \geq 0$, $g(x_n) = x_{n+1}$. It is a *pseudo-orbit* of a map g if for every $n \geq 0$,

$$|g(x_n), x_{n+1}| < \delta \qquad (11.2)$$

The first definition of approximability of a map f assumes knowledge of orbits of the approximants (perhaps contaminated by noise) and derives knowledge of the behavior of true orbits of f to any degree of accuracy.

Definition 11.12 *A δ-pseudo-orbit $x_0, x_1, \cdots$ is ε-shadowed by f if there exists a $z \in X$ so that for all $n \geq 0$,*

$$|x_n, f^n(z)| < \varepsilon. \qquad (11.3)$$

The map g is said to be shadowed, *or* traced, *by f on X if for every $\varepsilon > 0$ there exists some $\delta > 0$ such that f ε-traces every δ-pseudo-orbit of g. Given a family of maps $\mathcal{D} \subseteq \mathcal{F}(X)$, f is* sp-approximated *by $\mathcal{D}$ if inequality (11.2) holds for all n and for all δ-pseudo-orbits of some $g = g_\varepsilon \in \mathcal{D}$. The set of all functions shadowing $\mathcal{D}$ is denoted $\mathcal{S}(\mathcal{D})$, and is referred to as the* shadowing hull *of $\mathcal{D}$.*

In actual computational practice, it might be desirable to require only tracing of finite pseudo-orbits in the definition of sp-approximation. Two reasons can be given for the choice. First, for continuous maps the two definitions are easily seen to be equivalent. Secondly, Definition 11.12 clearly implies traceability of finite pseudo-orbits and so is a stronger condition.

Note that according to Definition 11.12, f shadows itself on X iff for every $\varepsilon > 0$ there exists a $\delta > 0$ such that every δ-pseudo-orbit of f is ε-shadowed (-traced) by one of its orbits. This is the shadowing property of Chapter 9. $\mathbf{SP}(X)$ (respectively, $\mathbf{SP}$) will denote the family of maps of X (respectively, the unit interval I of $\mathbf{R}$) satisfying the shadowing property, i.e.,

$$\mathbf{SP}(X) := \{f \in \mathbf{F}(X) : f \in \mathcal{S}(f)\}.$$

In other words, a map with the shadowing property is *observable* in the sense that even if errors are randomly introduced in the input or propagated upon iterated evaluation, the resulting pseudo-orbits reveal the true behavior of the map to arbitrary degrees of accuracy. Analogously, the main idea behind sp-approximation is that the observation of appropriate pseudo-orbits of maps in $\mathcal{D}$ reveal the true dynamical behavior of (random orbits of) f, i.e., that f is observable *through* the maps in the family $\mathcal{D}$.

The second way to approximability of a function f dually requires that the behavior of arbitrary true orbits of f be pseudo-orbits of the approximants. Despite the superficial analogy, it will turn out to be approximation of a different nature.

Definition 11.13 *The map f is said to be* orbitally approximated *by a family $\mathcal{D}$ of maps on X if for every $\varepsilon > 0$ there exists some $g \in \mathcal{D}$ such that every orbit of f is an ε-pseudo-orbit of g. The set of all maps that are orbitally approximated by the family $\mathcal{D}$ is denoted $\mathcal{S}_u(\mathcal{D})$.*

Example 11.14 Over the real numbers,

1. Only discontinuous functions can be sp-approximated by the identity map. In particular, in the presence of noise, the identity cannot sp-approximate itself, but it does approximate itself orbitally. In fact, $\mathcal{S}_u(\mathbf{id}) = \{\mathbf{id}\}$.

2. The family $\mathcal{S}(\mathbf{c})$, where $\mathbf{c}$ denotes a constant function, consists of those maps f that remain ε-close to $\mathbf{c}$ on an ε-dense subset E_ε of their domain upon iteration, for each $\varepsilon > 0$. By contrast, $\mathcal{S}_u(\mathbf{c}) = \{\mathbf{c}\}$.

3. More generally, if $\mathcal{D}$ is a finite family of continuous maps, nothing is gained from the cooperation of the elements in the family, i.e., $\mathcal{S}(\mathcal{D}) = \mathcal{D} \cap \mathbf{SP}$. They can only orbitally approximate themselves, i.e., $\mathcal{S}_u(\mathcal{D}) = \mathcal{D}$. However, an infinite family $\mathcal{D}$ can capture in cooperation many more continuous functions.

Of particular interest for a base family $\mathcal{D}$ are global dynamics induced by neural nets, either infinite boolean or finite analog. Two more families will eventually become involved in the problem. The linear maps (which are well known in the continuum and were defined in Chapter 3 for configuration space) and simple maps. Simple maps are a well known type for approximation in classical analysis but they require a careful definition in configuration space.

Definition 11.15 *A map f on euclidean spaces $\mathbf{R}^n$ is* simple *if it has finite range, only finitely many discontinuities, and for each valued c in the range, $f^{-1}(c)$ is a disjoint union of finitely many intervals of positive length. A map f on a configuration space is* simple *if it is continuous, it has finite range, each value in the range has finite support (i.e., each configuration in the range is nonzero only at finitely many cells) and, furthermore, $f^{-1}(c)$ contains an open set for every c in the range of f. Let $\mathbf{SF}(X)$ (respectively, $\mathbf{SF}(X)^c$) denote the family of (continuous) simple maps of X, or of $\mathbf{R}$, if the X is omitted.*

These two classes of maps provide a handle on the general question of characterizing the types of dynamical systems that can be approximated by neural networks. The rest of this section gives a number of properties useful in finding out just what maps are approximable by neural nets.

First, simple maps with the shadowing property can be characterized precisely.

Proposition 11.16 *A simple function has the shadowing property iff it is continuous at every point of its range.*

Proof. Say f is discontinuous at p. There exists an open interval J where f is constant of value p, say centered at x and of radius $\varepsilon_0 > 0$. Let $\varepsilon :=$ $\frac{1}{2}\min\{\varepsilon_0, |p_i - p_j| \ (i \neq j)\}$, where p_i are all the distinct values of f. Consider p_* such that $|p_* - f(x)| < \delta$ but $f(p_*) \neq f(p)$, where $p = f(x)$. For no $\delta > 0$ is the δ-pseudo-orbit

$$x \to p_* \to f(p_*) \to f^2(p_*) \to \cdots$$

ε-traceable by f since for every $z \in J$, $f(z) = f(x) = p$, $f^2(z) \neq f(p_*)$ and $|f^2(z) - f(p_*)| > \varepsilon$. Therefore, f does not have the shadowing property.

Conversely, given $\varepsilon > 0$, let

$$\delta := \min\{\delta_i : f(]p_i - \delta_i, p_i + \delta_i[) = p_j\}.$$

Any δ-pseudo-orbit $\{x_n\}$ satisfies $|f(x_i) - x_{i+1}| < \delta$. In particular, $|f(x_0) - x_1| < \delta$. Therefore $f(x_1) = f^2(x_0)$. Inductively, if $f(x_n) = f^{n+1}(x_0)$, then $f^2(x_n) = f(x_{n+1})$ and $f(x_{n+1}) = f^{n+2}(x_n)$. On the other hand, $|f(x_n) - x_{n+1}| < \delta$, which implies that $|f^{n+1}(x_0) - x_{n+1}| < \delta$. □

Remark. In configuration space, every simple map is actually a neural net, and hence it is continuous and observable (see Problem 5).

It is easy to see that the only observable continuous map that can be (sp- or orbitally) approximated by a *finite* set $\mathcal{D}$ of observable continuous maps are the maps in $\mathcal{D}$ themselves. In this case, it is easy to see that it suffices to check approximability using actual orbits instead of pseudo-orbits.

Proposition 11.17 *If every function in $\mathcal{D}$ has the shadowing property, then*

1. *$f \in \mathcal{S}(\mathcal{D})$ if and only if for every $\varepsilon > 0$, there exists $g \in \mathcal{D}$ such that every orbit of g is ε-traced by f.*

2. *$f \in \mathcal{S}_u(\mathcal{D})$ if and only if f is the uniform limit of functions in $\mathcal{D}$.*

Proof. For the first part, it is obvious that the condition is necessary for $f \in \mathcal{S}(\mathcal{D})$ since orbits are δ-pseudo-orbits for any $\delta > 0$. Conversely, because g has the shadowing property, every δ-pseudo-orbit of $g \in \mathcal{D}$ can be traced by an orbit of g, and hence of f.

For the second part, given any $\varepsilon > 0$, one can find $g \in \mathcal{D}$ such that every orbit of f is an ε-pseudo-orbit of g. Therefore $|f, g|_u < \varepsilon$. Conversely, if $\varepsilon > 0$ and an orbit $x, f(x), f^2(x), \ldots$, and $g \in \mathcal{D}$ is ε-uniformly close to f, then $|f^{n+1}(x), g(f^n(x))| < \varepsilon$ for all $n \geq 0$. $\square$

Now we can give a fairly satisfactory characterization of the maps that are accessible through neural networks by orbital approximation.

Theorem 11.18 *On a compact space X, a map is orbitally approximated by a family $\mathcal{D}$ if and only if it is a uniform limit of a sequence of functions in $\mathcal{D}$. In particular, in either euclidean or configuration space,*

1. $S_u(\mathcal{L}) = \mathcal{L}$

2. $S_u(\mathcal{L}^c) = \mathcal{L}^c$

Proof. The inclusions $\supseteq$ in parts 1 and 2 are obvious from the definitions. The reverse inclusions follow from Proposition 11.17 and the fact that a uniform limit of linear and continuous maps is linear and continuous. $\square$

This result yields a characterization of orbital approximation by neural nets in configuration space.

Theorem 11.19 *All continuous maps in configuration space can be orbitally approximated by boolean neural nets, or even neural nets of a very special type (simple maps), i.e.,*

$$S_u(\mathbf{SF(C)}) = S_u(\mathbf{NN(C)}) = \mathbf{C}$$

Proof. The inclusions $\subseteq$ follow from monotony of the operator S – see Problem 4 – and the fact that a uniform limit of continuous maps is continuous. Finally, the second $\supseteq$ follows Proposition 9.40. $\square$

By contrast, not all continuous maps are captured by analog networks with identity or hard-threshold transfer functions in the real line or unit interval; see Problems 8 and 9.

Three important questions remain. First, what maps can be sp-approximated by neural nets in configuration space. Secondly, what maps in the continuum can be sp-computed by boolean neural nets via some type of encoding. And third, what maps in the continuum can be sp-computed using analog nets. These questions turn out to be surprisingly difficult despite the similarity of the two definitions of approximability. What little is known about them is presented next.

11.3.2 Shadowing Bases

This section presents some results that might prove be useful in finding some indication of the sp-approximation power of neural nets (boolean or analog) in the continuum.

The maps in the shadowing hull of $\mathcal{D}$ are generated by relative shadowing in the sense that their orbits can be approximated by pseudo-orbits of the family $\mathcal{D}$. Naturally, the smallest such set is of interest and is defined as follows.

Definition 11.20 *Assume there exists a nonempty set $\mathcal{D}$ such that $f \in \mathcal{S}(\mathcal{D})$. A shadowing basis for a map f (or more generally, a family $\mathcal{M}$) is a family of maps of minimum cardinality with the property that $f \in \mathcal{S}(\mathcal{D})$ (respectively, $\mathcal{M} \subseteq \mathcal{S}(\mathcal{D})$). The shadowing dimension of a map $f : X \to X$ is the minimum cardinality of one such family $\mathcal{D}$.*

Let $\overline{\mathcal{D}}$ denote the closure of the family of continuous maps $\mathcal{D}$ of X with the uniform topology, and $\mathbf{SP}(X)$ the set of maps on X with the shadowing property.

Proposition 11.21 *If $\mathcal{D}$ is a family of continuous maps with $\overline{\mathcal{D}}$ compact in $\mathcal{F}(X)$, then*

$$\mathcal{S}(\mathcal{D}) \subseteq \overline{\mathcal{D}}$$

In particular, if $\mathcal{D}$ is a finite family of continuous functions, then

$$\mathcal{S}(\mathcal{D}) \cap \mathbf{SP}(X) = \mathcal{D} \cap \mathbf{SP}(X).$$

Proof. Assume $f \notin \overline{\mathcal{D}}$ and put $2\varepsilon := |f, \overline{\mathcal{D}}| > 0$. It is easy to see that for every $g \in \mathcal{D}$ and every $\delta > 0$, no orbit of f can ε-trace a δ-pseudo-orbit of g. $\square$

This inclusion may be strict, as the following example shows. Consider the sequence $\mathcal{D}$ of functions $f_n : I \to I$ given by

$$f_n(x) = \left(1 - \frac{1}{n}\right)x.$$

$f_n(x) \to \mathbf{id} \notin \mathcal{S}(\mathcal{D})$. Each f_n is a contraction so their orbits converge to 0 hence they can't be traced by the identity $\mathbf{id}$. However, $\mathbf{id} \in \overline{\mathcal{D}}$.

Entropy (denoted *ent*) is a measure of dynamical complexity introduced in Chapter 8. Roughly speaking the entropy of a map measures the asymptotic growth rate of the number of its distinct orbits. The following relation indicates the relevance of relative shadowing to dynamical properties.

Theorem 11.22 *If $\mathcal{D} \subseteq C(X)$, $f \in \mathcal{S}(\mathcal{D})$ and f is continuous, then*

$$ent(f) \le \sup_{g \in \mathcal{D}} ent(g).$$

Proof. Fix $\varepsilon > 0$ and $n > 0$. Since f^i $(0 \leq i \leq n-1)$ are uniformly continuous, there exists $0 < \varepsilon' < \varepsilon$ such that

$$|x - y| < \varepsilon' \Rightarrow |f^i(x) - f^i(y)| < \varepsilon \quad (0 \leq i \leq n-1).$$

Also, there exists $g \in \mathcal{D}$ and $\delta > 0$ such that any δ-pseudo-orbit of g can be ε'-traced by f. Let $E_{(n,\varepsilon')}$ be an (n, ε')-spanning set for g of minimum cardinality. Let

$$Y := \{z : \exists y \in X \; (\forall k \geq 0) \; (|g^k(y), f^k(z)| < \varepsilon') \}.$$

Clearly the set Y is ε'-dense, i.e., for all $x \in X$, there is $z \in Y$ such that $|x, z| < \varepsilon'$. For every $x \in E_{(n,\varepsilon')}$ there exists $z_x \in Y$ such that

$$|g^k(x), f^k(z_x)| < \varepsilon, \quad \forall k \geq 0.$$

Fix one such z_x for each such x and consider the set

$$E^f := \{z_x : x \in E_{(n,\varepsilon')}\} \; .$$

In order to show that E^f is $(n, 4\varepsilon)$-spanning for f, let $x \in X$. There exists $z \in Y$ such that

$$|f^i(x), f^i(z)| < \varepsilon \quad , i = 0, 1, \cdots n - 1.$$

Since $z \in Y$ there exists $y \in X$ such that

$$|f^k(z), g^k(y)| < \varepsilon', \quad \forall k \geq 0.$$

Because $E_{(n,\varepsilon')}$ is spanning for g, there exists $w \in E_{(n,\varepsilon')}$ such that

$$|g^i(y), g^i(w)| < \varepsilon', \quad i = 0, 1, \cdots, n - 1.$$

Now f must ε'-trace the orbit of w under g:

$$|g^k(w), f^k(z_w)| < \varepsilon' \quad \forall k \geq 0.$$

Therefore

$$|f^i(x), f^i(z_w)| < 4\varepsilon, \quad i = 0, 1, \cdots, n - 1.$$

This proves that E^f is an $(n, 4\varepsilon)$-spanning set for f. Thus

$$ent_{4\varepsilon}(f) \leq ent_{\varepsilon'}(g) \leq \sup_{g \in \mathcal{C}} ent(g).$$

In the limit as $\varepsilon \to 0$, the Theorem follows. $\qquad\qquad\qquad\qquad\qquad\qquad$ $\square$

Strict inequality may, in fact, occur in Theorem 11.22, even if $\mathcal{D}$ is a shadowing basis for f: a dynamically trivial map, namely 0, can shadow a set of maps of positive entropy, as the following example shows. Let f' be the map obtained from a given map f on I consisting of two copies scaled down to half-size in

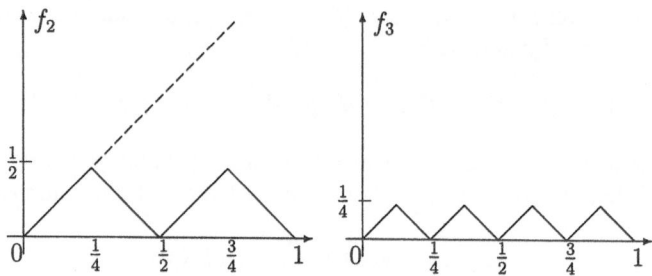

Fig. 11.1. Two functions in a shadowing basis of **0**

the subintervals $[0, \frac{1}{2}]$ and $[\frac{1}{2}, 1]$. Let $\mathcal{D}$ be the sequence given recursively by $f_{n+1} := f'_n$ $(n \geq 1)$ and $f_1 :=$ the tent map of height 1, say, $f_1(x) := x$ $(x \leq \frac{1}{2})$ and $f_1(x) := 1 - x$ $(x \geq \frac{1}{2})$. It is easy to see that $\mathbf{0} \in S(f_1, f_2, \ldots)$. Two of the maps, f_2 and f_3, are shown in Fig. 11.1. In this example, $ent(f_n) = \log 2$ for all $n \geq 2$, but $ent(\mathbf{0}) = 0$.

The next result proves that no countable set of maps (let alone π-computable) can generate all the maps of the interval via relative shadowing.

Theorem 11.23 *The set of all maps of a perfect metric space does not have a countable shadowing basis, even for finite shadowing.*

Proof. Let $\mathcal{D} = \{g_1, g_2, \cdots\}$ be a countable family of real- valued functions $g_i : I \to I$. Given $x_0 \in I$, consider

$$
\begin{array}{cccc}
x_0 & g_1(x_0) & g_1^2(x_0) & \cdots \\
x_0 & g_2(x_0) & g_2^2(x_0) & \cdots \\
x_0 & g_3(x_0) & g_3^2(x_0) & \cdots \\
& \vdots
\end{array}
$$

Let $\varepsilon := \frac{1}{8}$ and define f as follows for $y \in]x_0 - \varepsilon, x_0 + \varepsilon[$.

$f(y) := x_1,$ where $|x_1 - g_1(x_0)| \geq \varepsilon$ and $|x_1 - x_0| \geq \varepsilon;$

$f(x_1) := x_2,$ where $|x_2 - g_2^2(x_0)| \geq \frac{1}{4}$ and $x_2 \neq x_1$ and $|x_2 - x_0| \geq \varepsilon$

$f(x_2) := x_3,$ where $|x_3 - g_3^3(x_0)| \geq \frac{1}{4}$ and $x_3 \notin \{x_1, x_2\}$ and $|x_3 - x_0| \geq \varepsilon$

$$\vdots$$

The values of f can be defined arbitrarily at all other points of I. In order to prove that $f \notin S(\mathcal{D})$, let $\varepsilon := \frac{1}{8}$. For every $\delta > 0$, there exists a δ-pseudo-orbit of g_n, namely the orbit of x_0, that is not ε-traceable by f due to the n-th inequality in the definition of f. The same argument applies to finite shadowing. $\square$

There even exists a *universal tracer* that is sp-computed by an arbitrary families of maps of second countable spaces (i.e., those with a countable topological basis of open sets generating all sets through countable unions).

Theorem 11.24 *Let X be a second countable and perfect metric space. There exists a map u of X such that $u \in S(\mathcal{D})$, for every set of functions $\mathcal{D}$ of X. In particular, $S(\mathcal{D})$ is never empty.*

Proof. The proof requires use of Zorn's lemma. Consider the Hilbert cube $\Theta := I^\omega$ consisting of all sequences of real numbers in I with the uniform topology. It suffices to prove the existence of a suitable dense subset $\mathcal{P}$ of Θ since one can then define a self-map u of I such that each element $s \in \mathcal{P}$ is an orbit of u and $u(x) = 0$ elsewhere. Thus, given any pseudo-orbit of an arbitrary self-map f of I, one can ε-trace it by some orbit of u.

The components of all sequences in $\mathcal{P}$, however, must be all distinct from one another within the same and also for different sequences, in order for u to be well-defined. Toward this end, begin with the set of sequences of rational numbers, i.e., the product $\Pi_n Q_1$ (countably many copies of the rationals), which is dense in I^ω. Let x_1 denote a real number modulo 1. By Zorn's lemma, there exists an (uncountable) basis $\{1\} \cup B$ of $\mathbf{R}$ over $\mathbf{Q}$. For each $\lambda \in B$ put $A_\lambda := \{(n\lambda)_1; n \in \mathbf{Z}\}$. Since B is linearly independent it is easy to see that these sets are pairwise disjoint. Also, each set A_λ is uniformly distributed over I [N, Theorem 6.3], hence it is everywhere dense [N, Corollary 6.4]. Now, let $\mathcal{C}$ be the family of sequences A_λ with $\lambda \in B$. It is easy to see that $\mathcal{C}$ can be partitioned into a disjoint union of countably many uncountable subsets, say

$$\mathcal{C} = \bigcup_n \mathcal{C}_n \quad \text{and} \quad \mathcal{C}_n = \bigcup_j \mathcal{C}_n^j. \tag{11.4}$$

Let $\mathbf{Q}_1 = \{q_0, q_1, \cdots\}$ be an ordering of the rational numbers in the unit interval I. Each element $s \in \Pi_n Q_1$ is a sequence of rational numbers. For each $i = 1, 2, \ldots$, substitute s by the sequences obtained by replacing its i-th component s_i by each of the terms of a sequence $(s_n^i)_n$ converging to s_i and with entries in $\bigcup \mathcal{C}_i^{i_j}$, where q_{i_j} is the rational number corresponding to s_i in $\mathbf{Q}_1$ – see equation (11.4). Since the $\mathcal{C}_i^{i_j}$ are uncountable and disjoint for different i_j's, the sequences $(s_n^i)_n$ can be chosen with disjoint ranges even if $s_i = s_j$ for different s's and/or i's. Thus, the set

$$\mathcal{P} := \Pi_i \{s_n^i : n \in \mathbf{N}\}$$

can now be used to define u as originally intended.   □

Note that the foregoing proof establishes in fact the existence of uncountable many such u's.

It is known (but not proven here) that the continuous functions of the interval is not a shadowing basis of $\mathcal{F}(I)$. Thus, classical functions such as polynomials, etc., will *not* suffice to allow arbitrarily accurate simulations of *all* self-maps of the interval on a computer. This statement does not preclude, however, the existence of such a countable basis for restricted subfamilies of functions, e.g., the family of all continuous functions.

Proposition 11.25 *If either (i) the set of periodic points of a continuous self-map f of the unit interval I is dense, or (ii) its set of eventually periodic points is closed, then f has a countable shadowing basis consisting of simple functions.*

In particular, the dynamical behavior of certain *chaotic* maps of the interval (for example, tent and logistic maps) can be observed through analog neural nets ('chaotic' in the sense of Devaney [D], i.e., maps with a dense orbit, a dense set of periodic points, and sensitive dependence on initial conditions).

Proof. (*i*): Let $p_1, p_2, \cdots$ be a set of periodic points of f which is dense, and, without loss of generality, assume these points have disjoint orbits $O(p_i)$ under f. Let f_1 be a simple function satisfying

1. the range of f_1 is $O(p_1)$;

2. f_1 is continuous at every point of $O(p_1)$;

3. $\{f_1^n(p_1)\}_n = O(p_1)$

Now define f_2 satisfying

1. the range of f_2 is $O(p_1) \cup O(p_2)$;

2. f_2 is continuous at every point of $O(p_1) \cup O(p_2)$;

3. $\{f_2^n(p_1)\}_n = O(p_1)$ and $f_2^n(p_2) = O(p_2)$.

Likewise one can define f_n ($n \geq 2$) by considering $p_1, p_2, \cdots, p_n$.

For every $\varepsilon > 0$ there exists n such that $p_1, \cdots, p_n$ is ε-dense, i.e., the maximum distance between two of these consecutive points is less than ε. For a given n, for some $\delta > 0$ small enough, any pseudo-orbit of the corresponding f_n is ε-traced by the orbit of some p_i under f_n, which coincides with the orbit of p_i under f.

(*ii*): Assume that the set of eventually periodic points P of f is closed. Given $x, y \in P$, there exists n such that $f^n(x) = x$ and $f^n(y) = y$. Furthermore, for every $z \in]x, y[$, either $f^n(z) < z$ or $f^n(z) > z$. Hence, f has the shadowing property on $]x, y[$, and the same method as in part (*i*) applies.   □

Simple maps have the same approximation power as neural networks, but still there is an uncountable number of them. It would be desirable to trim the generating set of primitives to as small a family as possible. This is known to be possible for a fairly interesting uncountable subclass, although it is open for the whole class of discrete neural networks and will be given without proof.

Theorem 11.26 *The family of global maps of discrete neural networks over an abelian group whose order is square free with uniformly bounded neighborhoods including the center cell can be sp-approximated by a countable family D consisting of simple maps.*

Countable is the optimal number, since by Example 3, a finite family of networks (being continuous maps) will not suffice.

Finally, the set of all self-maps of the interval does not have a basis of any cardinality, i.e., there exist completely untraceable functions.

Theorem 11.27 *There are as many functions $f : I \to I$ such that $f \notin S(\mathcal{F}(I))$ as the cardinality of $\mathcal{F}(I)$.*

The proof of Theorem 11.27 requires the following ordering of the rationals in I and a lemma about it. Let $\mathbf{Q}_1 = \{q_0, q_1, \cdots\}$ be the sequence

$$q_0 = 0, q_1 := \frac{1}{2}, q_2 := 1 \quad q_3 := \tfrac{1}{3}, \tag{11.5}$$
$$q_4 := \frac{2}{3}, q_5 := \frac{1}{4}, \quad \cdots$$

of rational numbers in the unit interval I ordered lexicographically, without repetitions, by increasing denominators in the relatively prime expression, so that $\frac{s}{t} < \frac{s'}{t'}$ whenever $2 < t < t'$ or ($2 < t = t'$ and $s < s'$).

Lemma 11.28 *For every q_i $(i > 0)$, there exists k such that $|q_k - q_{i+k}| > \frac{1}{9}$.*

Proof. In fact $k = 0, 1$, or 2 will do. Let $q_i := \frac{s}{t}$ with s, t relatively prime. Assume neither $k = 0$ nor $k = 1$ satisfy the required inequality for q_i, so that $\frac{s}{t} < \frac{1}{9}$ and, if $q_{i+1} := \frac{p}{t}$ and $q_{i+2} := \frac{q}{t}$,

$$\left|1 - \frac{p}{t}\right| < \frac{1}{9}.$$

Therefore,

$$\left|\frac{q}{t} - \frac{1}{2}\right| > \frac{1}{9},$$

which proves that $k = 2$. $\qquad\square$

Now define the map f so that sequence (11.5) is the orbit of 0 under f and $f(x) := 0$ for irrational $x \in I$. By mapping x to q_0 for $x \in A$, an arbitrary set of irrational numbers, and to q_1 outside A, we can obtain more than uncountable many maps for which the following argument applies.

Proof. (Theorem 11.27.) It suffices to show that for $\varepsilon := \frac{1}{18}$, for every $g \in \mathcal{F}$ and for every $\delta > 0$, some δ-pseudo-orbit is not ε-traceable by f. Let $g \in \mathcal{F}$. Without loss of generality, assume that there are two distinct pseudo-orbits (in fact, the orbits of x and y) of g, $\min\{\frac{\delta}{3}, \frac{1}{18}\}$-traced by f, say by the orbits of z_x and z_y. Since the orbits of z_x and z_y under f are dense, there exists $n_0 > 4$ such that $|f^{n_0}(z_x) - z_y| < \delta/3$. Thus,

$$x, g(x), \cdots, g^{n_0-1}(x), y, g(y), g^2(y), \cdots$$

is a δ-pseudo-orbit of g since

$$|g^{n_0}(x) - y| \le |g^{n_0}(x) - f^{n_0}(z_x)| + |f^{n_0}(z_x) - z_y| + |z_y - y| < \delta.$$

If this pseudo-orbit were traced by some orbit of f, say that of w, then

$$
\begin{aligned}
|w - x| &< \varepsilon, \cdots, \\
|g^{n_0-2}(x) - f^{n_0-2}(w)| &< \varepsilon \\
|g^{n_0-1}(x) - f^{n_0-1}(w)| &< \varepsilon \\
|y - f^{n_0}(w)| &< \varepsilon \cdots,
\end{aligned}
$$

and so

$$
\begin{aligned}
|w - z_x| &< \frac{1}{9}, \cdots, \\
|f^{n_0-1}(w) - f^{n_0-1}(z_x)| &< \frac{1}{9},
\end{aligned}
$$

which, by the previous lemma, implies that $w = z_x$. Hence

$$
\begin{aligned}
|z_y - f^{n_0}(w)| &< \frac{1}{9} \\
|f(z_y) - f^{n_0+1}(w)| &< \frac{1}{9} \\
|f^2(z_y) - f^{n_0+2}(w)| &< \frac{1}{9}, \cdots
\end{aligned}
$$

which contradicts the previous lemma. $\qquad\qquad\qquad\qquad\qquad\qquad\square$

11.4 Problems

The problems marked * may need to be looked up.

EXACT COMPUTATION

1. Find a truly 2D representation (i.e., not an encoding of an expansion) of real numbers in which addition can be performed efficiently (possibly in constant time). Likewise for multiplication.

2. Complete the details of the proof of Theorem 11.7.

3. Show that there exist complete maps in the class of π-computable maps which are ω-piecewise linear on the complement of the Cantor set. (A map is ω-piecewise linear if its domain can be partitioned in countably many disjoint intervals in the interior of each of which f is linear.)

APPROXIMATE COMPUTATION

4. Show that the operator S preserves inclusion, that is $S(\mathcal{D}) \subseteq S(\mathcal{E})$ whenever $\mathcal{D} \subseteq \mathcal{E}$.

5. Prove that every simple map is the global dynamics of a neural network in configuration space.

6. Show that every simple function in the unit interval is the global map of an analog neural network with linear or threshold units, i.e.,

$$\mathbf{SF} \subseteq \mathbf{NN}_{\text{id,ht}} \cdot$$

7. Verify that $S_u(\mathbf{c}) = \{\mathbf{c}\}$ in Example 11.14.

8. Show that over the reals, $S_u(\mathbf{NN_{ht}}) = S_u(\mathbf{SF}) = {}^\omega\mathbf{PC}$, the class of ω-piecewise continuous functions. * (A map $f : \mathbf{R} \to \mathbf{R}$ is called ω-piecewise continuous if the following two conditions hold: $(i)\,f$ is continuous except at most at countably many points; and (ii) at every point of discontinuity (including both $\pm\infty$), f is one-sided continuous (from the left or from the right) and the other one-sided limit also exists (except at $\pm\infty$). In particular, such a function is bounded and has two horizontal asymptotes.)

9. Show that over the reals, $S_u(\mathbf{NN_{id,ht}}) = \mathcal{L} + {}^\omega\mathbf{PC}$ (a sum of a linear map and an ω-piecewise continuous map) and $S_u(\mathbf{NN_s}) = \mathbf{NN_s}$, the class of analog neural nets with sigmoidal activation functions. *

10. Show that a homeomorphism of the unit interval has a countable basis of simple functions, and therefore countable shadowing dimension. [Eventually periodic points of a homeomorphism are periodic, in fact, fixed points (if increasing) or 2-cycles (if decreasing, since their second iterations are increasing), hence a closed set.]

11. Prove that the set of funtions sp-approximated by neural networks in configuration space have a countable shadowing basis. *

11.5 Notes

A detailed treatment of recursive analysis in terms of classical computability and complexity can be found in the literature in several places, for example, Aberth [A, Ko, PE-R], Ko [A, Ko, PE-R], and Pour-El–Richards[PE-R]. Weihrauch [W] studies higher-order analysis in the same context. The notion of BSS-computability and complexity of real-valued functions was introduced in Blum-Shub-Smale [B-S-S] and can be regarded as an attempt to overcome the shortcomings of the classical notion of 'computable function' from the point of view of the continuum. Computer arithmetic is a cultivated branch of study and several accounts can be found – see for example [Mu1, Ka]. An introductory overview of dynamical systems can be found in [D].

It is well established that arbitrary continuous functions can be approximated to any degree of accuracy by virtually every type of artificial neural nets. Early results include, among others, Cybenko [C], Funahashi [Fu] and Hornik–Stinchcombe–White [H-S-W]. More recently, Koiran [Ko] has proven similar results with a contructive proof which allows the use of piecewise continuous transfer functions such as linear-threshold units. Moreover, Leshno et al. [L-L-P-S] have shown that multilayer feedforward neural networks with a common activation function are universal approximators for continuous functions if and only if their activation function is nonpolynomial. Since continuity is preserved under composition, a similar result holds for any fixed finite number of iterations of a given map. Furthermore, Gallant–White [G-W] have shown that one can simultaneously approximate a differentiable function and any given number of its derivatives with the same type of network. A powerful aspect of recurrent neural networks, however, is their asymptotic behavior as dynamical systems through unbounded time. The notions introduced in Sect. 11.3 attempt to capitalize on this strength.

Nontrivial distributed representations for numeric computation are suggested by Takeda–Goodman [T-G]. Real-valued computation by cellular automata has been explored in Garzon–Botelho [Ga-Bo1]. The concept of iterated approximability of a map by elements in a family of maps has been introduced by Botelho–Garzon [Bo-Ga2, Ga-Bo2] in a general context. The notion of π-computability originates with Garzon–Botelho [Bo-Ga3]. Most of the results in Sect 11.3 can be found in these two papers. Results similar to Proposition 11.9 about recursively enumerable sets are established by Siegelman–Sontag [S-S] with analog neural nets.

Evidence for an information-theoretical approach to physical processes lending credit to the study of distributed representations can be found by Stonier [S]. The most fundamental problem in characterizing maps of the continuum computable by boolean and analog neural nets is perhaps determining the "right" *encodings*. It is clear that an encoding can partly or wholly perform work for the computer. Distributed representations have been used in applications of cellular automata with an entirely different semantic content. Definition 11.8 is given with respect to arbitrary distributed representations that we are barely beginning to understand. That such representations may be much closer to physical reality is the fundamental hypothesis of an emerging branch of physics, as described by Stonier [S]. In view of the results in this chapter, it is at least conceivable that in higher dimensions there exist distributed representations for which real arithmetic may be π- computed exactly, at least for a dense set of real numbers. It is even quite conceivable that a systematic study of distributed representation will give rise to a realistic approach to modeling of natural phenomena through more than just real number computation as we know it.

The second basic problem is the *halting criterion*, i.e., when to look at the net to read off a function value. We have chosen an approach that appears almost necessary from a practical point of view for the user. However, it is easy to imagine other possibilities. For instance, one could use a clock (Turing computable function or a clock as defined in Chapter 10) to stop the automaton and obtain the answer. Or one could simply demand convergence toward the desired value (and hence, in general, one only obtains approximations at finite times). Although acceptable on a purely computational basis, these alternatives are somewhat undesirable from the point of view of a complex parallel computer performing the calculation under minimum supervision and intervention on the part of the user, the more likely situation in eventual applications.

References

[A]   O. Aberth: Computable analysis. McGraw-Hill, New York, 1980

[B-S-S]   L. Blum, M. Shub, S. Smale: On a theory of computation over the real numbers: **NP**-completeness, recursive functions and universal machines. Bull. Amer. Math. Society **21** (1989) 1–46

[Bo-Ga1]   F. Botelho, M. Garzon: On dynamical properties of neural networks. Complex Systems **5**:4 (1992) 401–413

[Bo-Ga2]   F. Botelho, M. Garzon: Generalized shadowing properties. J. Random and Comput. Dynamics **2**:2 (1994) 145–164

[Bo-Ga3]   F. Botelho, M. Garzon: Boolean neural nets are observable. Theoret. Comput. Sci. **134** (1994) 51–61. Corrigendum, forthcoming

[C]   G. Cybenko: Approximation by superpositions of a sigmoidal function. Math. Control, Signals and Syst. **2** (1989) 303-314

[D]   R.L. Devaney: An introduction to chaotic dynamical systems. Addison-Wesley, Reading MA, 1986

[Fu]   K. Funahashi: On the approximate realization of continuous mappings by neural networks. Neural Networks **2** (1989) 183–192

[G-W]   A.R. Gallant, H. White: Universal approximation of an unknown mapping and its derivatives using multilayer feedforward networks. Neural Networks **3** (1990) 551–560

[Ga1]   M. Garzon: Cellular automata and discrete neural networks. Physica D **45** (1990) 431–440

[G-F]   M. Garzon, S. Franklin: Neural computability II (Extended Abstract). Proc. 3rd Int. Joint Conf. on Neural Networks IJCNN (1989) I 631–637.

[GFBBD]   M. Garzon, S.P. Franklin, W. Baggett, W. Boyd, and D. Dickerson: Design and testing of a general-purpose neurocomputer. J. Par. Distr. Proc. **14**(1992) 203–220

[Ga-Bo1]   M. Garzon, F. Botelho: Real computation with cellular automata. In: Proc. Workshop on Cellular Automata and Cooperative Systems, Les Houches, N. Boccara et al. (eds.). Kluwer, Amsterdam, 1993, pp. 191–202

[Ga-Bo2]   M. Garzon, F. Botelho: Dynamical approximation by neural networks. Second Swedish Conference on Connectionism. Lawrence Erlbaum, pp. 57–66

[H-S-W]   K. Hornik, M. Stinchcombe, H. White: Multilayer feedforward networks are universal approximators. Neural networks **2**(1989) 359–366

[Ka]   H. Kai: Computer arithmetic: principles, architecture, and design. John Wiley & Sons, New York, 1979

[Ko]   K.I. Ko: Complexity theory of real functions, Birkhäusser, Basel, 1991

[Ko]   P. Koiran: On the complexity of approximating mappings using feedforward networks. Neural Networks **6** (1993), 649–653

[K-C-G]   P. Koiran, M. Cosnard, M. Garzon: Computability with low-dimensional dynamical systems. In: STACS'93. Lecture Notes in Computer Science 665, P. Enjalbert et al. (eds.) Springer-Verlag, Berlin, 1993 pp 365–373

[L-L-P-S] M. Leshno, V.Y. Lin, A. Pinkus, S. Schocken: Multilayer feedforward networks with a nonpolynomial activation function can approximate any function. Neural Networks **6** (1993) 861–867

[M-M-M]   C. Mazenc, X. Merrheim, J.M. Muller: High-Radix on-line arithmetic for very long precision computations on a parallel computer, Rapport de Recherche 92-28, Ecole Normale Supérieure, Lyon, 1992

[Mu1]   J.M. Muller: Arithmétique des ordinateurs. Etudes et recherches en informatique, Mason, 1989

[Mu2]   J.M. Muller: Some characterizations of functions computable in on-line arithmetic, Rapport de Recherche 90-15, Ecole Normale Supérieure, Lyon.

[N]   I. Niven: Rational and irrational numbers. Carus Mathematical Monographs AMS, Providence RI, 1956

[PE-R]   M.B. Pour-El, J.I. Richards: Computability in analysis and physics. Springer-Verlag, Berlin-New York, 1989

[S-S]   H. Siegelman, E. Sontag: Neural nets are universal computing devices. Appl. Math. Letters (1992) 77-80

[S]   T. Stonier: Information and the internal structure of the universe. Springer-Verlag, Berlin, 1990

[T-G]   M. Takeda, J. Goodman: Neural networks for computation: number representations and programming complexity. J. Applied Optics **25**:18 (1986) 3033–3046

[W]   K. Weihrauch: Computability. Springer-Verlag, Berlin, 1987

12. A Bibliography of Applications

He more rightly deserves the name of happy who knows how to use the god's gifts wisely.

Horace

Although this volume has been chiefly concerned with analysis of various aspects of the evolution of cellular automata and generalizations, their applications have played a catalytic role in their genesis and study. A survey would not be fairly complete without a description, even if superficial, of the major applications of cellular models (including probabilistic cellular automata).

The Wolfram collection [Wo5, Sects. 6–9] contains an annotated bibliography of applications in various fields up to its publication (1986), to which the reader is referred. Most of the sections below cite later references.

12.1 Physics

It is rumored that the cellular automaton model was introduced by von Neumann as an attempt to provide models for meteorological forecasting. The idea has expanded even to the extreme that the entire universe is just a giant all-encompassing cellular automaton. Efforts in this direction have developed models of: *Digital mechanics* by Fredkin [Fr]; *spin systems* by Vichniac [V], Creutz [C], Domany [D], Pomeau [Po]; *condensed matter* by Wolfram [Wo4]; *statistical mechanics* by Hartman et al. [H-T-K]; a lot in *hydrodynamics*: general theory in Wolfram [Wo1], the Navier–Stokes equation by Frisch–Hasslacher–Pomeau [F-H-P], the Burger's equation by Beghosian–Levermore [B-L], the Boltzmann equation by Hersbach [Her], soliton behavior by Park–Steiglitz–Thurston [P-S-T], and other models by d'Humiéres–Lallemand [H-L], Wolf–Nasilowski–Vogeler [WG-N-V], Boghosian–Taylor, Rothman [Be]; *representation* of general analog quantities and physical variables by Smith [S]; galaxy formation in *astronomy* by Vicsek–Szalay [V-S]; and *quantum mechanics* by Feynman [Fe], Rujàn [Ru], Drescher [D], and Grössing–Zeilinger [G-Z].

12.2 Chemistry

The earliest application in chemistry is perhaps the discrete model for the Beluzhov–Zhabotinsky reaction by [G-H, M-F]. Models of *reaction-diffusion* have generated lots of work, from Greenberg–Greene–Hastings [G-G-H] through Allouche–Reder [A-R], Oono–Kohmoto [O-M], Winfree–Winfree–Seifert [W-W-S], to Hartman–Tamayo [H-T]. Packard [Pa] uses them in models of *solidification and aggregation*.

12.3 Biology

Cellular automata were first suggested by Ulam [U] as models of *growth, self-reproduction* and *biological evolution*. The idea of brain modeling using cellular automata appears to have been on von Neumann's mind while working on his [VN], in addition to the more obvious use as a logical model of self-reproduction (at a time when the the biological mechanism, the double-helixed DNA, was yet unknown). Biological motivations are patent in virtually all early works – see those in Burks [Bu]. Conway's LIFE itself is inspired by a biological analogy, though asexual, that provides a simple mechanism for *social evolution*. Sex was later introduced into the model by Vitanyi [Vi]. Kitagawa [Ki] surveys possible applications of 2D cellular automata to biomathematics. There seems to be a renewed interest in biomodelling in the 1990s through several conferences, e.g., [INBS].

Victor [V] has more recently examined the possibilities of what automata studies can tell us about the *brain*. Langton [La3] has used cellular automaton models to locate order on the edge of chaos as an essential logical *property of life*; Tamayo–Hartman [T-M] have used them as a starting model on clay particles in attempting to provide a plausible mechanism for the *origin of life*.

More specific applications have been shown by Sieburg et al. [Sa1] and Agur [A] in simulation of *artificial immune systems* for HIV infection; by Boccara–Cheong [B-C] in *epidemic models*; by Brown [Br] as models of *competition and selection*; by Sieburg–Clay [S-C] for general systems for modeling *biology on a computer*; by Cosnard et al. [Co1] for neural models with *bounded memory*; by Hilke et al. [HNRL] for modeling nucleotide interactions and nonenzymatic transcription in DNA; and by Stark–Pedersen [S-P] for modeling biological oscillators and tissue.

12.4 Computer Science

Much of the work in other areas has reverted into applications in computer science. Cellular automata have been applied to a variety of areas as parallel models, in addition to the applications mentioned in Chapter 10.

A systematic study of *image processing* applications can be found in the books by Preston–Duff [D-F] and Duff–Fountain [D-F]. Some algorithms are also presented in Goles–Martinez [Go-Ma]. Cellular automata proved useful in *hardware design* of the CM-2 by Hillis [Her, CM], Califano–Margolus–Toffoli [Ca]; *fault-tolerant computing* has proved possible with cellular automata by the works of Lee–Frieder [L-F], Pippenger [Pi1, Pi2], Gács [Ga1, Ga2, Ga3], Gács–Reif [G-R], Tóth [Tot], and Wang [Wa]; *random-number generation* in

Wolfram [Wo3]; *fractal generation* in Martin [M]; *cryptographic ciphers* in Guan [Gu]; *full cryptographic systems* in Kari [K] and Gutowitz [Gut2, Gut3, Gut4]; *numerical solutions* to differential equations in [Me-Fe].

12.5 Artificial Intelligence and Cognitive Science

John von Neumann's work [VN] on the design of a universal computer and constructor (see the references in Chapter 4) marked the beginning of a whole new era of applications of cellular automata to artificial intelligence. More recently, neural networks have only reinforced that trend.

Applications of automata networks in *artificial intelligence* are surveyed by Fogelman–Soulié [FS2]. *Knowledge representation* in automata networks is explored by Fogelman–Soulié [FS3]. Much of the motivation and current work in *artificial life* came from studies in cellular automata or their application in modeling, as seen in the works of Langton [La1, La2, La3], Mitchell–Crutchfield–Hraber [M-C-H], Tamayo–Hartman [T-M], Balbi [B]. Chess [Ch] has used cellular models to simulate the *evolution of behavior*. Gharavi–Anantharam [G-A] have examined the effect of noise in *long-term memory* using asynchronous delays between cells. Caianiello et al. [C] use a linear model for *associative memories*. Gutowitz [Gut5, Gut6] shows a way to solve the inverse problem of *learning and recognition* with cellular automata.

Wolfram [Wo2] describes a possible foundation to a new complex systems engineering. A fundamental difficulty in this direction is the fact that, until recently, there was little research in the direction of learning and adaptation. A promising field of *evolutionary engineering* seems to be emerging from several reserach efforts. Prime examples are the object-oriented Swarm project at the Santa Fe Institute for experimentation and research in artificial life, and the effort at building a billion neuron artificial brain which grows/evolves at electronic speeds in a cellular automata machine by at least one group [DG] using the CAM-8 architecture.

12.6 Miscellaneous

1. **Geography and Cartography.** Models for geographical organization: Tobler [To]; Appel–Evangelisti–Stein [A-E-S].

2. **Geophysics.** Seismic wave propagation: Rothman [R].

3. **Management.** Models of management of complex organizations: Gelfand-Walker [G-W].

4. **Sociological Models.** Society evolution under symmetric local influences: Polzak–Sura [P-S]; Zamperoni [Z].

References

[A]   Z. Agur: Fixed points of majority rule cellular automata with applications
         to plasticity and precision of the immune system. Complex Systems **5**:3
         (1991) 351–357

[A-R]   J.P. Allouche, C. Reder: Oscillations spatio-temporelles engendreés par un
         automate cellulaire. Disc. Appl. Math. **8** (1984) 215

[A-S]   A. Appel, A.J. Stein: Cellular automata for mixing colors. IBM Tech. Disc.
         Bull. **24** (1981) 2032

[A-E-S] A. Appel, C.J. Evangelisti, A.J. Stein: Animating quantitative maps with
         cellular automata. IBM Tech. Disc. Bull. **26** (1983) 953

[B]   P.P. Balbi de Oliviera: Methological issues within a framework to support
         a class of artificial-life worlds in cellular automata. TR-School of Cognitive
         Science, the U. of Sussex, 1993

[Be]   B.M. Beghosian, W. Taylor IV, D.H. Rothman: A cellular automata simu-
         lation of two-phase flow on the CM-2 connection machine system. Thinking
         Machines Co. TR series, 1988

[B-L]   B.M. Beghosian, C.D. Levermore: A cellular automaton for Burger's equa-
         tion. Complex Systems **1**:1 (1987) 17–30

[B-C]   N. Boccara, K. Cheong: Automata networks of epidemic models. In:
         [BGMP], pp. 29–44

[BGMP]   N. Boccara, E. Goles, S. Martinez, P. Picco (eds.): Cellular automata and
         cooperative systems. Kluwer, Dordrecht, 1993

[Br]   D.B. Brown: Competition of cellular automata rules. Complex Systems **1**:1
         (1987) 169–180

[Bu]   A.E. Burks: Essays on Cellular Automata. U. of Illinois Press, Chicago,
         1972

[C]   E.R. Caianiello, A. Esposito, M. Marinaro, R. Tagliaferri: The behavior
         and learning of a deterministic neural net. Complex Systems **6**:6 (1992)
         507–517

[Ca]   A. Califano, N. Margolus, T. Toffoli: CAM-6: a high-performance cellu-
         lar automata machine (Users guide and hardware manual). MIT lab for
         computer science, Cambridge MA, 1987

[Ch]   D. Chess: Simulating the evolution of behavior: the iterated prisoner's
         dilemma problem. Complex Systems **2**:6 (1988) 663–670

[Cho]   C. Choffrut (ed.): Automata networks. Lecture Notes in Computer Science,
         Vol. 316, Springer-Verlag, Berlin, 1987

[CM]   Connection machine models, CM-2 technical summary. Thinking machines
         co., Cambridge MA, 1989

[Co1]   M. Cosnard, M. Tchuente, G. Tindo: Sequences generated by neuronal
         automata with memory. Complex Systems **6**:1 (1987) 13–20

[C]   M Creutz: Deterministic Ising systems. Ann. Phys. **67** (1986) 62

[D-G]   H. de Garis: The "CAM-BRAIN" project, Parts I (fundamentals) and II
         (a billion neuron artifical brain). New Generation Computing **12**:2(1994).
         Also announced on ALife Digest No. 117 (January, 1994) and the Internet
         (June 1994, February 1995)

[D-G-T]   J. Demongeot, E. Goles, M. Tchuente: Dynamical systems and cellular au-
            tomata. Proc. Journeés de la Société Mathématique de France. Academic
            Press, New York, 1985

[D]   E. Domany: Exact results for two- and three-dimensional Ising and Potts
            models. Phys. Rev. Lett. **52** (1984) 871

[D]   G.L. Drescher: Demystifying quantum mechanics. Complex Systems **5**:2
            (1991) 207–237

[D-F]   M.J.B. Duff, T.J. Fountain: Cellular logic image processing. Academic
            Press, New York, 1986 .

[F-T-W]   D. Farmer, T. Toffoli, S. Wolfram (eds.): Cellular Automata. Proc. of an
            interdisciplinary workshop at Los Alamos. North-Holland, New York, 1984

[Fe]   R. Feynman: Simulating physics with computers. Int. J. Theor. Phys.
            21:6/7 (1982) 467–488

[F-R-T]   F. Fogelman-Soulié, Y. Robert, M. Tchuente: Automata networks in com-
            puter science: theory and applications. Princeton University Press, Prince-
            ton NJ, 1987

[FS2]   F. Fogelman-Soulié: Automata networks and artifical intelligence. In
            [F-R-T], pp. 133-186

[FS3]   F. Fogelman-Soulié: Representation of knowledge and learning in automata
            networks. In [Cho], pp. 95–16

[Fr]   E. Fredkin: Digital mechanics: an informational process based on reversible
            universal cellular automata. In [Gut1], pp. 254–270

[F-H-P]   U. Frisch, B. Hasslacher, Y. Pomeau: Lattice gas automata for the Navier-
            Stokes equation. Phys. Rev. Lett. **56** (1986) 1505

[Ga1]   P. Gács: Reliable computation with cellular automata. In: Proc. 15th
            Symp. Theory of Computing STOC (1983) 32–41

[Ga2]   P. Gács: Reliable computation with cellular automata. J. Comput. Syst.
            Sci. **32** (1986) 15–78

[Ga3]   P. Gács: Self-correcting two-dimensional arrays. Preprint, Bellcore, 1987

[G-R]   P. Gács, J. Reif: A simple three-dimensional real-time reliable cellular ar-
            ray. J. Comput. Syst. Sci. **36** (1988) 125–147

[G-A]   R. Gharavi, V. Anantharam: Effect of noise in long-term memory in cellular
            automata with asynchronous delays between processors. Complex Systems
            6:3 (1992) 287–300

[Go-Ma]   E. Goles, S. Martinez: Neural and automata networks. Kluwer Academic
            Publishers, Dordrecht, 1990

[G-H]   J. Greenberg, S. Hastings: Spatial patterns for discrete models of diffusion
            in excitable media. SIAM J. Appl. Math. **34** (1978) 515

[G-G-H]   J. Greenberg, C. Greene, S. Hastings: A combinatorial problem arising in
            the study of reaction-diffusion equations. SIAM J. Alg. Disc. Meth. **1**:1
            (1980) 34–42

[G-W]   A.E. Gelfand, C.C. Walker: A systems theoretic approach to the manage-
            ment of complex organizations: management by consensus and its interac-
            tion with other management strategies. Behavioral Sci. **25** (1980) 250–260

[G-Z]   G. Grössing, A. Zeilinger: Quantum cellular automata. Complex Systems
            2:2 (1988) 197–209. A corrigendum **2**:5 (1988) 611–623

[Gu]   P. Guan: Cellular automaton public-key cryptography. Complex Systems
            1:1 (1987) 51–56

[Gut1]   H. Gutowitz (ed.): Cellular automata: theory and applications. Proc. 3rd. Int. Conf. Cellular Automata, Los Alamos, 1991. Also issued as a separate book by MIT Press, 1992

[Gut2]   H. Gutowitz: Cryptography with dynamical systems. In: [Gut1], pp. 237–274

[Gut3]   H. Gutowitz: A cellular automaton cryptosystem; specification and call for attack. Preprint, Laboratoire d'Electronique, ESPCI, Paris, France, 1992

[Gut4]   H. Gutowitz: Method and apparatus for encryption, decryption and authentication using dynamical systems. Preprint, Laboratoire d'Electronique, ESPCI, Paris, France, 1992

[Gut5]   H. Gutowitz: Statistical properties of cellular automata in the context of learning and recognition, Part I: introduction. In: Learning and Recognition – A Modern Approach, K.H. Zhao (ed.). World Scientific Publishing. Singapore (1989), pp. 233–255

[Gut6]   H.A. Gutowitz: Statistical properties of cellular automata in the context of learning and recognition, Part II: inverting local structure theory equations to find cellular automata with specified properties. In: Learning and Recognition – A Modern Approach, K.H. Zhao. (ed.) World Scientific Publishing. Singapore (1989), pp. 256–280

[HNRL]   J. Hilke, R. Navarro-González, J. Reggia, J. Lohn: A modified cellular automata model of nucleotide interactions and nonenzymatic transcription of DNA. In: Proc. Int. IEEE Symposium in Intelligence in Neural and Biological Systems, Washington DC, 1995. In press.

[H-L]   D. d'Humiéres, P.Lallemand: Numerical simulations of hydrodynamics with lattice gas in two dimensions. Complex Systems 1:2 (1987) 599–630

[H-M]   B. Hasslacher, D.A. Meyer: Knot invariants and cellular automata. In: [Gut1], pp. 328–344

[Her]   H. Hersbach: A cellular automaton for a solvable Boltzmann equation. Complex Systems 4:3 (1990) 251–268

[Hi]   W.D. Hillis: The connection machine. MIT Press, Cambridge MA, 1985

[H-T-K]   H. Hartman, P. Tamayo, W. Klein: Inhomogeneous cellular automata and statistical mechanics. Complex Systems 1:2 (1987) 245–256

[H-T]   H. Hartman, P. Tamayo: Reversible cellular automata and chemical turbulence. In [Gut1], 293–306

[INBS]   First Int. IEEE symposium on Intelligence in neural and biological systems, Washington DC, 1995.

[K]   J. Kari: Cryptosystems based on reversible cellular automata. Technical Report, U. of Turku, Finland, 1992

[Ki]   T. Kitagawa: Cell space approaches in biomathematics. Math. Biosciences 19 (1974) 27

[La1]   C. Langton: Virtual state machines in cellular automata. Complex Systems 1:2 (1987) 257-271

[La2]   C. Langton: Computation at the edge of chaos: phase transition and emergent computation. Physica D 42 (1990) 12–37

[La3]   C. Langton (ed.): Artificial life. Santa Fe Institute Studies, Vol. 6. Addison-Wesley, Reading MA, 1989

[L-F]   M.S. Lee, G. Frieder: Self-configuration of defective cellular arrays. Complex Systems 1:1 (1987) 81–106

[LPVW]   T. Legendi, D. Parkinson, R. Vollmar, G. Wolf: Parallel processing by cellular automata arrays. North-Holland, Amsterdam, 1987

[M]   B. Martin: Self-similar fractals can be generated by cellular automata. In: [BGMP], pp. 463–471

[M-F]   B. Madore, W. Freedman: Computer simulations of the Belousov-Zhabotinsky rection. Science **222** (1983) 615

[Me-Fe] A.J. Meade Jr., A.A. Fernandez: Solutions of nonlinear differential equations by feedforward neural networks. Mathematical and Computer Modelling, 1994. In press

[M-C-H] M. Mitchell, P.T Hraber, J.P. Crutchfield: Revisiting the edge of chaos: evolving cellular automata to perform computations. Complex Systems **7**:2 (1993) 89–130

[O-M]   Y. Oono, M. Kohmoto: A discrete model of chemical turbulence. Phys. Rev. Lett. **55** (1985) 2927

[Pa]   N. Packard: Deterministic lattice models for solidification and aggregation. Proc. First Int. Symposium for Science on Form, 1985

[Pi1]   N. Pippenger: Symmetry in self-correcting cellular automata. J. Comput. Syst. Sci. **49**:1 (1994) 83–95

[Pi2]   N. Pippenger: Developments in the synthesis of reliable organisms from unreliable components. In: The legacy of von Neumann, J. Glimm, J. Impagliazzo, I. Singer (eds.). Amer. Math. Soc. Providence RI, 1990

[Po]   Y. Pomeau: Invariant in cellular automata. J. Phys. A **117** (1984) L415

[P-D]   K. Preston, Jr., M.J.B. Duff: Modern cellular automata (Theory and applications). Plenum Press, New York, 1984

[P-S]   S. Polzak, M. Sura: On periodical behavior in societies with symmetric influences. Combinatorica **3**:1 (1983) 119–121

[R]   D.H. Rothman: Modeling seismic P-waves with cellular automata. Geophys. Research Lett. **14**:1 (1987) 17-20

[Ru]   P. Rujàn: Cellular automata and statistical mechanical models. J. of Statist. Physics **49**:1/2 (1987) 139–222

[P-S-T] J. Park, K. Steiglitz, W. Thurston: Soliton-like behavior in cellular automata. Physica D **19** (1986)

[S-C]   H.B. Sieburg, O. Clay: The cellular device machine development system for modeling biology on the computer. Complex Systems **5**:6 (1991) 575–602

[Sal]   H.B. Sieburg, M. McCutchan, O. Clay, L. Caballero, J. Oslund: Simulation of HIV-infection in artifical immune systems. In: [Gut1], pp. 208–227

[S]   F. Smith: Representations of geometrical and topological quantities in cellular automata. In: [Gut1], pp. 271–277

[S-P]   W.R. Stark, J.F. Pedersen: Understanding global behavior of distributed systems with irregular asynchronous communication. In: Proc. Int. IEEE Symposium in Intelligence in Neural and Biological Systems, Washington DC, 1995. In press.

[Sto]   T. Stonier: Information and the Internal Structure of the Universe. Springer-Verlag, Berlin, 1990.

[T-M]   P. Tamayo, H. Hartman: Cellular automata, reaction-difussion systems and the origin of life. In: [La3], pp. 105–124

[T-M]   T. Toffoli, N. Margolus, Cellular automata machines (A new environment for modeling). MIT Press, Cambridge MA, 1987

[To]   W.R. Tobler: Cellular geography. In: Philosophy in geography, S. Gale and
           G. Olsson (eds.). Reidel, 1979

[Tot]   N. Tóth: Fault-tolerant programming of a two-layer cellular array. In:
           [LPVW], pp. 101–108

[U]   S. Ulam: On some mathematical problems connected with patterns of
           growth of figures. In [Bu], pp. 219–231

[V]   G. Vichniac: Simulating physics with cellular automata. Physica D **10**
           (1984) 96

[Vi]   P. Vitanyi: Sexually reproducing cellular automata. Math Biosciences **18**
           (1973) 23

[VN]   J. von Neumann: Theory of self-reproducing automata. U. of Illinois Press,
           Chicago, 1966

[V-S]   T. Vicsek, A.S. Szalay: Fractal distribution of galaxies modeled by a
           cellular-automaton type stochastic process. Phys. Rev. Lett. **58** (1987)
           2828

[Wa]   W. Wang: An asynchronous two-dimensional self-correcting cellular au-
           tomaton. In: Proc. 32nd Symp. Found. Comput. Sci. FOCS (1991) 278–285

[WG-N-V]   D.A. Wolf-Gladrow, R. Nasilowski, A. Vogeler: Numerical simulations of
           hydrodynamics with a pair interaction automaton in two dimensions. Com-
           plex Systems **5**:1 (1991) 89–100

[W-W-S]   A. Winfree, E. Winfree, H. Seifert: Organizing centers in a cellular excitable
           medium. Physica D **17** (1985) 109

[Wo1]   S. Wolfram: Cellular automata fluids I: basic theory. Institute for Advanced
           Study preprint, 1986

[Wo2]   S. Wolfram: Approaches to complexity engineering. In: [Wo5], pp. 400–413

[Wo3]   S. Wolfram: Random sequence generation by cellular automata. Adv. in
           Applied Math. **7** (1986) 123–169. Reprinted in [Wo3]

[Wo4]   S. Wolfram: Cellular automata and condensed matter physics. In: Scaling
           phenomena in disordered systems, R. Pynn, A. Skjeltorp (eds.). Plenum
           Press, 1985

[Wo5]   S. Wolfram: Theory and Applications of cellular automata. World Scientific
           Publishing, Singapore, 1986

[Z]   P. Zamperoni: Analysis of some migration processes in a two-dimensional
           cellular space. In: [LPVW], pp. 155–162

Author Index

Symbol Index

Subject Index

Texts in Theoretical Computer Science – An EATCS Series

J. L Balcázar, J. Díaz, J. Gabarró
Structural Complexity I
2nd ed. (see also overleaf, Vol. 22)

M. Garzon
Models of Massive Parallelism
Analysis of Cellular Automata and Neural
Networks

Monographs in Theoretical Computer Science – An EATCS Series

C. Calude
Information and Randomness
An Algorithmic Perspective

A. Nait Abdallah
The Logic of Partial Information

K. Jensen
Coloured Petri Nets
Basic Concepts, Analysis Methods
and Practical Use, Vol. 2
(see also overleaf)

Former volumes appeared as
EATCS Monographs on Theoretical Computer Science